Fourier Transform Methods in Finance

For other titles in the Wiley Finance series
please see www.wiley.com/finance

Fourier Transform Methods in Finance

Umberto Cherubini
Giovanni Della Lunga
Sabrina Mulinacci
Pietro Rossi

WILEY
A John Wiley and Sons, Ltd., Publication

This edition first published 2010

Registered office
John Wiley & Sons Ltd, The Atrium, Southern Gate, Chichester, West Sussex, PO19 8SQ, United Kingdom

For details of our global editorial offices, for customer services and for information about how to apply for permission to reuse the copyright material in this book please see our website at www.wiley.com.

Library of Congress Cataloging-in-Publication Data

Fourier transform methods in finance / Umberto Cherubini . . . [et al.].
p. cm.
Includes bibliographical references and index.
ISBN 978-0-470-99400-9 (cloth)
1. Options (Finance)–Mathematical models. 2. Securities–Prices–Mathematical models.
3. Finance–Mathematical models. 4. Fourier analysis. I. Cherubini, Umberto.
HG6024.A3F684 2010
332.63′2042–dc22

2009043688

A catalogue record for this book is available from the British Library.

ISBN 978-0-470-99400-9 (H/B)

Typeset in 10/12pt Times by Aptara Inc., New Delhi, India
Printed in Great Britain by CPI Antony Rowe, Chippenham, Wiltshire

Contents

Preface

For a trader or an expert in finance, call him Mr Hyde, it is quite clear that a call or put spread is the derivative of an option and that a butterfly spread is the derivative of a call or put spread. Perhaps, he thinks, it should be approximately so. In fact, he knows that when a client asks for a digital option, he actually approximates that by taking large positions of opposite sign in European options with strikes as close as possible. So, for him a digital payoff is the limit of a call or put spread. He may also imagine what happens to the payoff of the butterfly spread as he increases the size of the positions and moves the strike prices closer and closer. He would get a tall spike with a tiny base, and, by iterating the process to infinity, he would get the *Dirac delta function*. So, gluing all the pieces together, Mr Hyde concludes that it is quite obvious that a Dirac delta function is the derivative of a digital payoff, which he knows is called the *Heaviside unit step function*.

For a mathematician, whose name could be Dr Jekyll, this conclusion is not so obvious, and for sure it is not rigorous. The digital payoff is a singular function, for which the derivative is not defined almost everywhere. In particular, it is not defined when it is most needed – that is, when the payoff jumps from zero to one, which is exactly where all the mass of the other singular function, Dirac's delta, is concentrated. Anyway, after a first sense of natural disgust, Dr Jekyll recalls that there is a special setting in which this holds exactly true, and that is the theory of generalized functions. Then, disgust may leave the way to a sort of admiration for the trader, and will of cooperation. The mathematician proposes that one could actually consider to recover the price in the framework of generalized functions. In this setting, the Fourier transform of the payoff of a digital option is well defined. Working out the convolution of that with the density is not straightforward, but something can be done. One could then retrieve the price of the digital options for general densities, under very weak conditions, and in a totally consistent and, why not, elegant framework.

In this book we arranged a meeting and thorough discussion between Mr Hyde and Dr Jekyll. The idea is to deal with Fourier transform analysis in the framework of generalized functions. To the best of our knowledge, this is the first application of the idea to finance, and it delivers an original viewpoint on the subject, even though it reaches consistent results with the literature on the subject. The book is entirely devoted to the presentation of this idea, and it is not its ambition to provide a comprehensive and complete review of the literature, nor to address all the issues that may arise in the use of Fourier transform analysis in finance. The task is instead to develop the Fourier transform methodology in a setting that, in our judgement, may be the most appropriate for several reasons: not least, because there the intuition of Mr Hyde meets the rigor and elegance of Dr Jekyll.

For this reason, we also chose a non-standard structure for the book, which would have not been appropriate for a textbook or a review monograph. So, just as in many police stories, we decided to start from the murder scene, and then to develop the whole story in a flashback explaining how we got to that. We may reassure the reader that in this case the murder is a happy ending, and does not involve either Dr Jekyll or Mr Hyde, who are both alive and kicking and get along very well.

Chapter 1 collects the main results of the approach, along with frontier issues in the modelling of asset prices consistently with both time series dynamics and option prices. Expert readers are advised to read this chapter first. However, remember that even the authors had to go to the chapters written by the others to find out more. Chapter 2 proposes a review of the stochastic models applied to the dynamics of asset prices within the general assumption of market efficiency: the chapter opens with Bachelier at the beginning of the twentieth century and closes with CGMY at the beginning of the twenty-first. From the chapter, it clearly emerges why the concept of characteristic function has substituted that of density, shedding attention to Fourier transform methods. Chapter 3 extends the analysis to allow for non-stationary returns, introducing additive processes on one side, and time change techniques (based both on stochastic volatility and subordinators) on the other. Chapter 4 addresses the problem of pricing contingent claims in the most general setting, well suited to cases in which the dynamics of prices is represented in terms of characteristic functions. Chapter 5 introduces the theory of generalized functions, and shows how to compute distributions and convolutions of distributions in this setting; the chapter also specifies the setting that allows us to rigorously recover the original results presented in Chapter 1. Chapter 6 simply extends the analysis of the previous chapter to the case of Fourier transforms. Chapter 7 concludes by presenting a sensitivity analysis of option prices and smiles for the most famous models, and a calibration exercise is carried out in the current period of crisis.

That is the story of this book. Since it was born from the discussion between Dr Jekyll and Mr Hyde, the book is naturally targeted to two opposite kinds of audience. Necessarily, some reader will find parts of the book too basic and some will find them too complex, but we hope that in the balance the reader will enjoy going through it and will find an original presentation of the topic. Coming to the conclusions, we would like to thank, without implicating, Prof. Marc Yor for agreeing to read and discuss the first draft of the text. We conclude with warmest thanks to our families for their infinite patience while we were writing this book, and (not necessarily warm) thanks from each author to the other three for their finite patience. And, needless to say, Mr Edward Hyde is thankful to his master, Dr Henry Jekyll.

Bologna, 1 July 2009

U. Cherubini
G. Della Lunga
S. Mulinacci
P. Rossi

List of Symbols

Symbol	Description
c.f.	characteristic function
$N(m, \sigma^2)$	Normal distribution
$B(n,)$	Binomial distribution
Poi(λ)	Poisson distribution
$\Gamma(\alpha, \lambda)$	Gamma distribution
$\mathcal{E}(\lambda)$	Exponential distribution
r	continuously compounded short rate
σ	scalar standard deviation
W_t	Brownian process
p.d.f.	Probability density function
p.d.e.	Partial differential equation
c.d.f.	Cumulative distribution function
SDE	Stochastic differential equation
$\mathbb{P}(x)$	c.d.f (objective measure)
$\mathbb{Q}(x)$	c.d.f (risk-neutral measure)
$B(t, T)$	Price at t of a risk-free coupon bond expiring at T
S_t	Price at t of a risky asset
$\mathcal{O}$	European option (call or put)
$\mathcal{C}$	European call option
$\mathcal{P}$	European put
CoN	Cash-or-Nothing (subscript)
AoN	Asset-or-Nothing (subscript)
$a \wedge b$	MIN(a, b)
$a \vee b$	MAX(a, b)
$\theta(x)$ or $H(x)$,	*Heaviside unit step function*
$\delta(x)$	*Dirac delta* function.
$\mathcal{F}f(x)$	Fourier transform
$\overline{\mathcal{F}}\hat{f}(x)$	Inverse Fourier transform
$\varphi(x)$	*Testing function*
$f \star g$	Convolution
$P\int_{-R}^{+R}(f(x)/x)\,\mathrm{d}x$ or p.v. $f(x)$	Principal value of function $f(x)$

1
Fourier Pricing Methods

1.1 INTRODUCTION

In recent years, Fourier transform methods have emerged as some of the major methodologies for the evaluation of derivative contracts. The main reason has been the need to strike a balance between the extension of existing pricing models beyond the traditional Black and Scholes setting and a parsimonious stance for the evaluation of prices consistently with the market quotes.

On the one hand, the end of the Black–Scholes world spurred more research on new models of the dynamics of asset prices and risk factors, beyond the traditional framework of normally distributed returns. On the other, restricting the search to the set of processes with independent increments pointed to the use of Fourier transform as a natural tool, mainly because it was directly linked to the characteristic functions identifying such processes.

This book is devoted to the use of Fourier transform methods in option pricing. With respect to the rest of the literature on this topic, we propose a new approach, based on generalized functions. The main idea is that the price of the fundamental securities in an economy – that is, digital options and Arrow–Debreu securities – may be represented as the convolution of two generalized functions, one representing the payoff and the other the pricing kernel.

In this chapter we present the main results of the book. The remaining chapters will then lead the reader through a sort of flashback story over the main steps needed to understand the rationale of Fourier transform pricing methods and the tools needed for implementation.

1.2 A GENERAL REPRESENTATION OF OPTION PRICES

The market crash of 19 October 1987 may be taken as the date marking the end of the Black–Scholes era. Even though the debate on evidence that market returns were not normally distributed can be traced back much further in the past, from the end of the 1980s departures from normality have become the usual market environments, and exploiting these departures has even suggested new business ideas for traders. Strategies bound to gain from changes in the skew or higher moments have become the usual tools in every dealing room, and concerns about exposures to changes in volatility and correlation have become a major focus for risk managers.

On the one hand, the need to address the issue of non-Gaussian returns started the quest for new models that could provide a better representation of asset price dynamics; and, on the other, that same need led to the rediscovery of an old idea. According to a model going back to Breeden and Litzenberger (1978), one may recover the risk-neutral probability from the prices of options quoted in the market. Notice that this finding only depends on the requirement to rule out arbitrage opportunities and must hold in full generality for all risk-neutral probability distributions. The idea is that the risk-neutral density can be computed as the second derivative of the price of options with respect to the strike. More precisely, we have that

$$B(t, T)f_{t,T}(K) \equiv B(t, T)\mathbb{Q}_t(S_T \in \mathrm{d}K) = \frac{\partial^2 \mathcal{P}(S_t; K, T)}{\partial K^2}$$

where $\mathcal{P}(S_t; K, T)$ denotes the put option and $B(t, T)$ is the risk-free discount factor – that is, the value at time t of earning a unit of cash for sure at future time T. This is true of all option pricing models. Notice that the no-arbitrage condition immediately leads to characterize $f_{t,T}(x)$ as a density. First, if one assumes to have bought a product paying a unit of cash if $(S_T \in dx)$ and zero otherwise, the price of this product cannot be negative. Second, if one assumes to have bought a set of products paying one unit of cash if $(S_T \in dx)$ in such a way as to cover the all-positive real line $[0, \infty]$, then one must earn one unit of cash for sure, so that we have

$$\int_0^\infty f_{t,T}(x)\,\mathrm{d}x = 1$$

Computing option prices amounts to an evaluation of the integrals of the density above, when it exists. Namely, consider the price of an option paying 1 unit of cash if the value of the underlying asset is lower than K at time T. The price of this option, which is called a digital *cash-or-nothing put* option, is

$$\mathcal{P}_{\mathrm{CoN}} = B(t, T)\int_0^K f_{t,T}(x)\,\mathrm{d}x = B(t, T)\mathbb{Q}_t(S_T \le K)$$

Now consider a similar product delivering one unit of asset S in the event $S_T \le K$. This product is called an *asset-or-nothing put* option. Likewise, its price will be

$$\mathcal{P}_{\mathrm{AoN}} = B(t, T)\int_0^K x f_{t,T}(x)\,\mathrm{d}x = B(t, T)\mathbb{E}_t^Q(S_T \mathbf{1}_{[S_T \le K]})$$

where $\mathbb{E}_t^Q(x)$ denotes the conditional expectation taken under probability measure $\mathbb{Q}$ with respect to the information available at time t. Consider now the portfolio of a short position on an *asset-or-nothing put* and a long position in K *cash-or-nothing put* options, with same strike price K and same maturity T. Then, at time T the value of such a portfolio will be

$$K\mathbf{1}_{[S_T \le K]} - S_T\mathbf{1}_{[S_T \le K]} = \max(K - S_T, 0)$$

which is the payoff of a European put option. The no-arbitrage assumption then requires that the value of the put option at any time $t < T$ should be equal to

$$\mathcal{P}(S_t; K, T) = B(t, T)\left(K\mathbb{Q}_t(S_T \le K) - \mathbb{E}_t^Q(S_T\mathbf{1}_{[S_T \le K]})\right)$$

It is easy to check that the no-arbitrage assumption requires that a digital option paying one unit of cash if, at time T, the underlying asset is worth more than K (*cash-or-nothing call*) must have the same value as that of a long position in the risk-free asset and a short position in a *cash-or-nothing put* option. Namely, we must have

$$\mathcal{C}_{\mathrm{CoN}} = B(t, T) - \mathcal{P}_{\mathrm{CoN}} = B(t, T)(1 - \mathbb{Q}_t(S_T \le K)$$

where $\mathcal{C}_{\mathrm{CoN}}$ denotes the *cash-or-nothing call* option. By the same token, and *asset-or-nothing call* option can be replicated by buying a unit of the underlying asset spot while going short the *asset-or-nothing put*

$$\mathcal{C}_{\text{AoN}} = S_t - B(t,T)\mathbb{E}_t^Q(S_T \mathbf{1}_{[S_T \le K]})$$

Notice that the value of an *asset-or-nothing call* option must also be equal to

$$\mathcal{C}_{\text{AoN}} = B(t,T)\mathbb{E}_t^Q(S_T \mathbf{1}_{[S_T > K]})$$

so that we have

$$\mathcal{C}_{\text{AoN}} + \mathcal{P}_{\text{AoN}} = B(t,T)\mathbb{E}_t^Q(S_T) = S_t$$

This defines the main property of the probability measure $\mathbb{Q}$. Under this measure, the asset S, and every other asset in the economy, is expected to earn the risk-free rate. For this reason, this measure is called *risk-neutral*. Alternatively, if one defines a new variable $Z_t \equiv S_t/B(t,T)$, it is evident that under measure $\mathbb{Q}$ we have

$$Z_t = \mathbb{E}_t^Q(Z_T)$$

and the price of the asset S, and every other asset, turns out to be a martingale when measured using the risk-free asset as the numeraire. For this reason, this measure is also called an *equivalent martingale measure* (EMM), where equivalent means that it gives zero measure to the events that have zero measure under the historical measure, and only to those.

Notice that just as for the put option, the price of a call option can be written as a long position in an *asset-or-nothing call* option and a short position in K *cash-or-nothing call* options. Formally,

$$\mathcal{C}(S_t : K, T) = B(t,T)\mathbb{E}_t^Q(S_T \mathbf{1}_{[S_T > K]}) - KB(t,T)(1 - \mathbb{Q}_t(S_T \le K))$$

Notice that by applying a change of numeraire, namely using S_t, we can rewrite the *asset-or-nothing* option in the form

$$\mathcal{C}_{\text{AoN}} = B(t,T)\mathbb{E}_t^Q(S_T \mathbf{1}_{[S_T > K]}) = S_t \mathbb{Q}_t^*(S_T > K)$$

where $\mathbb{Q}^*$ is a new probability measure. So, European options can be written in full generality as a function of two probability measures, one denoting the price of a *cash-or-nothing* option and the other pricing the *asset-or-nothing* one. For call options we have then

$$\mathcal{C}(S_t : K, T) = S_t(1 - \mathbb{Q}_t^*(S_T \le K)) - KB(t,T)(1 - \mathbb{Q}_t(S_T \le K))$$

and for put options

$$\mathcal{P}(S_t : K, T) = -S_t \mathbb{Q}_t^*(S_T \le K) + KB(t,T)\mathbb{Q}_t(S_T \le K)$$

So, the risk-neutral density completely specifies the price of options for all strikes and maturities.

1.3 THE DYNAMICS OF ASSET PRICES

From the discussion above, pricing derivatives in an arbitrage-free setting amounts to selecting a measure endowed with the martingale property. In a complete market, only one measure is sufficient to fit all prices exactly. This implies that all financial products can be exactly replicated by a dynamic trading strategy (all assets are *attainable*). In incomplete markets, the measure must be chosen according to auxiliary concepts, such as mean-variance optimization or the expected utility framework. Concerning this choice, the current presence of liquid option markets with different strike prices and maturities has added more opportunities to replicate derivative contracts and, at the same time, more information on the shape of the risk-neutral distribution. This has brought about the problem of selection and comparison of the models with the whole set of prices observed on the market – that is, the issue of calibration to market data.

By and large, two main strategies are available. One could try models with a limited number of parameters, but a sufficient number of degrees of freedom to represent the dynamics of assets as consistently as possible with the prices of options. The advantage of this route is that it allows a parsimonious arbitrage-free representation of financial prices and it directly provides dynamic replication strategies for contingent claims. This has to be weighted against the risk of model mis-specification. On the other hand, one could try to give a non-parametric representation of the dynamics, based on portfolios of cash positions and derivative contracts held to maturity. This approach is known as *static replication* and it has the advantage of providing the best possible fit to observed prices. The risk is that some products used for static replication may be illiquid, and their prices inconsistent with the no-arbitrage requirement.

This book is devoted to the first strategy, that is the selection of a convenient fully specified dynamics for the prices of assets. The models reviewed in this book are based on two assumptions that jointly determine what is called the *Efficient Market Hypothesis*. The first is that prices are Markovian, meaning that all information needed to predict future price changes is included in the price currently observed, so that past information cannot produce any improvement in the forecast. The second assumption is that such forecasts are centred around zero, so that price changes are not predictable.

The above framework directly leads to modelling the dynamics of asset prices as processes with *independent increments*. The price, or more precisely the logarithm of it, is assumed to move according to a sequence of shocks such that no shock can be predicted from a previous shock. If one adds that all these shocks have the same distribution – that is, are identically distributed, and finite variance – the standard result, called, the central limit theorem, predicts that these log-changes, when aggregated over a reasonable number of shocks, should be normally distributed, so that the prices should be log-normally distributed. This is the standard model used throughout most of the last century, and named the Black–Scholes model after the famous option pricing formula that is recovered under this assumption.

In the Black–Scholes setting, the logarithm of each asset is then assumed to be driven by a Brownian motion with constant diffusion and drift parameters. Formally, if we denote $X_t \equiv \ln(S_t)$ we have

$$dX_t = \left(r - \frac{1}{2}\sigma^2\right) dt + \sigma \, dW_t$$

where σ is the diffusion parameter, r is the instantaneous risk-free rate of return and W_t is a *Wiener process*. The dynamics of price S is then represented by a *geometric Brownian motion*. Notice that this model predicts that all options traded on the market should be consistent with the same volatility figure σ, for all strike and maturity dates. As discussed before, this prediction is clearly at odds with the empirical evidence gathered from option market prices. In many option markets, prices of *at-the-money* options are consistent with volatility levels different from those implied by *out-of-the-money* and *in-the-money* option prices. Namely, in markets such as foreign exchange and interest rate options, the volatility of both *in* and *out of* the money options is higher than that of *at-the-money* options, producing a phenomenon called the *smile effect*, after the scatter of the relationship between volatility and moneyness that resembles the image of a smiling mouth. In other markets, such as that of equity options, this relationship is instead generally negative, and it is called *skew*, recalling the empirical regularity that volatility tends to increase in low price scenarios. Moreover, volatility also tends to vary across maturities, generating *term structures of volatility* typical of every market.

The quest for a more flexible representation of the asset price dynamics, consistent with smiles and term structures of volatility, has brought us to dropping either of the two assumptions underlying the Black–Scholes framework. The first is that the assets follow a diffusion process, and the second is the stationarity of the increments of log-prices. So, more general models could be constructed allowing for the presence of jumps in asset price dynamics and for changes in the volatility and the probability of such jumps – that is, intensity. If we stick to processes with independent stationary increments, this defines a class of processes called *Lévy processes*. An effective way to describe these processes is to resort to their *characteristic function*. We recall that the characteristic function of a variable X_t is defined as

$$\phi_{X_t}(\lambda) = \mathbb{E}\left(e^{i\lambda X_t}\right)$$

A general result holding for all Lévy processes is that this characteristic function may be written as

$$\phi_{X_t}(\lambda) = e^{-t\psi(\lambda)}$$

where the function $\psi(\lambda)$ is called the *characteristic exponent* of the process. Notice that stationarity of increments implies that the characteristic exponent is multiplied by the time t so that increments of the process over time intervals of the same length have the same characteristic function and the same distribution. A fundamental result is that such a characteristic exponent can be represented in full generality using the so-called *Lévy–Khintchine formula*.

$$\psi(\lambda) = -ia\lambda + \frac{1}{2}\sigma^2\lambda^2 - \int_{-\infty}^{+\infty} \left(e^{i\lambda x} - 1 - i\lambda x \mathbb{I}_{\{|x|\leq 1\}}\right) \nu(dx) \qquad \lambda \in \mathbb{R}$$

Every Lévy process can then be represented by a triplet $\{a, \sigma, \nu\}$, which uniquely defines the characteristic exponent. The first two parameters define the diffusion part of the dynamics, namely drift and diffusion. The last parameter is called the *Lévy measure* and refers to jumps in the process. Loosely speaking, the Lévy measure provides a synthetic representation of the contribution of jumps by the product of the instantaneous probability of such jumps, the intensity, and the probability density function of the dimension of jumps. Intuitively, keeping this measure finite requires that relatively large jumps must have finite intensity, while jumps with infinite intensity must have infinitesimal length. The former kind of jumps are denoted as *finite activity*, while the latter are called *infinite activity* and describe a kind of dynamics similar to that of diffusion processes. For further generalization, positive and negative jumps may also be endowed with different specifications.

Stationarity may be a limit for Lévy processes. As a matter of fact, this would imply that the distribution of log-returns on assets over holding periods of the same length should be the same, while in the market we usually see changes in their distribution: typically, we see periods of very huge movements followed by periods of relative calm, a phenomenon which is known as *clustering* of volatility. An intuitive way of moving beyond stationary increments is to assume that both the volatility of the diffusive part and the intensity of jumps change randomly as time elapses. Even the economic rationale for that goes back to a very old stream of literature of the 1970s. Clark (1973) proposed a model to explain the joint dynamics of trading volume and asset prices using subordinated processes. In the field of probability theory, Monroe (1978) proved that all semi-martingale processes can be represented as Brownian processes evaluated at stochastic times. Heuristically, this means that one can always represent any general process by sampling a Brownian motion at random times. Several *stochastic clocks* may be used to switch from the non-Gaussian process observed at *calendar time* to a Brownian motion. If the stochastic clock is taken to be a continuous process, then the required change of time is its quadratic variation. As an alternative, a stochastic clock can be constructed by any strictly increasing Lévy process: these processes are called *subordinators*. One could also use other variables as proxies for this level of activity of the market. The main idea is in fact to model the process of information arrival to the market: in periods in which the market is hectic and plenty of information flows to the market, business time is moving more quickly, but when the market is illiquid or closed, the pace of time slows down.

In the time change approach, the characteristic function is obtained by a composition of the characteristic exponent of the stochastic clock process and that of the subordinated process. The result follows directly from the assumption that the subordinator is independent of the time-changed process. As an alternative approach, it is possible to remain within the realm of stochastic processes with independent increments by extending the Lévy–Khintchine representation. In this case, the characteristic function becomes

$$\phi_{X_t}(\lambda) = \exp(-\psi_t(\lambda))$$

with characteristic exponent

$$\psi_t(\lambda) = ia_t\lambda - \frac{1}{2}\sigma_t^2\lambda^2 + \int_{-\infty}^{+\infty} \left(e^{i\lambda x} - 1 - i\lambda x\mathbb{I}_{\{|x|\leq 1\}}\right)\nu_t(dx) \qquad \lambda \in \mathbb{R}$$

Notice that, unlike the case of Lévy processes, $\psi_t(\lambda)$ is no longer linear in t. Technical requirements must be imposed on the process governing volatility and the Lévy measure (heuristically, they must not decrease with the time horizon).

1.4 A GENERALIZED FUNCTION APPROACH TO FOURIER PRICING

From what we have seen above, a pricing system can be completely represented by a *pricing kernel*, which is the price of a set of digital options at each time t. We now formally define the payoff of such options, for all maturities $T > t$. We start by denoting $m \equiv (B(t, T)K)/S_t$ the scaled value of the strike price, where the forward price is used as the scaling variable. This is a natural measure of moneyness of the option. Now, define $k \equiv \ln(m)$ as our key variable representing the strike. We omit the subscript t to the strike for ease of convenience, but notice that at time T, $k = \ln(K/S_T)$. Let $X_t = \ln(S_t/B(t, T))$. Then, the *Heaviside function* $\theta(\omega(X_T - X_t - k))$, where $\omega = -1$, defines the event $\{S_T \leq K\}$ and $\omega = 1$ refers to the complementary event. So, in what follows we will refer to the probability measure of the variable $X_T - X_t$, that is, the increment of the process between time t and time T, rather than its level at the terminal date. Anyway, since we are concerned with pricing a set of contingent claims at time t, when X_t is observed, this will only amount to a rescaling by a known constant.

As for the function $\theta(x)$, we recall its formal definition as

$$\theta(x) = \begin{cases} 1 & x > 0 \\ 0 & x < 0 \end{cases}$$

1.4.1 Digital payoffs and the Dirac delta function

In financial terms, the *cash-or-nothing* product can be considered as the limit of a sequence of bull/bear spreads. This limit leads to the derivative of the call option pricing formula with respect to the strike price. It is also easy to check that – in financial terms – just as the digital option is the limit of a sequence of call spreads, the derivative of this option is the limit of a sequence of *butterfly spreads*. In fact, it may be verified by heuristic arguments that the payoff of such a product is a *Dirac delta* function assigning infinite value to the case $S_T = K$ and zero to all other events. Not surprisingly, the price of such a limit product, computed as the expected value under the equivalent martingale measure, is the density, when it exists, of the *pricing kernel*, and it is considered to be the equivalent of Arrow–Debreu prices for asset prices that are continuous variables.

Then, from a financial viewpoint, it is quite natural to consider the *Dirac delta* function as the derivative of the *Heaviside step* function. It is not so from a mathematical viewpoint, unless we introduce the concept of *generalized functions*. Loosely speaking, a generalized function may be defined as a linear functional from an assigned set of functions, called *testing functions* to the set of complex numbers. This set of functions is chosen to be infinitely smooth and with compact support, or with some particular regularity condition on their speed of descent. Formally, if we denote $\varphi(x)$ to be a testing function, a generalized function $f(x)$ is defined through the operator assigning a complex number to the function

$$\langle f, \varphi \rangle \equiv \int_{\mathbf{R}} f(x)\varphi(x)\,\mathrm{d}x$$

Notice that by the main property of the *Dirac delta* function we have that

$$\langle \delta, \varphi \rangle = \varphi(0)$$

Furthermore, by a straightforward application of integration by parts, one may prove that the derivative of the distribution $f(x)$ is

$$\langle f', \varphi \rangle = \int_{\mathbf{R}} f'(x)\varphi(x)\,\mathrm{d}x = -\int_{\mathbf{R}} f(x)\varphi'(x)\,\mathrm{d}x = -\langle f, \varphi' \rangle$$

Now notice what happens if we compute the derivative of the *Heaviside step function* $\theta(x)$. We have

$$\langle \theta', \varphi \rangle = -\langle \theta, \varphi' \rangle = -\int_{\mathbf{R}} \theta(x)\varphi'(x)\,\mathrm{d}x = \phi(0) - \varphi(\infty) = \varphi(0)$$

where we have used bounded support or the rapid descent property of the testing functions. We have then that

$$\langle \theta', \varphi \rangle = \langle \delta, \varphi \rangle$$

and the conjecture based on financial arguments is rigorously proved: in the realm of generalized functions, the derivative of the *Heaviside step function* is actually the *Dirac delta function*.

The strategy followed throughout this book is to remain in the realm of a generalized function to consistently recover the price of options in terms of Fourier transforms.

1.4.2 The Fourier transform of digital payoffs

The starting point of our approach is to recover the Fourier transform of the payoff of digital options. This is clearly not defined if the Fourier transform is applied to functions, but it is well defined in the setting of generalized functions.

For a start, we will denote by $\mathcal{F}$ the Fourier transform operator, and by $\overline{\mathcal{F}}$ its inverse, and write

$$\hat{f} = \mathcal{F} f, \quad f = \overline{\mathcal{F}} \hat{f}$$

following the convention:

$$\mathcal{F} f(v) \equiv \int \mathrm{d}u \, \mathrm{e}^{i2\pi uv} f(u)$$
$$\overline{\mathcal{F}} g(v) \equiv \int \mathrm{d}u \, \mathrm{e}^{-i2\pi uv} g(u)$$

We report here the main result concerning the Fourier transform of the digital option that is fully developed and explained in Chapter 5. Let us introduce

$$\delta^+(x) \equiv \frac{i}{2\pi} g^+(x)$$

where

$$g^+(x) = \lim_{\epsilon \to 0^+} \frac{1}{x + i\epsilon}$$

We are now going to show that $\overline{\mathcal{F}}[\delta^+] = \theta$, from which $\mathcal{F}[\theta] = \delta^+$. Since

$$\langle \overline{\mathcal{F}}[\delta^+], \varphi \rangle = \langle \delta^+, \overline{\mathcal{F}}[\varphi] \rangle$$

$$\begin{aligned}
\langle \delta^+, \overline{\mathcal{F}}[\varphi] \rangle &= \frac{i}{2\pi} \lim_{\epsilon \to 0^+} \int \mathrm{d}x \, \mathrm{d}\lambda \frac{\varphi(\lambda)}{x + i\epsilon} \mathrm{e}^{-2\pi i \lambda x} \\
&= \frac{i}{2\pi} \lim_{\epsilon \to 0^+} \int_{-\infty}^{0} \mathrm{d}\lambda \, \varphi(\lambda) \int \mathrm{d}x \frac{1}{x + i\epsilon} \mathrm{e}^{2\pi i |\lambda| x} \\
&\quad + \frac{i}{2\pi} \lim_{\epsilon \to 0^+} \int_{0}^{+\infty} \mathrm{d}\lambda \, \varphi(\lambda) \int \mathrm{d}x \frac{1}{x + i\epsilon} \mathrm{e}^{-2\pi i |\lambda| x} \\
&= \lim_{\epsilon \to 0^+} \int_0^{\infty} \mathrm{d}\lambda \, \varphi(\lambda) \mathrm{e}^{-2\pi \epsilon \lambda} = \int_0^{\infty} \mathrm{d}\lambda \, \varphi(\lambda)
\end{aligned}$$

it follows that:

$$\overline{\mathcal{F}}[\delta^+] = \theta$$

Now, it is possible to compute that the distributional value of $g^+(x)$ is p.v. $1/x - i\pi\delta(x)$ (see Example 5.4.3), so that we conclude

$$\begin{aligned}
\mathcal{F}[\theta](v) = \delta^+(v) &= \frac{i}{2\pi} \left(\text{p.v.} \frac{1}{v} - i\pi\delta(v) \right) \\
&= \frac{1}{2} \delta(v) + \frac{i}{2\pi} \text{p.v.} \left(\frac{1}{v} \right)
\end{aligned}$$

where p.v. denotes the principal value and δ is the *Dirac delta* function.

1.4.3 The cash-or-nothing option

We are now going to recover the price of digital *cash-or-nothing* options. We shall treat both the probability distribution $\mathbb{Q}$ and the payoff as generalized functions, and the pricing formula as a convolution of distributions. In this setting, we have already computed the Fourier transform of the payoff. As for the distribution, we assume that we only know its *characteristic function*, which we redefine in a slightly different way, which is useful for computational purposes:

$$\phi_X(v) \equiv \mathbb{E}\left[e^{i2\pi v X_T}\right] = \int \mathbb{Q}(du)\, e^{i2\pi vu} = \int \mathcal{F}\, d\mathbb{Q} \tag{1.1}$$

Notice that with respect to the usual definition we have simply multiplied the exponent by 2π.

The maths concerning these assumptions is thoroughly discussed in the main body of this book, namely Chapters 5 and 6, so here we stick to essential definitions for the reader who is already familiar with the technique.

Let 'f', and 'g' be two generalized functions. The convolution will be denoted as:

$$f \star g \equiv \int du\, f(u)g(y-u)$$

If $\mathbb{Q}$ is a (probability) measure, we shall write:

$$(\mathbb{Q} \star g)(y) \equiv \int \mathbb{Q}(du)g(y-u)$$

We are interested in the convolution, in a generalized function sense, of the density and the digital payoff function $\theta(x)$.

$$\mathbb{Q}(k) = \mathbb{Q} \star \theta(k) \equiv \int \mathbb{Q}(du)\theta(k-u) \tag{1.2}$$

Notice that the main pillar of our approach is the requirement that this convolution of generalized functions be well defined. In Chapter 5, section 8, we give a proof under very weak conditions, which amount to the existence of the first moment of the probability distribution. We now apply the Fourier transform to the convolution and obtain:

$$f \star g = \overline{\mathcal{F}}[(\mathcal{F}f)(\mathcal{F}g)] \tag{1.3}$$

and

$$\langle \mathcal{F}f, \phi \rangle = \langle f, \mathcal{F}\phi \rangle$$

We now use equation (1.3) to compute (1.2):

$$\mathbb{Q}(k) = \int du\, e^{-2\pi iku} \phi_X(u)\delta^+(u) \tag{1.4}$$

Replacing the value for δ^+ in equation (1.4) and applying a result that may be found in Chapter 5, Example 5.4.2, we end up with

$$\mathbb{Q}(k) = \frac{1}{2} + \frac{i}{2\pi} \int \frac{du}{u} \left[\phi_X(u)\, e^{-2\pi iuk} - 1\right] \tag{1.5}$$

The above formula is certainly not new (see, for example, Kendall and Stuart, 1977, vol. III). It provides the relationship between the characteristic function and the cumulative probability distribution, which in our case is the pricing kernel of the economy.

The value of a *cash-or-nothing put* option is then given by

$$\mathcal{P}_{\mathrm{CoN}}(k) = B(t,T)\left\{\frac{1}{2} + \frac{i}{2\pi}\int \frac{\mathrm{d}u}{u}\left[\phi_X(u)\,\mathrm{e}^{-2\pi iuk} - 1\right]\right\} \tag{1.6}$$

It is now immediate to obtain the price of the corresponding cash-or-nothing call option. Namely, we have

$$\begin{aligned} 1 - \mathbb{Q}(k) &\equiv 1 - \int \mathbb{Q}(\mathrm{d}u)\theta(k-u) \\ &= 1 - \left\{\frac{1}{2} + \frac{i}{2\pi}\int \frac{\mathrm{d}u}{u}\left[\phi_X(u)\,\mathrm{e}^{-2\pi iuk} - 1\right]\right\} \end{aligned} \tag{1.7}$$

and we immediately obtain

$$\mathcal{C}_{\mathrm{CoN}} = B(t,T)\left\{\frac{1}{2} - \frac{i}{2\pi}\int \frac{\mathrm{d}u}{u}\left[\phi_X(u)\,\mathrm{e}^{-2\pi iuk} - 1\right]\right\} \tag{1.8}$$

1.4.4 The asset-or-nothing option

We now extend the analysis to *asset-or-nothing* options. The whole analysis above would of course lead to a result analogous to that obtained for *cash-or-nothing* options. As a matter of fact, we saw before that the two prices are linked by a change of measure. Namely,

$$B(t,T)\mathbb{E}(S_T \mathbf{1}_{[S_T \le K]}) = S_t Q^*(k)$$

Under our notation, which is based on the forward price rescaled with respect to the price at time t (that is, $S_t = 1$), the Radon–Nikodym derivative linking the two measures is S_T, so that we may write

$$\int \mathbb{Q}^*(\mathrm{d}u) = 1, \quad \mathbb{Q}^*(\mathrm{d}u) = \mathbb{Q}(\mathrm{d}u)\,\mathrm{e}^u \tag{1.9}$$

We may now denote the characteristic function of measure $\mathbb{Q}^*$ as

$$\phi_X^*(k) \equiv \int \mathbb{Q}^*(\mathrm{d}u)\,\mathrm{e}^{2\pi iku} \tag{1.10}$$

and a straightforward computation gives the relationship between the characteristic function of measure $\mathbb{Q}^*$ and that of measure $\mathbb{Q}$:

$$\phi_X^*(k) = \int \mathbb{Q}(\mathrm{d}x)\exp\left[2\pi i\left(k - \frac{i}{2\pi}\right)x\right] = \phi_X\left(k - \frac{i}{2\pi}\right) \tag{1.11}$$

Asset-or-nothing options may then be computed using the same formalism as *cash-or-nothing* options. Namely, we have

$$\mathcal{C}_{\mathrm{AoN}} = S_t\left\{\frac{1}{2} - \frac{i}{2\pi}\int \frac{\mathrm{d}u}{u}\left[\phi_X\left(u - \frac{i}{2\pi}\right)\mathrm{e}^{-2\pi iuk} - 1\right]\right\} \tag{1.12}$$

for call options and

$$\mathcal{P}_{\text{AoN}} = S_t \left[\frac{1}{2} + \frac{i}{2\pi} \int \frac{\mathrm{d}u}{u} \left\{ \phi_X \left(u - \frac{i}{2\pi} \right) \mathrm{e}^{-2\pi iuk} - 1 \right] \right\} \tag{1.13}$$

for put options.

1.4.5 European options: the general pricing formula

It is now possible to derive a general pricing formula for European options that will be used to calibrate pricing models to market data. Notice that all the information content concerning the dynamics of the risk factor S, the underlying asset of our options, is summarized in the function

$$\mathrm{d}(k, \alpha) \equiv \int \frac{\mathrm{d}u}{u} \left(\mathrm{e}^{-2\pi iuk} \phi_X (u - \alpha) - 1 \right) \tag{1.14}$$

We call this function the **characteristic integral** of asset S. The probability distribution used in the pricing of all *cash-or-nothing* and *asset-or-nothing* options for all maturities can be synthetically reported with the common notation:

$$D(k, \alpha, \omega) = \frac{1}{2} - \omega \frac{i}{2\pi} \mathrm{d}(k, \alpha) \tag{1.15}$$

Clearly the *cash-or-nothing* case corresponds to $\alpha = 0$ while the *asset-or-nothing* case is covered by $\alpha = i/2\pi$. Furthermore, as stated before, $\omega = 1$ denotes call options, while $\omega = -1$ denotes put.

Adopting this notation for European options, the prices for call or put can be written as:

$$\mathcal{O}(S_t; K, T, \omega) = \omega \left[S_t D \left(k, \frac{i}{2\pi}, \omega \right) - B(t, T) K\, D(k, 0, \omega) \right] \tag{1.16}$$

which only depends on the **characteristic integral**. In order to highlight that, the European option pricing formula can be rewritten as

$$\mathcal{O}(S_t; K, T, \omega) = \frac{1}{2} \omega S_t (1 - m) + S_t \frac{i}{2\pi} \left[\mathrm{d}(k, 0) m - \mathrm{d} \left(k, \frac{i}{2\pi} \right) \right] \tag{1.17}$$

where we recall that $m \equiv B(t, T)K/S_t$ denotes *moneyness* (in the forward price sense). Notice that the **characteristic integral** enters the formula with the same sign for both call and put options. The shape of the smile could then be recovered by using the statistics

$$\frac{\mathcal{C}(m) + \mathcal{P}(m)}{S_t} = \frac{i}{\pi} \left[\mathrm{d}(k, 0) m - \mathrm{d} \left(k, \frac{i}{2\pi} \right) \right] \tag{1.18}$$

where $\mathcal{C}$ and $\mathcal{P}$ denote call and put options as usual.

Finally, notice that for the *at-the-money forward* option ($m = 1$) we have

$$\mathcal{O}_{\text{AtM}}(S, t; K, T) = S_t \frac{i}{2\pi} \left[d(0, 0) - \mathrm{d} \left(0, \frac{i}{2\pi} \right) \right] \tag{1.19}$$

which may be useful to calibrate the term structure of volatility around the most liquid option quotes.

With this general structure we are then ready not only to price options but also to use option prices to back out in a synthetic way all relevant information concerning the dynamics of the underlying assets.

1.5 HILBERT TRANSFORM

We are now going to show that the **characteristic integral** defined above can be represented in an alternative way, resorting to what is known as the *Hilbert transform*. This technique was recently applied to the option pricing problem by Feng and Linetsky (2008).

The Hilbert transform $\mathcal{H}f$ of a function f is obtained by performing the convolution of the function with the distribution p.v.$1/x$, in formula:

$$[\mathcal{H}f](y) = \frac{1}{\pi}\int \mathrm{d}x\, f(x)\,\text{p.v.}\left[\frac{1}{y-x}\right]$$

If we call $h(x)$ the tempered distribution:

$$h(x) = \text{p.v.}\frac{1}{\pi x}$$

we may define the Hilbert transform by the alternative notation:

$$\mathcal{H}f = h \star f$$

We can immediately see that the characteristic integral defined above, and yielding the prices of options, can be written in terms of the Hilbert transform

$$\begin{aligned}
\mathbb{Q}(k) &= \int \mathrm{d}u\, \mathrm{e}^{-2\pi iku}\phi_X(u-\alpha)\delta^+(u) \\
&= \int \mathrm{d}u\, \mathrm{e}^{-2\pi iku}\phi_X(u-\alpha)\left(\frac{1}{2}\delta(u) + \frac{i}{2\pi}\text{p.v.}\left(\frac{1}{u}\right)\right) \\
&= \frac{1}{2} + \frac{i}{2\pi}\int \mathrm{d}u\, \mathrm{e}^{-2\pi iku}\phi_X(u-\alpha)\text{p.v.}\left(\frac{1}{u}\right) \\
&= \frac{1}{2} + \frac{1}{2i}[\mathcal{H}f_k](0), \quad \text{where } f_k : u \to \mathrm{e}^{-i2\pi ku}\phi_X(u-\alpha).
\end{aligned} \tag{1.20}$$

In order to compute Hilbert transforms of the quantities in which we are interested in the development of this chapter, we anticipate some relations that will be presented in Chapter 5:

$$\text{p.v.}\left(\frac{1}{x}\right) = \frac{1}{x-i\epsilon} - i\pi\delta(x) = \frac{1}{x+i\epsilon} + i\pi\delta(x)$$

We then get:

$$[\mathcal{H}f](y) = \frac{1}{\pi}\int \mathrm{d}x\, \frac{f(x)}{y-x-i\epsilon} - if(y)$$

as a general rule to compute the Hilbert transform.

Adopting the usual *hat* notation for the Fourier transform we can write:

$$[\mathcal{H}f](y) = \overline{\mathcal{F}}\left(\hat{h}\hat{f}\right) \tag{1.21}$$

where

$$\hat{h} := \mathcal{F}\left(\text{p.v.}\frac{1}{u}\right)$$

A result that will be needed in the development is the Fourier transform of p.v.$(1/u)$.

Example 1.5.1 *From the definition of $\hat{h}$ we get:*

$$\begin{aligned}
\hat{h}(k) &= \frac{1}{\pi}\int \mathrm{d}x\, \frac{\mathrm{e}^{i2\pi kx}}{x - i\epsilon} - i \\
&= 2i\theta(k) - i \\
&= i\,\mathrm{sign}(k)
\end{aligned}$$

where the "signum" function "sign" is defined by:

$$\mathrm{sign}(x) = \begin{cases} 1 & x > 0 \\ -1 & x < 0 \end{cases}$$

We now provide a set of examples that should (a) illustrate how to compute the Hilbert transform of functions and (b) lead to a formula that will be paramount in the development of the numerical implementation.

Example 1.5.2 *Consider the function $\mathrm{e}_\beta : x \to \mathrm{e}^{i2\pi\beta x}$, following the definition we have:*

$$\begin{aligned}
[\mathcal{H}\mathrm{e}_\beta](y) &= -i\mathrm{e}^{i2\pi\beta y} + \frac{1}{\pi}\int \mathrm{d}x\, \frac{\mathrm{e}^{i2\pi\beta x}}{y - x - i\epsilon} \\
&= -i\mathrm{e}^{i2\pi\beta y} + 2i\,\mathrm{e}^{i2\pi\beta y}\theta(-\beta) \\
&= -i\mathrm{e}^{i2\pi\beta y}\mathrm{sign}(\beta)
\end{aligned}$$

There is also a second method available (as in most cases) to get to the result. The method exploits equation (1.21). *We observe that*

$$\hat{h}(u) = i\,\mathrm{sign}(u), \quad \hat{e}_\beta(u) = \delta(u + \beta)$$

therefore:

$$\begin{aligned}
[\mathcal{H}e_\beta](y) &= \overline{\mathcal{F}}\left(\hat{h}\hat{e}_\beta\right) \\
&= i\int \mathrm{d}u\, e^{-i2\pi uy}\mathrm{sign}(u)\delta(u+\beta) \\
&= -i\mathrm{e}^{i2\pi\beta y}\,\mathrm{sign}(\beta)
\end{aligned}$$

We can exploit the linearity of the Hilbert transform and the result in the example above to recover the transform of trigonometric functions.

Example 1.5.3 *Let $s_m : x \to \sin(mx)$. If we set $\mu = m/2\pi$, from the definition we get:*

$$\begin{aligned}
[\mathcal{H}s_m](y) &= \frac{1}{2i}[\mathcal{H}e_\mu](y) - \frac{1}{2i}[\mathcal{H}e_{-\mu}](y) \\
&= -\frac{1}{2}\,\mathrm{e}^{imy}\,\mathrm{sign}(m) + \frac{1}{2}\,\mathrm{e}^{-imy}\,\mathrm{sign}(-m) \\
&= -\cos(my)\,\mathrm{sign}(m)
\end{aligned}$$

We are now ready to compute the Hilbert transform of a function that will be crucial in the numerical applications below.

Example 1.5.4 *Let's consider the function:*

$$\mathrm{sinc}_m : x \to \frac{\sin(mx)}{x}$$

then:

$$[\mathcal{H}\mathrm{sinc}_m](y) = -i\frac{\sin(my)}{y} - \frac{1}{\pi}\int \mathrm{d}x \, \frac{\sin(mx)}{x[x-(y-i\epsilon)]}$$

Exploiting the relation:

$$\frac{1}{x}\frac{1}{x-(y-i\epsilon)} = \frac{1}{y}\left[\frac{1}{x-(y-i\epsilon)} - \frac{1}{x}\right]$$

we have:

$$[\mathcal{H}\mathrm{sinc}_m](y) = -i\frac{\sin(my)}{y} - \frac{1}{\pi y}\int \mathrm{d}x \, \sin(mx)\left[\frac{1}{x-(y-i\epsilon)} - \frac{1}{x}\right]$$

The integral

$$\int \mathrm{d}x \, \frac{\sin(mx)}{x}$$

is finite, so among the different ways to compute it, one particularly convenient for us is to replace it with its principal value (since it is finite, its value must coincide with its principal value). With this understanding we get:

$$\begin{aligned}[\mathcal{H}\mathrm{sinc}_m](y) &= \frac{1}{y}[\mathcal{H}\sin](y) - \frac{1}{y}[\mathcal{H}\sin](0) \\ &= \mathrm{sign}(m)\frac{1-\cos(my)}{y}\end{aligned}$$

1.6 PRICING VIA FFT

We are now going to address the numerical issues involved in the application of Fourier pricing methods to market data.

It is quite clear that all the numerical work needed to compute prices for vanilla options consists in performing the *characteristic integral* as defined in equation (1.14). For later convenience we will introduce now a change in notation:

$$\mathrm{d}(k,\alpha) = \int_{-\infty}^{+\infty} \mathrm{d}u \, \frac{f(u,k,\alpha)-1}{u} \tag{1.22}$$

or equivalently, in terms of the Hilbert transform,

$$\mathrm{d}(k,\alpha) = \int_{-\infty}^{+\infty} \mathrm{d}u \; f(u,k,\alpha)\,\mathrm{p.v.}\left(\frac{1}{u}\right) \tag{1.23}$$

where

$$f(u, k, \alpha) = e^{-2\pi iuk}\phi_X(u - \alpha) \tag{1.24}$$

As we shall see in the following sections, there are powerful numerical methods to compute, with great accuracy, the Hilbert transform of a characteristic function.

Having devised a method to compute the integral, the problem is how to compute many of such integrals in a run. This problem is particularly relevant in finance. In fact, nowadays we have plenty of information concerning not only the historical dynamics of market data, but also the *forward-looking* dynamics of the distribution implied by market data, even though the two sources of information are referred to different probability measures and so are not directly comparable. There are many instances, both in time series and cross-section analysis, in which it is required to compute many prices by inversion of the Fourier transform. In this case, a well known technique, called *Fast Fourier Transform* (FFT), is typically applied. At the end of the section we will address how to cast the computation of the characteristic integral in a FFT setting.

1.6.1 The sampling theorem

We start now to develop the theory concerning the numerical integration of the *characteristic integral*.

We recall the fundamental relation between a function $p(x)$ and its Fourier transform.

$$p(x) = \int_{-\infty}^{+\infty} du\ \hat{p}(u)\, e^{-i2\pi ux}, \qquad \hat{p}(u) = \int_{-\infty}^{+\infty} dx\ p(x)\, e^{i2\pi ux}$$

In the language of the previous section, and later as well, $p(x)$ is the p.d.f. of some stochastic process at a given time t and $\hat{p}(u)$ is its characteristic function. More precisely, the variable x would represent the log return of the process. A first step to be taken while landing from the realm of theory to applications is that for any practical purpose we are required to restrict the support of this variable, which is typically taken to be unbounded, to a bounded subset. This justifies the change of notation from $\phi_X(x)$, the characteristic function of the process, to $\hat{p}(x)$ as a characteristic function of the density defined on a bounded support. We want to use the latter as an approximation for the former, so that outside the support of $p(x)$ the value of the true probability distribution function is so close to zero that it can be considered zero for any practical purpose. In other words, we are saying that there exists a value X_c such that:

$$p(x) < \epsilon, \qquad |x| > X_c$$

So, if the value of the asset is normalized with respect to its price today, even a modest value of X_c such as 4 means that we give a negligible probability to moves beyond 140% or below 60%, and this may be large enough, particularly if we are not looking at extremely long time intervals and at times of normal volatility.

From this point on, we then substitute $\phi_X(x)$ with a function $p(x)$ such that $p(x) = 0$ for $|x| > X_c$. Therefore the characteristic function is given by:

$$\hat{p}(u) = \int_{-X_c}^{+X_c} dx\ p(x)\, e^{i2\pi ux} \tag{1.25}$$

Example 1.6.1 *To gain some insight into what happens when the p.d.f. is (nearly) zero outside its bounded domain, as described above, we make things very simple and assume that*

$$p(x) = 1, \quad |x| < X_c, \quad p(x) = 0, \quad |x| > X_c$$

Despite its simplicity, it will turn out that this example is extremely useful, so the reader is well advised to work through it until a good grasp is achieved. Performing the simple integral we obtain:

$$\hat{p}(u) := \int_{-X_c}^{+X_c} \mathrm{d}x \, \mathrm{e}^{i2\pi ux} = 2X_c \, \mathrm{sinc}(2\pi X_c u)$$

where we have adopted the definition for the "sinc" functions as:

$$\mathrm{sinc}(x) = \frac{\sin(x)}{x}$$

Let us now define

$$\Delta := \frac{1}{2X_c}, \qquad u_n := n\Delta, \qquad \hat{p}_n := \hat{p}(u_n)$$

then we can compute the l.h.s. in equation (1.25) only at the (sampling) values u_n

$$\hat{p}_n = \int_{-X_c}^{+X_c} \mathrm{d}x \; p(x) \, \mathrm{e}^{i2\pi nx\Delta}$$

From the theory of Fourier series, we know that we can use the sequence $\{\hat{p}_n\}$ to get back the function $p(x)$, and the inversion formula is given by:

$$p(x) = \frac{1}{2X_c} \sum_{n=-\infty}^{+\infty} \hat{p}_n \, \mathrm{e}^{-i2\pi nx\Delta}, \qquad |x| < X_c$$

We can get rid of the explicit constraint $|x| < X_c$ resorting to the indicator function and write:

$$p(x) = \frac{1}{2X_c} \mathbf{1}_{[|x|<X_c]} \sum_{n=-\infty}^{+\infty} \hat{p}_n \, \mathrm{e}^{-i2\pi \Delta nx}$$

The original function $\hat{p}(u)$ can be recovered by applying Fourier transform to $p(x)$:

$$\hat{p}(u) = \frac{1}{2X_c} \sum_{n=-\infty}^{+\infty} \hat{p}_n \int_{-\infty}^{+\infty} \mathrm{d}x \, \mathbf{1}_{[|x|<X_c]} \, \mathrm{e}^{i2\pi x(u-n\Delta)}$$

The remaining integral

$$\int_{-\infty}^{+\infty} \mathrm{d}x \, \mathbf{1}_{[|x|<X_c]} \, \mathrm{e}^{i2\pi x(u-n\Delta)}$$

is nothing but the integral performed in Example 1.6.1 and the result is:

$$\frac{\sin[2\pi X_c(u - n\Delta)]}{\pi(u - n\Delta)}$$

Then, we can conclude that the whole Fourier spectrum of the function $p(x)$ with bounded domain is given by:

$$\hat{p}(u) = \frac{1}{2X_c} \sum_{n=-\infty}^{+\infty} \hat{p}_n \frac{\sin[2\pi X_c(u - n\Delta)]}{\pi(u - n\Delta)} \tag{1.26}$$

This remarkable result, also known as the sampling theorem, shows that the Fourier transform $\hat{p}(u)$ of a function with bounded domain can be fully known provided it is known at discrete sampling points.

1.6.2 The truncated sampling theorem

Numerical approximations will be introduced, replacing the infinite sum with a finite one.

$$\hat{p}^N(u) = \frac{1}{2X_c} \sum_{n=-N}^{+N} \hat{p}_n \frac{\sin[2\pi X_c(u - n\Delta)]}{\pi(u - n\Delta)} \tag{1.27}$$

We will discuss the type of error introduced by this truncation in the next section, after presenting the final result for the computation of the *characteristic integral*. Presently we limit ourselves to a numerical verification of the accuracy of the truncated sampling theorem. The whole foundation of the approach adopted in this book is that, for many interesting models, the characteristic function is easy to compute. Accordingly, we know the exact form of the l.h.s. of equation (1.26) and we are in a position to check the accuracy of the approximation produced by the r.h.s. of equation (1.27) when we select different values for the bound X_c and different values for N.

The measure that we propose to represent the error in the representation of characteristic function consists in looking at the quantity:

$$\mathrm{d}_N(X_c) := \sqrt{\sum_{i=0}^{n} |\phi_X(-x_{\min} + si) - \hat{p}_N(-x_{\min} + si)|^2}$$

where n is the number of points in which the distance between the two functions is computed, $x_{\min}$ is the lowest value of x where the comparison is made, and s is the increment of x from one point to the latter. This distance is computed for fixed X_c as a function of N and for fixed N as a function of X_c. In the former case, provided we select X_c large enough, this will give us insight on the number of Fourier modes needed to achieve the desired accuracy. In the latter case, provided we take N large enough, we may gauge the values of X_c for which the p.d.f. can be considered negligibly small when $|x| > X_c$.

As an example, in Figure 1.1 we look at a simple diffusion model with $\sigma = 0.4423$, $X_c = 4.0$. We see that we reach machine precision with as little as 60 Fourier modes, while in Figure 1.2 we look at the same model but keep fixed the number of Fourier modes at $N = 64$. We see that we can consider negligible the p.d.f. for values $|x| > 4.0$.

In Figures 1.3 and 1.4 we present the same model but with $\sigma = 0.1$. Smaller volatility means a narrower distribution, so we do expect to be able to use a much lower cutoff X_c. As we can see, in fact we reach machine precision for $N \geq 50$ and $X_c < 1.0$.

The reader is warmly invited to run the same test for the case $\sigma = 0.4423$, keeping the spatial cutoff at $X_c = 1.0$. It should not come as a surprise that no amount of Fourier modes will be able to reduce the error to acceptable values.

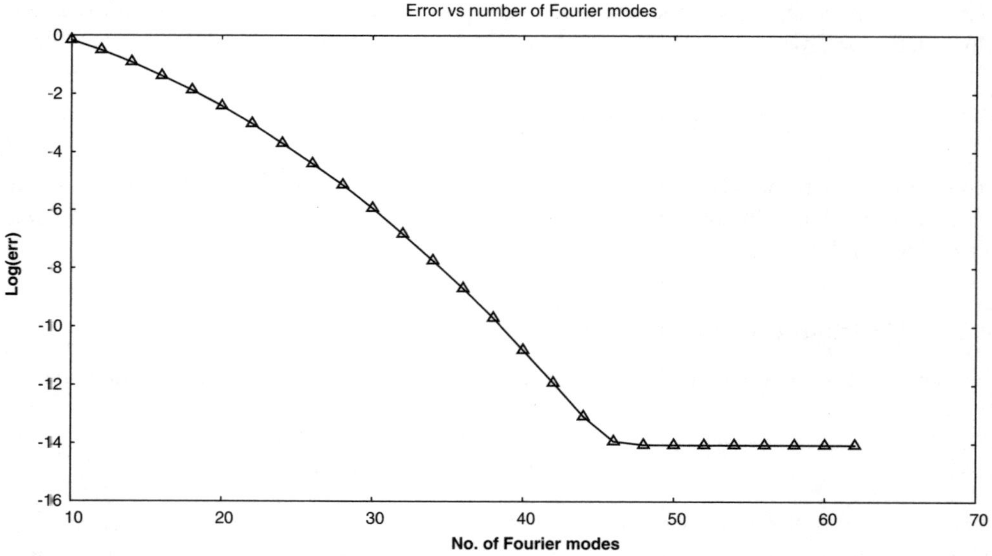

Figure 1.1 The dependency of the error on the number of Fourier modes for the truncated sampling theorem. The model used is simple diffusion with $\sigma = 0.4423,\ T = 1$ year, the spatial cutoff is $X_c = 4.0$

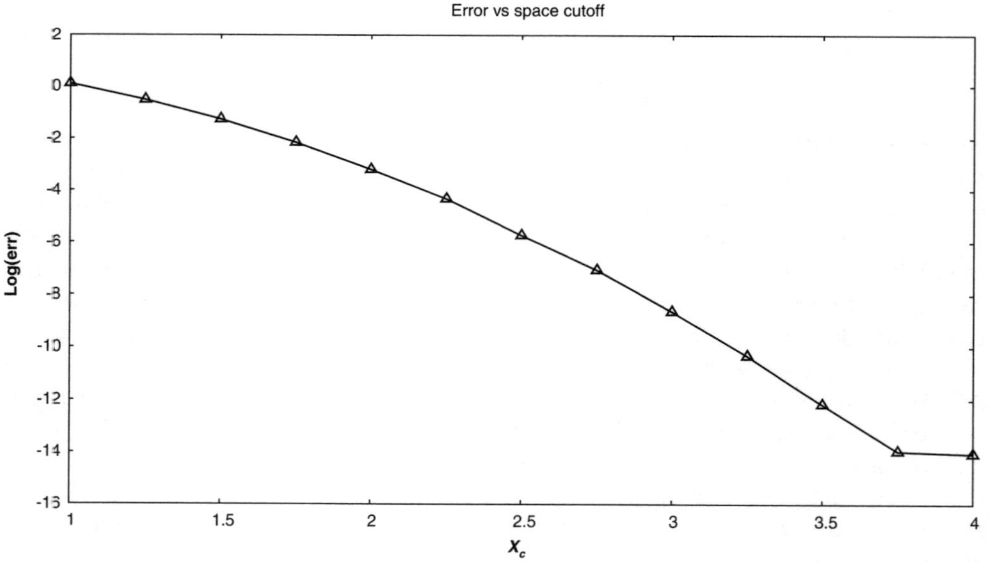

Figure 1.2 The dependency of the error on the spatial cutoff for the truncated sampling theorem. The model used is simple diffusion with $\sigma = 0.4423,\ T = 1$ year. The number of Fourier modes used is $N = 64$

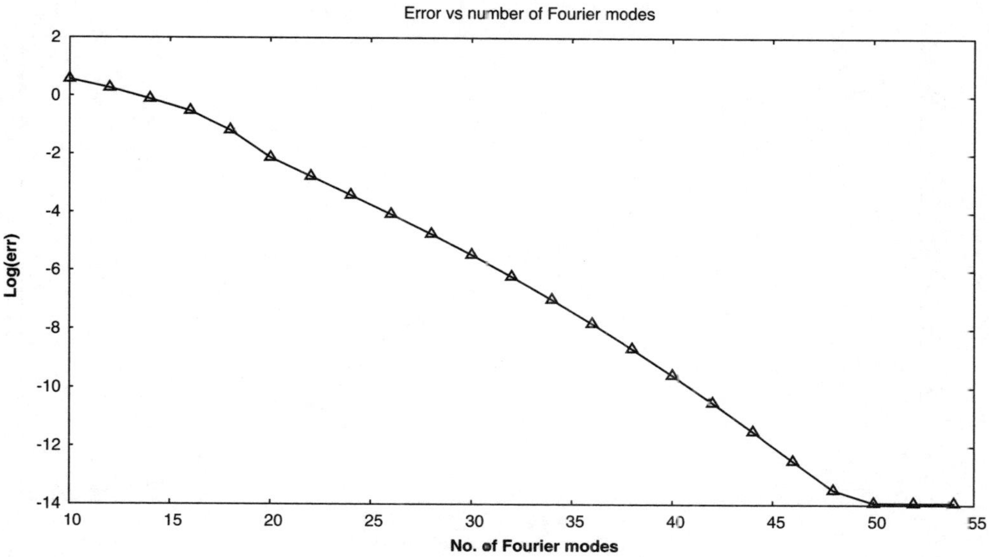

Figure 1.3 The dependency of the error on the number of Fourier modes for the truncated sampling theorem. The model used is simple diffusion with $\sigma = 0.1$, $T = 1$ year, the spatial cutoff is $X_c = 1.0$

As a final example we run the same test on the Heston model. The results are presented in Figures 1.5 and 1.6. The parameters of the model are detailed in the figure captions. Also in this case we see that with a judicious choice of the spacial cutoff we can reach machine accuracy.

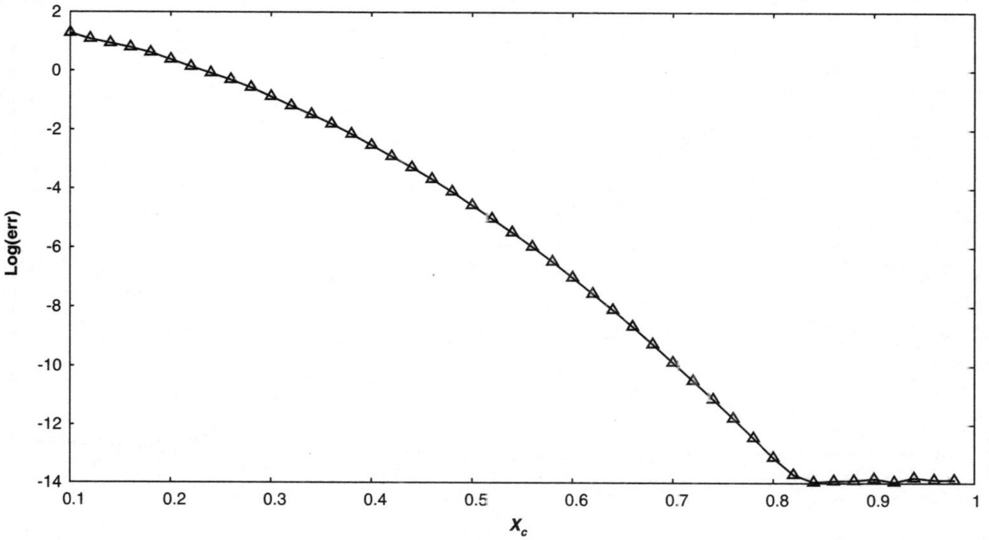

Figure 1.4 The dependency of the error on the spatial cutoff for the truncated sampling theorem. The model used is simple diffusion with $\sigma = 0.1$, $T = 1$ year. The number of Fourier modes used is $N = 56$

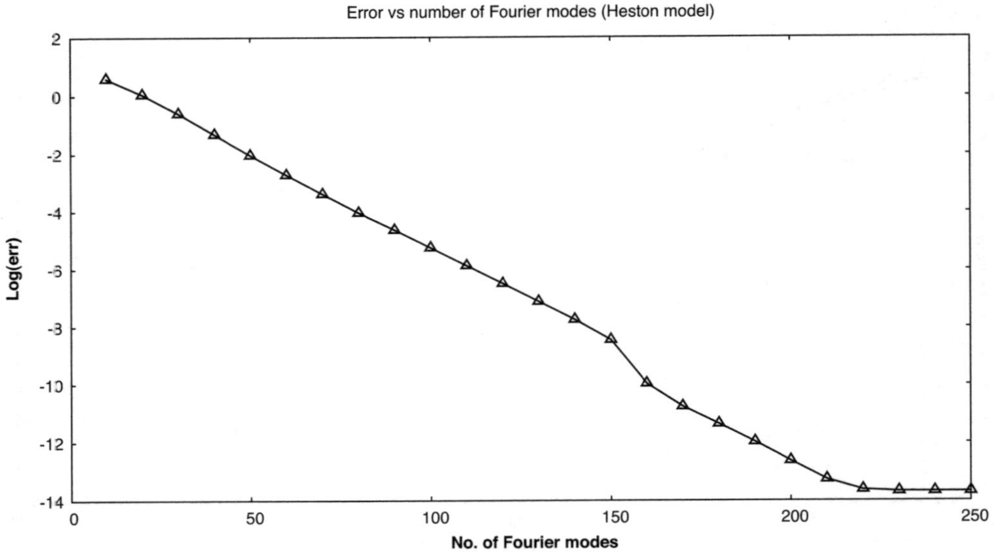

Figure 1.5 The dependency of the error on the number of Fourier modes for the truncated sampling theorem. The model used is the Heston model with $\eta = 0.256$, $\lambda = 1.481$, $\nu_0 = 0.2104$, $\overline{\nu} = 0.1575$, $\rho = -0.8941$, $T = 1$ year. The spatial cutoff is $X_c = 8.0$

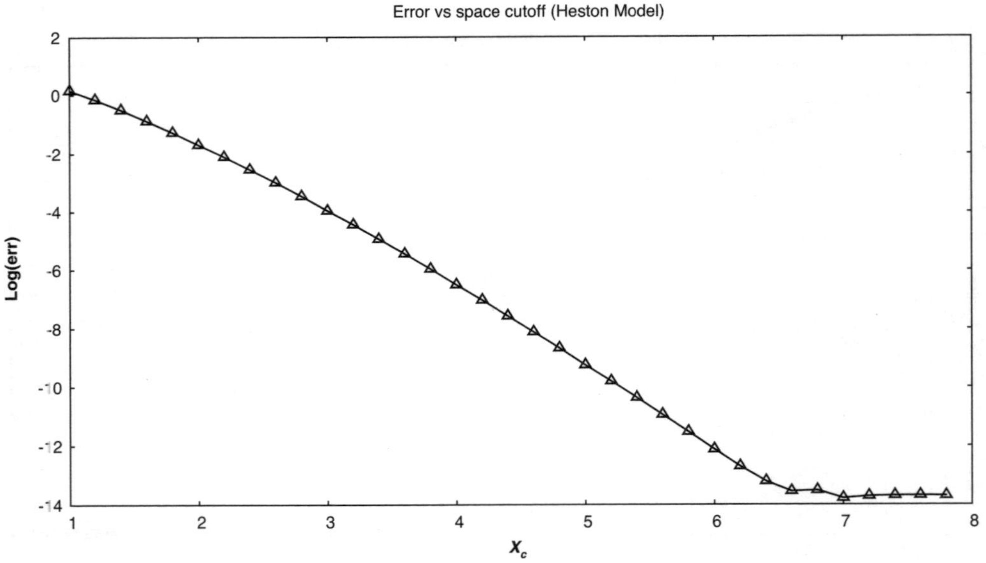

Figure 1.6 The dependency of the error on the spatial cutoff for the truncated sampling theorem. The model used is the Heston model with $\eta = 0.256$, $\lambda = 1.481$, $\nu_0 = 0.2104$, $\overline{\nu} = 0.1575$, $\rho = -0.8941$, $T = 1$ year. The number of Fourier modes used is $N = 256$

1.6.3 Why bother?

The wily reader might have nursed a cunning question. If we can get "exactly" the characteristic function of the process under examination, why bother to compute it via its representation given by the sampling theorem? The answer rests on the fact that, for pricing purposes, we need to compute the convolution of the characteristic function with the distribution p.v.$1/u$. In general this cannot be computed in closed form for the characteristic function of most models. A naive numerical integration of that convolution would prove to be highly delicate due to the oscillatory nature of the characteristic function itself, and is emphatically *something that should not be done*.

On the contrary, the sampling theorem gives us a nice and exact representation in term of the "sinc" function, and the convolution of this function with p.v.$1/u$ is something we can compute. In equation (1.24) the function $\phi_X(u-\alpha)$ is the characteristic function of some p.d.f. $p(x)$ with bounded (approximately bounded) support. An immediate result from Fourier transform theory is that

$$\mathrm{e}^{-i2\pi ku}\phi_X(u-\alpha)$$

is the characteristic function of the p.d.f. $p(x-k)$ that is again a p.d.f. with approximately bounded support, so we can resort to the sampling theorem to represent it.

From equations (1.24) and (1.26), and the considerations expressed above, we see that the *characteristic integral* can be written as:

$$\mathrm{d}(k,\alpha) = \frac{1}{2X_c}\sum_{n=-\infty}^{+\infty} \mathrm{e}^{-i2\pi n\Delta k}\phi_X(n\Delta-\alpha)\int_{-\infty}^{+\infty} du\, \frac{\sin[\,2\pi X_c(u-n\Delta)\,]}{\pi(u-n\Delta)}\text{p.v.}\left(\frac{1}{u}\right) \tag{1.28}$$

The integral can now be performed. It is recognized as the Hilbert transform of the "sinc" function, and having done this we have disposed of the most delicate part of the numerical integration and are left with an infinite sum over discretely sampled values. Since this sum will be related in a straightforward manner with the sum coming from Fourier series, we will have at our disposal all of the tools to control the accuracy of the approximation introduced in replacing the infinite sum with a finite sum.

1.6.4 The pricing formula

The discussion above lead us to the conclusion that the numerical integration of the *characteristic integral* is equivalent to the computation of the r.h.s. of equation (1.28).

Let us concentrate on the integral on the r.h.s.

$$I = \int_{-\infty}^{+\infty} du\, \frac{\sin[\,2\pi X_c(u-n\Delta)\,]}{\pi(u-n\Delta)}\left(\text{p.v.}\frac{1}{u}\right)$$

performing the change of variables $u := -v/2\pi X_c + n\Delta$ we get:

$$I = \int_{-\infty}^{+\infty} dv\, \frac{\sin(\,v\,)}{\pi v}\text{p.v.}\left(\frac{1}{n\Delta - v/2\pi XC}\right)$$

$$= \left(\frac{\pi}{\Delta}\right)\frac{1}{\pi}\int_{-\infty}^{+\infty} dv\, \frac{\sin(\,v\,)}{v}\text{p.v.}\left(\frac{1}{\pi n - v}\right)$$

where in the last equality we have made use of the definition $\Delta = 1/2X_c$, and we recognize the integral on the r.h.s. as the Hilbert transform of the "sinc" function:

$$I = \frac{\pi}{\Delta}[\mathcal{H}\,\mathrm{sinc}_1](n\pi).$$

This is a result that we know from Example 1.5.4 and:

$$I = \frac{\pi}{\Delta}\frac{1-\cos(n\pi)}{n\pi} = \frac{1-(-1)^n}{n\Delta}.$$

Having done this we have achieved an amazingly accurate formula by which to compute the *characteristic integral* numerically

$$\mathrm{d}(k,\alpha) = \frac{1}{2X_c}\sum_{n=-\infty}^{+\infty} \mathrm{e}^{-i2\pi n\Delta k}\phi_X(n\Delta-\alpha)\frac{1-(-1)^n}{n\Delta} \tag{1.29}$$

It is worth stressing once more that, apart from the assumption that the p.d.f. of the process under examination has bounded domain, this is an exact integration formula.

Some approximation arises when we decide to introduce a cutoff in the number of Fourier modes that we use. The terms

$$\mathrm{e}^{-i2\pi n\Delta k}\phi_X(n\Delta-\alpha)$$

are the Fourier coefficients for a function with approximately bounded support. Without proof we will quote the following well-known theorem:

Theorem 1.6.1 *Let $p(x)$ a function with support in the interval $I_c = [-X_c, X_c]$.*

$$p(x) = 0, \quad x \notin I_c$$

Let

$$c_k = \int_{-X_c}^{X_c} \mathrm{d}x\,\mathrm{e}^{i2\pi xk}p(x)$$

If $p(x) \in C^q$ then

$$\sum_{n=-\infty}^{+\infty} |n^q c_n| < \infty$$

in particular

$$\lim_{n\to\infty} n^q c_n = 0$$

The meaning of this theorem is that the truncation error we are going to incur depends on the smoothness property of the p.d.f. of the process under examination (actually, it depends on the smoothness at $x = 0$).

The general smoothness property of a generic model cannot be assessed in advance without knowledge of the model, so the issue concerning truncation errors has to be addressed from case to case relative to each individual model.

For the time being we replace the infinite sum with a truncated sum,

$$d_N(k, \alpha) = \sum_{n=-N/2}^{+N/2} e^{-2\pi i \alpha k \Delta} \phi_X(n\Delta - \alpha) \frac{1-(-1)^n}{n\Delta} \tag{1.30}$$

and (7.1), the fundamental pricing equation, is modified accordingly:

$$\mathcal{O}(S_t; K, T, \omega) = \frac{1}{2}\omega S_t(1-m) + S_t \frac{i}{2\pi}\left(d_N(k, 0)m - d_N\left(k, \frac{i}{2\pi}\right)\right) \tag{1.31}$$

To steer clear of any form of circular reasoning, the assessment of the quality of the numerical approximation embedded in equation (1.31) can be performed with arbitrary accuracy only for models that admit an analytical solution that is NOT obtained by performing a Fourier integral.

For the sake of providing a simple example, we report some results for the Black–Scholes model. More precisely, Figures 1.7 and 1.8 report results for a volatility of $\sigma = 0.4423$, and Figures 1.9 and 1.10 for a volatility of $\sigma = 0.1$.

1.6.5 Application of the FFT

The next issue to address concerns the best way to perform the finite sum in equation (1.30). If all we need is just one value of the option at fixed strike, the issue is non-existent, and we simply sum the terms exactly as they are described in equation (1.30). Whenever we need to extract a larger set of results – quite a common situation when calibrating a model – it might be

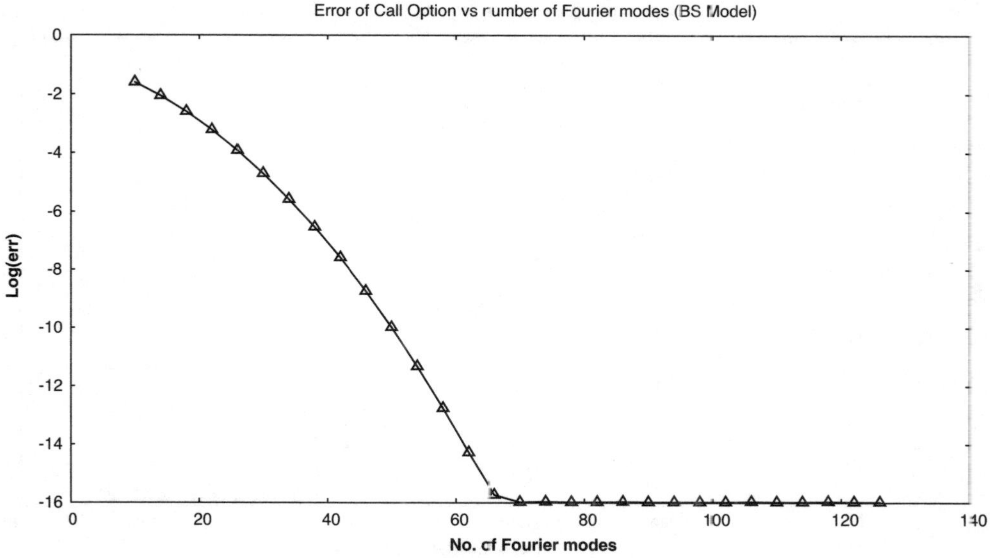

Figure 1.7 The dependency of the error on the number of Fourier modes used to compute a call option. Parameters are $T = 1$, $K = 1.0$, $\sigma = 0.4423$, $r = 0.05$. The spatial cutoff is $X_c = 6.0$

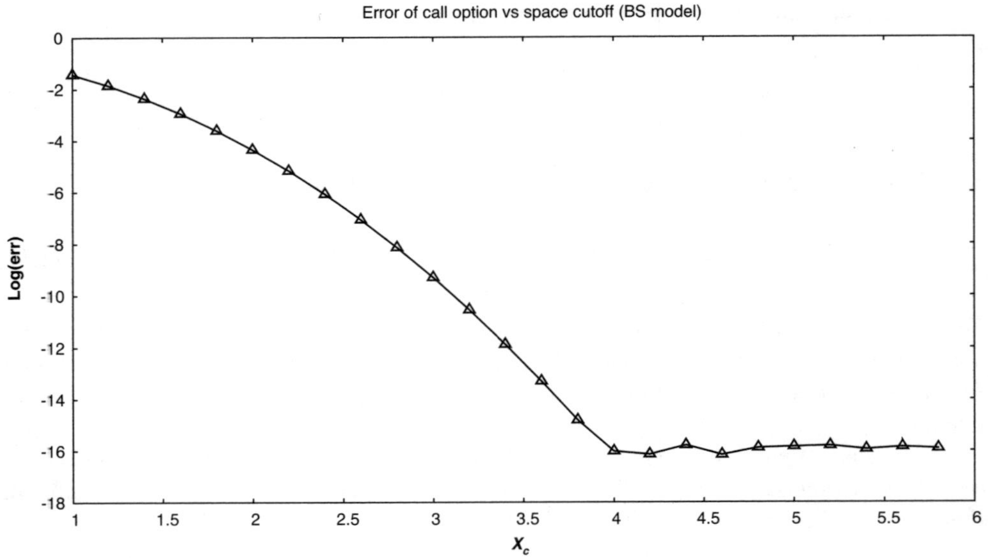

Figure 1.8 The dependency of the error on the spatial cutoff used to compute a call option. Parameters are $T = 1$, $K = 1.0$, $\sigma = 0.4423$, $r = 0.05$. The number of Fourier modes is 128

convenient to resort to the fast Fourier transform (FFT). The FFT can compute in $o(N \log(N))$ operation the sum in (1.30) for a set on N values $k_1, \ldots, k_N$ of the strike k, provided that:

- N can be written as 2^l.
- We confine the computation to the values $k_q = q/N\Delta$.

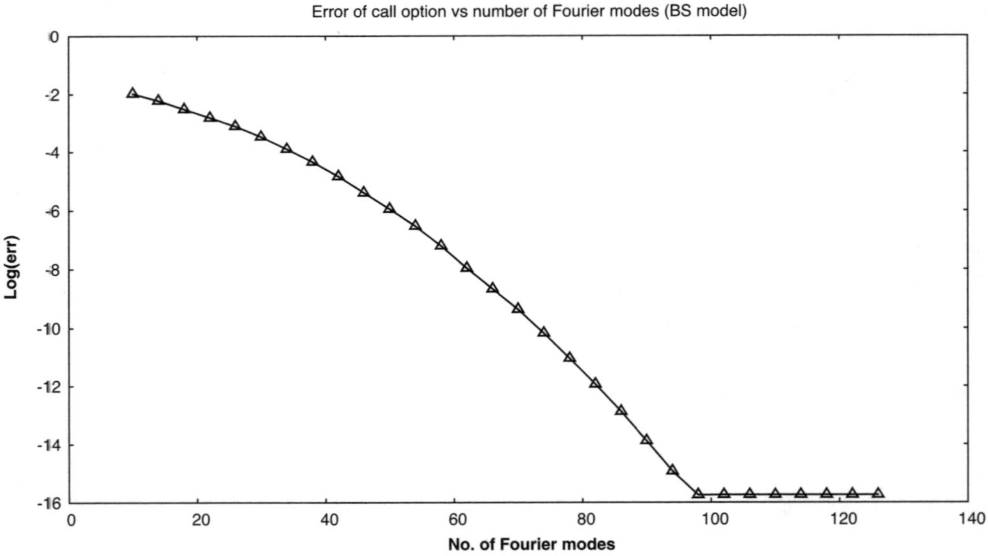

Figure 1.9 The dependency of the error on the number of Fourier modes used to compute a call option. Parameters are $T = 1$, $K = 1.0$, $\sigma = 0.1$, $r = 0.05$. The spatial cutoff is $X_c = 2.0$

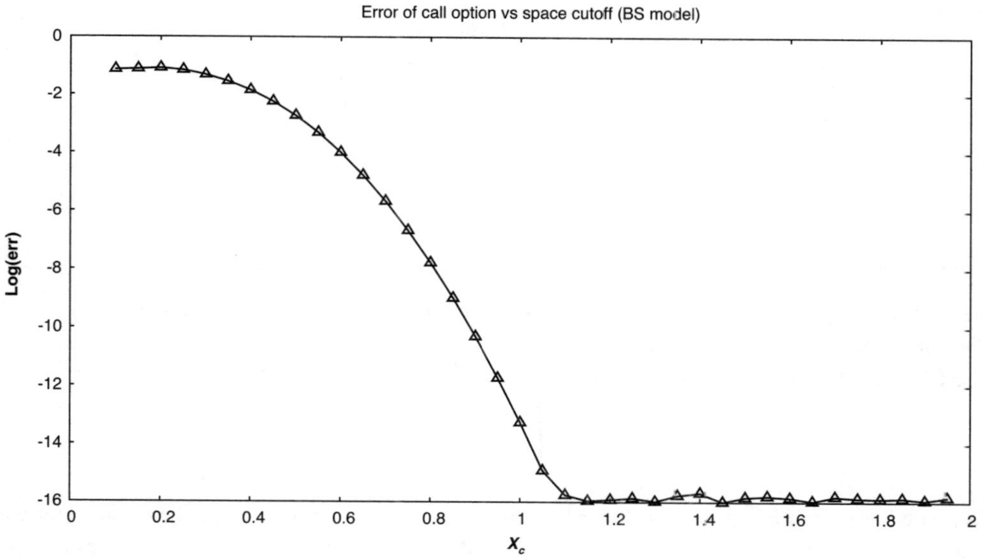

Figure 1.10 The dependency of the error on the spatial cutoff used to compute a call option. Parameters are $T = 1$, $K = 1.0$, $\sigma = 0.1$, $r = 0.05$. The number of Fourier modes is 128

Let

$$\zeta_n := \phi_X(n\Delta)\frac{1-(-1)^n}{n\Delta}$$

if we consider only FFT compliant strikes we can write:

$$\mathrm{d}_N\left(\frac{q}{N\Delta}, \alpha\right) = \sum_{n=-N/2}^{+N/2} \mathrm{e}^{-\frac{2\pi i n q}{N}} \zeta_n$$

then we separate positive and negative frequencies:

$$\mathrm{d}_N\left(\frac{q}{N\Delta}, \alpha\right) = \sum_{n=-N/2}^{1} \mathrm{e}^{-\frac{2\pi i n q}{N}} \zeta_n + \sum_{n=0}^{N/2-1} \mathrm{e}^{-\frac{2\pi i n q}{N}} \zeta_n + \mathrm{e}^{-\pi i q} \zeta_{N/2}$$

The last term is clearly zero ($N/2$ is even, in the working hypothesis), and the first term can be written as

$$\sum_{n=-N/2}^{1} \mathrm{e}^{-\frac{2\pi i n q}{N}} \zeta_n = \sum_{n=N/2}^{N-1} \mathrm{e}^{-\frac{2\pi i (n-N) q}{N}} \zeta_{n-N} = \sum_{n=N/2}^{N-1} \mathrm{e}^{-\frac{2\pi i n q}{N}} \zeta_{n-N}$$

Finally,

$$\mathrm{d}_N\left(\frac{q}{N\Delta}, \alpha\right) = \sum_{n=0}^{N-1} \mathrm{e}^{-\frac{2\pi i n q}{N}} \tilde{\zeta}_n \tag{1.32}$$

where

$$\tilde{\zeta}_n = \begin{cases} \zeta_n & 0 \le n < N/2 \\ \zeta_{n-N} & N/2 \le n < N \end{cases}$$

$$\zeta_n := \phi_X(n\Delta)\frac{1-(-1)^n}{n\Delta}$$

Small practical matters

The periodicity of the sum on the r.h.s. of equation (1.32) implies a periodicity of the l.h.s. of the same equation. This observation is what is required to extract the correct frequencies from the FFT sum that in fact turn out to be:

$$k_q = \begin{cases} \frac{q}{N\Delta} & k_q > 0 \\ \frac{q-N}{N\Delta} & k_q < 0 \end{cases}$$

When we use the FFT algorithm for calibration, usually we cannot choose the strikes that we want to calibrate. Our procedure is to compute the array of strikes at the FFT values k_q and interpolate linearly for the desired ones. This opens up the question of whether the FFT strikes are dense enough to populate reasonably the range needed. The best resolution we can achieve is given by:

$$\delta k = \frac{2X_c}{N}$$

The smallest X_c is dictated by the "boundedness" of the domain of the p.d.f. so we cannot use that as a free parameter; therefore if, for a given N, the resolution turns out to be too coarse, we have only two options:

- increase the number of Fourier modes even though the selected N is large enough for the desired accuracy;
- switch to the "fractional FFT" (FFFT) that allows for a different discretization of the FFT strikes.

The fractional FFT turns out to be, on average, four times slower that the straight FFT, so alternatives must be weighted carefully if performance is an issue. The basic ideas underlying the fractional FFT are presented in Appendix F.

1.7 RELATED LITERATURE

We provide here a very general review of the literature on Fourier transform applications to option pricing problems. We stick to a mandatory reading list on the subject with a particular focus on aspects of this literature that are related to our approach.

To the best of our knowledge, the gold rush to Fourier transform pricing applications was initiated by Heston (1993). Our approach shares the same philosophy of searching a relationship between the characteristic function of the pricing kernel of an underlying asset. In this sense, our work is also in the line of literature of Bakshi and Madan (2000) and Duffie *etal.* (2000) both of which define a spanning structure of the pricing kernel based on Fourier transforms. We denote by Arrow–Debreu prices the discounted value of the density instead of the digital options, but that is a mere question of taste, to keep a similarity with the binomial model.

Carr and Madan (1999) proposed a technique to represent the price of a plain vanilla option in terms of Fourier transforms, in such a way as to have a model that was well suited for application of the FFT technique. For this purpose, they addressed the problem of performing the Fourier transform of the payoff function with respect to the strike. This is then substituted in the pricing integral and, by a change of order of integration, produces the price of the option as a function of the characteristic function of the density. Lewis (2001) addressed the problem of computing the Fourier transform of the payoff function in a more general setting. Differently from Carr and Madan (1999), the Fourier transform is computed with respect to the underlying asset. With this technique, Lewis provides a pricing formula that is valid for general payoffs, including the pricing kernels that had represented the focus of the first stream of literature.

Our approach blends most of the features of the literature that we have so brutally reviewed. For one thing, our attention is focused on the pricing kernel, as in the first stream of literature quoted above. For another, our focus is on disentangling the payoff of this digital option from the characteristic function in the pricing formula. Differently from Lewis (2001), we are only interested in the pricing kernel, because our task is to use the model for calibration. While this interest in calibration recalls the contribution by Carr and Madan (1999), our focus is on digital instead of European options, even though we finally obtain pricing formulas for European options that can be applied in a FFT procedure to perform calibration to market data. What we think is original with respect to the literature is that our approach is cast in the framework of generalized functions, in which the Fourier transform of singular functions, such as the payoff of digital options (which is the core of our approach) are well defined, and so is the convolution of these payoffs with pricing density.

2
The Dynamics of Asset Prices

2.1 INTRODUCTION

In 1900 Louis Bachelier discussed at the University of Paris-Sorbonne a thesis on "*The theory of speculation*" proposing a mathematical theory of the dynamics of asset prices. His work was rediscovered fifty years later upon suggestion by Leonard J. Savage. Since the 1960s the same theory has become the standard in the financial economics literature. Eugene Fama and Paul Samuelson provided the theoretical foundations of what is known as the "*efficient market hypothesis*". The bottom line of this theory is that price changes of assets cannot be predicted and the rate of return of the market as a whole cannot be outperformed by any economic agent, unless by a matter of mere luck. Surprisingly, while denying any hope to build a mathematical model to forecast asset price movements, this theory predicts a very neat and stringent, albeit simple, stochastic dynamics for them. This is the so-called *random walk* model by which the price evolves according to a sequence of unexpected innovation or *shocks*. Formally, the dynamics is written as

$$S_t = S_{t-1} + Z_t$$

where S_t denotes the price of the asset at time t and Z_t is the innovation. By definition, Z_t cannot be predicted exploiting the set of information available at time $t - 1$. If this set of information only includes the past history of S_t, or in the jargon of probability it coincides with the *natural* filtration generated by S_t, the market is said to be "*weakly efficient*". If the set of information includes other pieces of news from public sources, such as the dynamics of other assets or the result of fundamental research published by analysts, the market is said to display "*semi-strong efficiency*". Finally, the market is said to be "*strongly efficient*" if private information available to insiders is also included in the price.

To summarize, a market is said to be efficient if it produces prices that "fully reflect" available information. Notice that besides being independent the distribution of increments Z_t must have particular features. In its strongest form we may add that it should have zero mean, but this is not required: we will see that market efficiency may be consistent with positive expected returns representing the risk premium that is considered fair by the market. Beyond this restriction on the mean, the dynamics and the probability distribution of Z_t can be characterized according to many different choices and models.

In this chapter, we review the main choices available within the set of processes with independent and stationary increments. We will see that these models may be uniquely defined by a specific formula, known as Lévy–Khintchine, which completely describes the characteristic function. In the first part of the chapter we will review this class of processes from the point of view of central limit theorem: in other words, we will assume that a large number of innovations reaches the market in a unit of time, and we will derive the dynamics of prices right from general requirements imposed to the distribution of these new pieces of information. In the second part of the chapter we will specify the nature of these innovations in more detail, giving a taxonomy of possible shocks. This will take us close to the price discovery process studied in the market microstructure literature. Finally, in the third part of the chapter we will review the main properties of this set of processes.

2.2 EFFICIENT MARKETS AND LÉVY PROCESSES

2.2.1 Random walks and Brownian motions

Going back to Bachelier, we begin by reporting a formal and general definition of *random walk*.

Definition 2.2.1 *Let Z_k, $k \geq 1$, be i.i.d. Then*

$$S_n = \sum_{k=1}^{n} Z_k, \qquad n \in \mathbb{N}$$

is called a **random walk**.

Formally, the increments of random walks are *stationary and independent*, where stationarity means that

$$Z_k = S_k - S_{k-1}, \qquad k \geq 1$$

have identical distribution.

If we now refer to $S_{m+n} - S_n$ as an increment over m time units, $m \geq 1$, we may ask the question: Which kind of distribution may this increment have? In fact, while any distribution may be freely chosen for Z_j, increments over m time units are sums of m i.i.d. random variables, and for this reason must have a specific property. We will see below that this property will be called *infinite divisibility* in its discrete form. We will also see that all infinitely divisible distributions can be obtained as limits of sums of independent random variables.

If we look at infinite divisibility from a temporal viewpoint, i.e. we ask what happens if we let the time unit tend to zero and take a finer look at a random walk, we will also find that Lévy processes are limits of random walks. For a start, in this section we stick to the simplest case within this family, that is the case of finite variance innovations.

Formally, we focus at a fixed time, say 1, and partition the process so as to make n steps per time unit. Next, we can use the standard central limit theorem:

Theorem 2.2.1 (*Lindeberg–Lévy*) *If $\sigma^2 = \mathrm{Var}(Z_1) < +\infty$, then*

$$\frac{S_n - \mathbb{E}(S_n)}{\sigma\sqrt{n}} \xrightarrow{d} Z \stackrel{d}{=} N(0, 1).$$

It is then immediate to apply this result to show that, in this case, the limit of a random walk is the Brownian motion process that was first proposed by Bachelier. If we denote by $S_{[nt]}$ the value of the process at time $[nt]$, the integer part of nt, we have

Theorem 2.2.2 (*Donsker*) *If $\sigma^2 = \mathrm{Var}(Z_1) < +\infty$, then, for $t \geq 0$*

$$X_t^{(n)} = \frac{S_{[nt]} - \mathbb{E}(S_{[nt]})}{\sigma\sqrt{n}} \xrightarrow{d} X_t \stackrel{d}{=} N(0, t).$$

Furthermore, $X^{(n)} \to X$ where X is a Brownian motion.

We recall here the definition and main properties of Brownian motion.

Definition 2.2.2 (***Brownian motion***) *A real-valued stochastic process $X = (X_t)_{t \geq 0}$ is called Brownian motion if it has independent and stationary increments and*

1. *the paths $t \to X_t$ are continuous almost surely;*
2. $X_t \stackrel{d}{=} N(0, t)$, *for all $t \geq 0$ (Normal distribution).*

The paths of Brownian motion are continuous, but turn out to be nowhere differentiable. This way, they exhibit erratic movements at all scales. This dynamics is represented in terms of a stochastic differential equation

$$\mathrm{d}S_t = \mu \,\mathrm{d}t + \sigma \,\mathrm{d}X_t$$

where μ is the *drift* term denoting the average increase in S_t and σ is the *diffusion* term representing the volatility of the process.

2.2.2 Geometric Brownian motion

Under the original Bachelier framework, the price increments Z_k are assumed to follow a Brownian motion. It is easy to see that this representation may not be well suited to represent price changes. On the one hand, this may produce negative prices and, on the other, we are used to thinking of returns in terms of percentage changes in a continuous compounding regime. After all, a 1 dollar gain out of a capital invested of 1 million is not the same as 1 dollar earned with an investment of 10 dollars. The continuous compounding regime suggests that we change the prices to logs before taking first differences. The result is a dynamics called *Geometric Brownian Motion*, which is represented by the stochastic differential equation

$$\mathrm{d}S_t = \mu S_t \,\mathrm{d}t + \sigma S_t \,\mathrm{d}X_t$$

Notice that both the drift and the diffusion terms are now proportional to the price S_t. If we take the log of price and use Itô's lemma, we easily obtain

$$\mathrm{d}\ln(S_t) = \left(\mu - \frac{\sigma^2}{2}\right) \mathrm{d}t + \sigma \,\mathrm{d}X_t$$

and the log of price is an arithmetic Brownian motion.

For most of the past century, this representation of the dynamics of prices represented the dominating paradigm: normality of log-returns and constant volatility. However, in the 1960s some scholars pointed out that this hypothesis could have been too restrictive. Since the crisis of 19 October 1987, this argument has become common knowledge and most of the research has been devoted to a more flexible and realistic representation of the dynamics of log-prices.

2.2.3 Stable processes

A possible extension of the model, allowing for departures from normality, was first pointed out in the 1960s by Mandelbrot. Let us just go back to the central limit theorem above and ask the question: What happens if the condition $\mathrm{Var}(Z_1) < +\infty$ fails to hold? Actually, we may provide an extended version of the central limit theorem according to which the sum of a large number of i.i.d. disturbances converges to a *stable* distribution under milder conditions. We begin giving the definition of stable distributions.

Definition 2.2.3 *A random variable Y is said to have a stable distribution if, for all $n \geq 1$, it satisfies*

$$Y_1 + \ldots + Y_n \stackrel{d}{=} a_n Y + b_n$$

where $Y_1, \ldots, Y_n$ are i.i.d. copies of Y, $a_n > 0$ and $b_n \in \mathbb{R}$.

By subtracting b_n/n from each of the terms of the left-hand side and dividing by a_n, we can see that any stable distribution is *infinitely divisible* (see Definition 2.2.6 below). So, if we sum independent random variables with stable distributions we obtain a variable with the same distribution. Likewise, if we partition a sum of stable variables in sums of their subsets, the new variables obtained retain the same distribution. It is intuitive to guess that the normal distribution must be included in this family.

Let us now introduce a more general version of the central limit theorem (see Sato, 1999, Theorem 15.7):

Theorem 2.2.3 *Let S_n be a random walk. A random variable Z is said to have a stable distribution if and only if for every n there exist $b_n > 0$ and $c_n \in \mathbb{R}$ such that*

$$b_n S_n + c_n \stackrel{d}{\to} Z$$

In the same way as we did above, we may obtain the corresponding result for processes.

Theorem 2.2.4 *For every $t \geq 0$, let*

$$X_t^{(n)} = b_n S_{[nt]} + c_{t,n} \stackrel{d}{\to} X_t$$

Furthermore, $X^{(n)} \to X$ where X is called a ***stable Lévy process****.*

2.2.4 Characteristic functions

Having extended the set of distributions that we may use to describe the stochastic dynamics of log-prices, a natural question arises as to how to represent their shapes. From basic probability we are used to representing distributions by their density functions. Unfortunately, such density functions can seldom be written in closed form. Even for the class of stable distributions, the models with density functions in closed form are limited to the cases of Example 2.2.1 below.

Fortunately, even though the density is not known in closed form, it can be fully characterized by means of its Fourier transform, known as its *characteristic function*.

$$\phi_X(u) = \mathbb{E}[e^{iuX}] = \int_{-\infty}^{+\infty} e^{iux}\, dF(x), \qquad u \in \mathbb{R}$$

This exists and is uniquely defined for all distributions. Moreover, if X and Y are independent random variables, then $\phi_{X+Y}(u) = \phi_X(u)\phi_Y(u)$. In particular, for centred stable distributions – that is, with $b_n = 0$ in Definition 2.2.3 – if ϕ_Y is the characteristic function of Y we get an interesting result. Namely,

$$\phi_Y^n(u) = \mathbb{E}[e^{iu(a_n Y)}] = \phi_Y(a_n u)$$

from which

$$\begin{aligned}\phi_Y^{nm}(u) &= \left(\phi_Y^n(u)\right)^m \\ &= \phi_Y^m(a_n u) = \phi_Y(a_m a_n u)\end{aligned}$$

but

$$\phi_Y^{nm}(u) = \phi_Y(a_{nm}u)$$

that is

$$a_{nm} = a_n a_m$$

whose solution is of type

$$a_n = n^H$$

for a certain $H > 0$. So, we have

$$\phi_Y^n(u) = \phi_Y(n^H u) \tag{2.1}$$

Remark 2.2.1 Clearly, the centring sequence b_n of Theorem 2.2.3 coincides with $a_n^{-1} = n^{-H}$.

Remark 2.2.2 Relation (2.1) can be extended beyond natural numbers. In fact let $r = m/n > 0$. From (2.1), $\phi_Y(u) = \phi_Y^{1/n}(n^H u)$ and $\phi_Y^m(u) = \phi_Y^{m/n}(n^H u)$; if $v = n^H u$, we get

$$\phi_Y^m(\frac{1}{n^H}v) = \phi_Y^{m/n}(v)$$

but

$$\phi_Y^m(\frac{1}{n^H}v) = \phi_Y((m/n)^H v)$$

so $\phi_Y^r(u) = \phi_Y(r^H u)$. Now let r_n be a sequence of rational numbers converging to a real number $r > 0$. By the continuity of the characteristic function it follows that

$$\phi_Y^r(u) = \phi_Y(r^H u) \tag{2.2}$$

for every real number $r > 0$.

It can be proved,

$$H \geq \frac{1}{2} \tag{2.3}$$

The parameter $\alpha = 1/H$ is called the *index* of the random variable Y: we then say that Y has an α-stable distribution. From (2.3), it follows that $0 < \alpha \leq 2$.

Getting to the characteristic function, Y has a stable distribution if and only if (see Sato, 1999, Theorem 14.15, that)

$$\phi_Y(u) = e^{iu\eta - c|u|^\alpha(1-i\beta sgn(u)g(u))} \tag{2.4}$$

with

$$g(u) = \begin{cases} tan\,(\pi\alpha/2) & \text{for } \alpha \in (0,1) \cup (1,2] \\ 2/\pi \log|u| & \text{for } \alpha = 1 \end{cases} \tag{2.5}$$

with $\alpha \in (0, 2]$, $\beta \in [-1, 1]$, $c > 0$ and $\eta \in \mathbb{R}$. We shall indicate $Y \stackrel{d}{=} \text{Stable}_\alpha(c, \beta, \eta)$. In this representation c is the scale parameter (note that it has nothing to do with the Gaussian component if $\alpha < 2$), η is the location parameter, α determines the shape of the distribution (known as the *index* of the stable distribution) and β is the skewness parameter.

When $\beta = 0$ and $\eta = 0$, Y is said to have a symmetric stable distribution and the characteristic function is given by

$$\phi_Y(u) = \mathrm{e}^{-c|u|^\alpha}$$

Furthermore, when $\alpha = 2$, $Y \overset{d}{=} N(\eta, 2c)$. It can be proved that these distributions for $0 < \alpha < 2$ are heavy tailed (we will revisit this fact in Section 2.4.3 below):

- $\mathbb{E}[|Y|^p] < +\infty$, if $p \in (0, \alpha)$
- $\mathbb{E}[|Y|^p] = +\infty$, if $p \in [\alpha, 2]$

Example 2.2.1 *The probability density of an α stable distribution law is not known in closed form except in the following cases (plus the degenerate case of a constant random variable):*

1. $\alpha = 2$ *corresponds to the Normal distribution.*
2. $\alpha = 1$, $\beta = 0$ *and* $\eta = 0$ *corresponds to the Cauchy distribution whose density is given by*

$$\frac{c}{\pi} \frac{1}{(x^2 + c^2)}, \text{ for } x \in \mathbb{R}$$

3. $\alpha = \frac{1}{2}$, $\beta = 1$ *and* $\eta = 0$ *corresponds to inverse Gaussian distribution whose density is given by*

$$\frac{c}{\sqrt{2\pi x^3}} \mathrm{e}^{-\frac{c^2}{2x}}, \text{ for } x > 0$$

It is the distribution of the random variable.

$$\tau_c = \inf\{t > 0 : B_t > c\}$$

where B is a Brownian motion.

As a specific instance, we can write the characteristic function of a Brownian motion as follows.

Example 2.2.2 (*Brownian motion*) *The Fourier transform of a Brownian motion X_t is given by*

$$\phi_{X_t}(\gamma) = \mathbb{E}\left[\mathrm{e}^{i\gamma X_t}\right] = \mathrm{e}^{-\frac{\gamma^2 t}{2}}$$

It follows that

$$\phi_{\sqrt{c}X_{\frac{t}{c}}}(\gamma) = \mathbb{E}\left[\mathrm{e}^{i\gamma\sqrt{c}X_{\frac{t}{c}}}\right] = \mathrm{e}^{-\frac{\gamma^2 c \frac{t}{c}}{2}} = \mathrm{e}^{-\frac{\gamma^2 t}{2}}$$

Moreover, it can also be proved that Brownian motion has the scaling property $\left(\sqrt{c}X_{\frac{t}{c}}\right)_{t\geq 0} \overset{d}{=} X$.

2.2.5 Lévy processes

The extension of model choices beyond the Gaussian distribution to the wider class of stable distributions allows more flexibility in the representation of the dynamics of logs of prices. Nevertheless, even this generalization is not free from flaws. There is first a question of taste. Except the normal case, all the other distributions in this family do not have finite variance, and variance has been the most popular measure of risk in the finance literature. There is a second issue of consistency with the evidence of financial data. According to a very well-known empirical regularity, daily returns have different distributions from monthly returns, and departures from normality are more evident at higher data frequencies. This is clearly at odds with the stability property, which predicts that we would find the same distribution for returns at different frequencies.

For this reason, we need a model in which, if we partition a process in increments over periods of equal length, we obtain returns that have equal distribution, but this distribution may be allowed to change if we change the period length. This requirement leads to a larger class of processes, known as *Lévy processes.*

Definition 2.2.4 *A real-valued (or $\mathbb{R}^d$-valued) stochastic process $X = (X_t)_{t\geq 0}$ is called a Lévy process if*

1. *it has* ***independent increments****, i.e. the random variables $X_{t_0}, X_{t_1} - X_{t_0}, \ldots, X_{t_n} - X_{t_{n-1}}$ are independent for all $n \geq 1$ and $0 \leq t_0 < t_1 < \ldots < t_n$*
2. *it has* ***stationary increments****, i.e. $X_{t+h} - X_t$ has the same distribution as X_h for all $h, t \geq 0$.*
3. *it is stochastically continuous: for every $t \geq 0$ and $\epsilon > 0$*

$$\lim_{s\to t} \mathbb{P}\left[|X_s - X_t| > \epsilon\right] = 0$$

4. *the paths $t \to X_t$ are right-continuous with left limits with probability* 1.

Condition (2) implies that $\mathbb{P}(X_0 = 0) = 1$. Moreover, it is an immediate consequence of (1) that a Lévy process is a Markov process.

From a look at the definition of Lévy processes it should not come as a surprise that they could be obtained as the limit of a sum of independent variables with some features, and thus they may be the outcome of some more general form of the central limit theorem. Actually, the property that the process can be always be partitioned as a sum of independent variables – which means *infinite divisibility* – can show up if we derive the central limit theorem in full generality, resorting to the so-called *triangular arrays*.

Definition 2.2.5 *A double sequence of random variables $\left\{Y_k^{(n)} : k = 1, \ldots, r_n; n = 1, 2, 3, \ldots\right\}$ is called a "triangular null array" if, for each fixed n, $Y_1^{(n)}, Y_2^{(n)}, \ldots, Y_{r_n}^{(n)}$ are independent and if, for any $\epsilon > 0$,*

$$\lim_{n\to+\infty} \max_{1\leq k\leq r_n} \mathbb{P}\left[|Y_k^{(n)}| > \epsilon\right] = 0$$

Let $S_n = \sum_{k=1}^{r_n} Y_k^{(n)}$. We have that

Theorem 2.2.5 (*Khintchine*) *Let $\left\{Y_k^{(n)}\right\}$, be a null array. If for some $b_n \in \mathbb{R}$, $n = 1, 2, \ldots$ there exists a random variable Z such that*

$$S_n - b_n \xrightarrow{d} Z$$

then Z has an infinitely divisible distribution. (See Sato, 1999, Theorem 9.3.)

Having done this, we recover the whole class of Lévy processes, thanks to the following theorem:

Theorem 2.2.6 (*Skorohod*) *In the previous theorem, $X_t^{(n)} = S_{[nt]}^{(n)} - b_{t,n} \xrightarrow{d} X_t$. Furthermore, $X^{(n)} \to X$ where X is a* ***Lévy process****.*

Example 2.2.3 *Take the binomial process $B(n, p_n)$. We have $B(n, p_n) \to Poi(\lambda)$ for $np_n \to \lambda$ and this is a special case of Khintchine's theorem. In fact, assuming $Y_k^{(n)} \stackrel{d}{=} B(1, p_n)$, $S_n \stackrel{d}{=} B(n, p_n)$. But, as $n \to +\infty$*

$$\binom{n}{k} p_n^k (1-p_n)^{n-k} = \frac{n(n-1)\ldots(n-k+1)}{k!} \frac{(np_n)^k}{n^k} \frac{(1-np_n/n)^n}{(1-p_n)^k} \to \frac{\lambda^k}{k!} \mathrm{e}^{-\lambda}$$

Example 2.2.4 (***Poisson process***) *Skorohod's theorem turns out to give an approximation of the* ***Poisson process*** *by Bernoulli random walks.*

An $\mathbb{N}$-valued stochastic process $X = (X_t)_{t \geq 0}$ is called a Poisson process with rate $\lambda \in (0, +\infty)$ if X satisfies (1)–(4) *in Definition* 2.2.4 *and*

4. $\mathbb{P}(X_t = k) = \frac{(\lambda t)^k}{k!} \mathrm{e}^{-\lambda t}$, $k \geq 0$, $t \geq 0$ *(Poisson distribution).*

Poisson processes have jumps of size 1. *In fact, if*

$$T_n = \inf\{t \geq 0 : X_t \geq n\},$$

$$\Delta X_{T_n} = 1, \ a.s.$$

Moreover, it can be proved (see Billingsley, 1986 for the details) *that the random variables $T_{n+1} - T_n$, $n \in \mathbb{N}$, are i.i.d. with $T_{n+1} - T_n \stackrel{d}{=} \mathcal{E}(\lambda)$ (exponential distribution with parameter λ).*

The Fourier transform of the Poisson process is

$$\begin{aligned} \phi_{X_t}(\gamma) &= \sum_{h=0}^{+\infty} \mathrm{e}^{i\gamma h} \mathrm{e}^{-\lambda t} \frac{(\lambda t)^h}{h!} \\ &= \sum_{h=0}^{+\infty} \mathrm{e}^{-\lambda t} \frac{(\lambda t e^{i\gamma})^h}{h!} = \mathrm{e}^{\lambda t(\mathrm{e}^{i\gamma}-1)} \end{aligned} \tag{2.6}$$

2.2.6 Infinite divisibility

By definition, if X is a Lévy process, any X_s can be decomposed, for every $m \geq 1$,

$$X_s = \sum_{j=1}^{m} \left(X_{\frac{js}{m}} - X_{\frac{(j-1)s}{m}} \right)$$

into a sum of m i.i.d. random variables.

Definition 2.2.6 *Y is said to have an* ***infinitely divisible distribution*** *if for every $m \geq 1$,*

$$Y \stackrel{d}{=} Y_1^{(m)} + \ldots + Y_m^{(m)}$$

for some i.i.d. random variables $Y_1^{(m)}, \ldots, Y_m^{(m)}$.

We stress again that the distribution of $Y_j^{(m)}$ may vary as m varies, but not as j varies.

Remark 2.2.3 The argument just before the definition shows that increments of Lévy processes are infinitely divisible. In particular, if X_t is a Lévy process, the distribution of X_t

necessarily has to be of the infinitely divisible type. For $t = 1$, this implies

$$\phi_{X_1}(u) = \phi^n_{X_{1/n}}(u)$$

If $t = m/n$, $\phi_{X_t}(u) = \phi^m_{X_{1/n}}(u) = \phi^{m/n}_{X_1}(u)$. If $t > 0$ is irrational, let r_n be rational numbers such that $r_n \to t$. We have, by the stochastic continuity of the Lévy process, that $X_{r_n} \to X_t$ in probability, hence $\phi_{X_{r_n}}(u) \to \phi_{X_t}(u)$. Then $\phi_{X_t}(u) = \phi^t_{X_1}(u)$, that is the distribution of X_t is the one given by the characteristic function $\phi^t_{X_1}(u)$. More precisely, the following result holds (see Sato, 1999, Theorem 7.10, for a detailed proof).

Theorem 2.2.7 *If $(X_t)_{t\geq 0}$ is a Lévy process, then, for any $t \geq 0$, the distribution of X_t is infinitely divisible and, if $\phi(u) = \phi_{X_1}(u)$, we have $\phi_{X_t}(u) = \phi^t_{X_1}(u)$.*

Conversely, if $\phi(u)$ is the characteristic function of an infinitely divisible distribution, then there is a Lévy process $(X_t)_{t\geq 0}$ such that $\phi_{X_1}(u) = \phi(u)$.

Moreover, if $(X_t)_{t\geq 0}$ and $(X'_t)_{t\geq 0}$ are Lévy processes such that $\phi_{X_1}(u) = \phi_{X'_1}(u)$, then they are identical in law.

Many known distributions are infinitely divisible, some are not.

Example 2.2.5 *The Normal, Poisson, Gamma and geometric distributions are infinitely divisible. This follows from the well-known fact that sums of independent random variables distributed as each of the above distributions, are again of the same type with adequate parameters.*

Example 2.2.6 *The Bernoulli distribution with parameter $p \in (0, 1)$ is not infinitely divisible. In fact, assume that one can represent a Bernoulli random variable X with parameter p as $Y_1 + Y_2$ for independent identically distributed Y_1 and Y_2. Then*

$$\mathbb{P}\left(Y_1 > \frac{1}{2}\right) > 0 \Rightarrow 0 = \mathbb{P}(X > 1) \geq \mathbb{P}\left(Y_1 > \frac{1}{2}\right)\mathbb{P}\left(Y_2 > \frac{1}{2}\right) > 0$$

is a contradiction, so we must have $\mathbb{P}\left(Y_1 > \frac{1}{2}\right) = 0$, but then

$$\mathbb{P}\left(Y_1 > \frac{1}{2}\right) = 0 \Rightarrow p = \mathbb{P}(X = 1) = \mathbb{P}\left(Y_1 = \frac{1}{2}\right)\mathbb{P}\left(Y_2 = \frac{1}{2}\right) \Rightarrow \mathbb{P}\left(Y_1 = \frac{1}{2}\right) = \sqrt{p} > 0$$

Similarly,

$$\mathbb{P}(Y_1 < 0) > 0 \Rightarrow 0 = \mathbb{P}(X < 0) \geq \mathbb{P}(Y_1 < 0, Y_2 < 0) > 0$$

is a contradiction, so we must have $\mathbb{P}(Y_1 < 0) = 0$ and then

$$1 - p = \mathbb{P}(X = 0) = \mathbb{P}(Y_1 = 0, Y_2 = 0) \Rightarrow \mathbb{P}(Y_1 = 0) = \sqrt{1 - p} > 0$$

This is impossible: in fact

$$0 = \mathbb{P}\left(X = \frac{1}{2}\right) \geq \mathbb{P}(Y_1 = 0)\,\mathbb{P}\left(Y_2 = \frac{1}{2}\right) > 0$$

Infinite divisible distributions are characterized by the following key result (see Sato, 1999, Theorem 8.1):

Theorem 2.2.8 (*Lévy–Khintchine theorem*) *A real-valued random variable X has an infinitely divisible distribution if there are parameters $a \in \mathbb{R}$, $\sigma^2 \geq 0$ and a locally finite* *measure* ν *on* $\mathbb{R} \setminus \{0\}$ *with* $\int_{-\infty}^{+\infty}(1 \wedge x^2)\nu\,(\mathrm{d}x) < +\infty$ *such that*

$$\phi_X(\lambda) = \mathbb{E}\left(\mathrm{e}^{i\lambda X}\right) = \mathrm{e}^{-\psi(\lambda)}$$

where

$$\psi(\lambda) = -ia\lambda + \frac{1}{2}\sigma^2\lambda^2 - \int_{-\infty}^{+\infty}\left(\mathrm{e}^{i\lambda x} - 1 - i\lambda x\mathbb{I}_{\{|x|\le 1\}}\right)\nu\,(\mathrm{d}x) \qquad \lambda \in \mathbb{R} \tag{2.7}$$

Infinite divisible distributions are parameterized by their Lévy–Khintchine characteristics (a, σ^2, ν). $\psi(\lambda)$ is called the *characteristic exponent*, and ν the *Lévy measure*.

Example 2.2.7

1. *For the Normal distribution,* $\nu = 0$ *and* $a = 0$.
2. *For stable distributions it can be proved* (see Sato, 1999, Theorem 14.3 and Remark 14.4 or Samorodnitsky and Taqqu, 1994) *that if* $0 < \alpha < 2$ *the characteristics of an* α*-stable distribution are* $\sigma = 0$ *and the Lévy measure*

$$\nu(\mathrm{d}x) = \begin{cases} c_1 x^{-1-\alpha}\,\mathrm{d}x, & \text{for } x > 0 \\ c_2|x|^{-1-\alpha}\,\mathrm{d}x, & \text{for } x < 0 \end{cases} \tag{2.8}$$

   with $c_1, c_2 > 0$.
3. *For the Poisson distribution with parameter* $\mu > 0$, $a = 0$, $\sigma = 0$ *and* $\nu(\mathrm{d}x) = \mu\delta_1(\mathrm{d}x)$.

By Theorem 2.2.7 and Theorem 2.2.8, without additional work we can state the following fundamental result representing the characteristic function of a Lévy process in terms of its characteristic triplet (a, σ^2, ν).

Theorem 2.2.9 (***Lévy–Khintchine representation***) *Let* X_t *be a Lévy process, then there are parameters* $a \in \mathbb{R}$, $\sigma^2 \ge 0$ *and a locally finite measure* ν *on* $\mathbb{R} \setminus \{0\}$ *with* $\int_{-\infty}^{+\infty}(1 \wedge x^2)\nu(\mathrm{d}x) < +\infty$ *such that*

$$\phi_{X_t}(\lambda) = \mathbb{E}\left(\mathrm{e}^{i\lambda X_t}\right) = \mathrm{e}^{-t\psi(\lambda)}$$

where

$$\psi(\lambda) = -ia\lambda + \frac{1}{2}\sigma^2\lambda^2 - \int_{-\infty}^{+\infty}\left(\mathrm{e}^{i\lambda x} - 1 - i\lambda x\mathbb{I}_{\{|x|\le 1\}}\right)\nu(\mathrm{d}x) \qquad \lambda \in \mathbb{R}$$

Lévy processes are characterized by their Lévy–Khintchine characteristics (a, σ^2, ν), where we call a the *drift coefficient*, σ^2 the *Brownian coefficient* and ν the *Lévy measure* or *jump measure*. $\psi(\lambda)$ defined in (2.7) is called the *characteristic exponent* of the Lévy process.

Remark 2.2.4 (***Stable process***) Since it can be proved that in this case the characteristic exponent is

$$\psi(u) = iu\eta - c|u|^\alpha\left(1 - i\beta\,\mathrm{sgn}\,(u)g(u)\right)$$

with

$$g(u) = \begin{cases} \tan\frac{\pi\alpha}{2} & \text{for } \alpha \in (0,1) \cup (1,2] \\ \frac{2}{\pi}\log|u| & \text{for } \alpha = 1 \end{cases}$$

it is easy to verify that the *scaling property* shown for the Brownian motion in Example 2.2.2, extends to all α-stable processes with $\eta = 0$, that is

$$(X_{ct})_{t\geq 0} \stackrel{d}{=} \left(c^{\frac{1}{\alpha}} X_t\right)_{t\geq 0}$$

2.3 CONSTRUCTION OF LÉVY MARKETS

Up to this point we have shown that by using several versions of the central limit theorem we can characterize the dynamics of the price of a financial asset. Namely, if a market is efficient every new piece of information is immediately reflected into the price, so that future price movements are independent of the information available at the present time. We have seen that under the assumption of stationarity of such increments we recover Lévy processes as the most general representation of asset price dynamics. More stringent requirements may then allow us to specialize the model first to stable and then to Gaussian processes. A remark is needed to say that stationarity of the increments remains a strong restriction. Nevertheless, the model is able to generate some of the empirical features that we see in the market, namely the change of distribution when we change the length of holding period of the returns.

Notice that all these results have been achieved without any reference whatsoever to the stochastic process driving the rate at which new information flows into the market. In this section we are making the picture richer by adding the specification of some probability laws that may govern the arrival of information and then trigger price changes. In the end, of course, we cannot expect to find anything other than Lévy processes, but we will have learned more about the kind of shocks that are consistent with them, and, more importantly, we will learn that Lévy markets may be very different from each other, as they are collections of different kinds of shocks. From this point of view, even though they are within the realm of stationary processes, these processes may seem well suited to represent the dynamics of very different markets.

Before leaving for this journey into the entomology of shocks, credit must be given to the path-breaking model on this subject published by Clark in his 1973 *Econometrica* paper. He was the first to propose the modelling of financial prices as subordinated stochastic processes with the task of jointly determining the dynamics of prices, trading volume and volatility. In this Section we refer to Winkel, Lecture notes, for details.

2.3.1 The compound Poisson process

We start by describing the process of arrival of new pieces of information to the market in the most straightforward way. The most intuitive assumption is that the information flow is a discrete process of independent events, and that the probability of arrival of new pieces of information in the time unit is constant. So, arrival of new information is modelled as a *Poisson process*. Whenever new information reaches the market, the price changes and such price changes are independently and identically distributed. We start by defining a first generalization of the Poisson process, setting

$$C_t = \sum_{k=1}^{N_t} Z_k, \qquad t \geq 0 \tag{2.9}$$

for a Poisson process $(N_t)_{t\geq 0}$, with rate λ, and independent (also of N_t) identically distributed jumps Z_k, $k \geq 1$. Such processes are called *compound Poisson processes*. The characteristic function of the compound Poisson process is

$$
\begin{aligned}
\phi_{C_t}(\gamma) = \mathbb{E}\left[\mathrm{e}^{i\gamma \sum_{k=1}^{N_t} Z_k}\right] &= \sum_{h=0}^{+\infty} \mathbb{E}\left[\mathrm{e}^{i\gamma \sum_{k=1}^{h} Z_k} \,\middle|\, N_t = h\right] \mathbb{P}(N_t = h) \\
&= \sum_{h=0}^{+\infty} \prod_{k=1}^{h} \mathbb{E}\left[\mathrm{e}^{i\gamma Z_k}\right] \mathrm{e}^{-\lambda t} \frac{(\lambda t)^h}{h!} \\
&= \sum_{h=0}^{+\infty} \left(\phi_{Z_1}(\gamma)\right)^h e^{-\lambda t} \frac{(\lambda t)^h}{h!} \\
&= \sum_{h=0}^{+\infty} \mathrm{e}^{-\lambda t} \frac{(\lambda t \phi_{Z_1}(\gamma))^h}{h!} = \mathrm{e}^{\lambda t(\phi_{Z_1}(\gamma)-1)} \\
&= \mathrm{e}^{\lambda t \int_{-\infty}^{+\infty} (\mathrm{e}^{i\gamma x}-1)\mu(\mathrm{d}x)}
\end{aligned} \tag{2.10}
$$

where ϕ_{Z_1} is the Fourier transform of Z_k and μ is the distribution of Z_k for every k. As a particular case ($Z_k = 1$) we trivially get the characteristic function of the Poisson process N_t (see Example 2.2.4).

It is very easy to check that the compound Poisson process is a Lévy process. By writing

$$
C_t = C_s + \sum\nolimits_{k=N_s+1}^{N_t} Z_k
$$

it is clear that C_t is the sum of C_s and an independent copy of C_{t-s}. Right-continuity and left-limits of the process N ensure right-continuity and left-limits for C.

Notice that a compound Poisson process is a random walk with jumps spaced out with independent and exponentially distributed periods. Let us make one step further to evaluate the properties of this model in the representation of the dynamics of prices. Intuitively, let us partition both time and the range of jumps in subsets. So, for example, we are interested in studying the variable of the number of price movements between 50 and 100 basis points over the next week.

More formally we define the support of each Z_k in $D_0 \subset \mathbb{R} \setminus \{0\}$. We then define $(a, b] \subset [0, +\infty)$ and $A \subset D_0$ as a Borel set. In the example above, the subset $(a, b]$ is the week and the subset A is the interval between 50 and 100 basis points. The random variable that we want to study is

$$
N\left((a, b] \times A\right) = \#\left\{t \in (a, b] : \Delta C_t \in A\right\}.
$$

First notice that $N\left((a, b] \times D_0\right) = N_b - N_a$: trivially, the overall number of price movements in the coming week is the difference between the number of movements at the end of the week and the number at the beginning. In addition to this, our variable has two interesting properties.

Proposition 2.3.1 *The function N satisfies the following two properties.*

1. for *all $n \geq 1$ and disjoint Borel sets $A_1, A_2, \ldots A_n \subset [0, +\infty) \times D_0$, the random variables $N(A_1), \ldots, N(A_n)$ are independent.*
2. *$N((a, b] \times A)$ is a Poisson random variable with parameter $(b - a)\lambda\mathbb{P}(Z_1 \in A)$.*

Proof. First, we recall the **Thinning property of Poisson processes.** If each point of a Poisson process $(N_t)_{t\geq 0}$ with rate λ is of type 1 with probability p and of type 2 with probability $1-p$, independent of one another, then the process $X^{(1)}$ and $X^{(2)}$ counting points of types 1 and 2, respectively, are independent Poisson processes with rates $p\lambda$ and $(1-p)\lambda$, respectively.

Let $(c,d] \subset D_0$. Consider the thinning mechanism, where the jth jump is of types 1 if $Z_j \in (c,d]$. Then, the process counting jumps in $(c,d]$ is a Poisson process with rate $\lambda\mathbb{P}(Z_1 \in (c,d])$ and so

$$N((a,b] \times (c,d]) = X_b^{(1)} - X_a^{(1)} \stackrel{d}{=} \text{Poi}((b-a)\lambda\mathbb{P}(Z_1 \in (c,d])).$$

For the independence of counts in disjoint rectangles $A_1, \ldots, A_n$, we first cut them into smaller rectangles $B_i = (a_i, b_i] \times (c_i, d_i]$, $1 \leq i \leq m$ such that for any two B_i and B_j either $(c_i, d_i] = (c_j, d_j]$ or $(c_i, d_i] \cap (c_j, d_j] = \emptyset$. Denote by k the number of different disjoint $(c_i, d_i]$. Now a straightforward generalization of the thinning property to k types splits $(N_t)_{t\geq 0}$ into k independent Poisson processes $X^{(i)}$ with rates $\lambda\mathbb{P}(Z_1 \in (c_i, d_i])$, $1 \leq i \leq k$. Now $N(B_1), \ldots N(B_m)$ are independent as increments of independent Poisson processes or of the same Poisson process over disjoint intervals. This property naturally extends to Borel disjoint sets $A_1, \ldots, A_n$. □

Intuitively, the first property says that the counting Poisson process can be considered as a collection of independent Poisson processes, each one defined with respect to a specific range of jump. The second property says that the number of counts grows linearly with time. So, once we have estimated that there may be five price changes between 50 and 100 basis points in one week, there must be 10 over a two-week period.

2.3.2 The Poisson point process

The above analysis somewhat reminds us of the standard Brownian motion process, for which variance grows linearly with time. The analogy may be stretched even further if we think that, just like volatility in the standard Brownian motion, intensity is also constant in the compound Poisson process. We then immediately think of possible extensions in which intensity may vary between one set and another. This leads us a definition of the so-called *Poisson point processes*.

Definition 2.3.1 *Let ν be a locally finite measure on $D_0 \subset \mathbb{R}\setminus\{0\}$. A process $(\Delta_t)_{t\geq 0}$ in $D_0 \cup \{0\}$ such that*

$$N((a,b] \times A) = \#\{s \in (a,b] : \Delta_s \in A\}, \quad 0 \leq a < b, A \subset D_0 \ (\textit{measurable})$$

satisfies

1. *for all $n \geq 1$ and disjoint $A_1, A_2, \ldots A_n \subset [0,+\infty) \times D_0$, the random variables $N(A_1), \ldots, N(A_n)$ are independent;*
2. *$N((a,b] \times A)$ is a Poisson random variable with parameter $(b-a)\nu(A)$;*

is called a ***Poisson point process with intensity measure*** *ν.*

To see this consider that

$$N_t(A) = N((0,t] \times A) = \#\{s \in (0,t] : \Delta_s \in A\}, \qquad t \geq 0$$

is a stochastic process. Since

$$N_{t+h}(A) - N_t(A) = N((t, t+h] \times A) \stackrel{d}{=} \mathrm{Poi}\,(h\nu(A)), \qquad t \geq 0, h > 0$$

and, for $0 \leq t_0 < t_1 < \ldots < t_n$

$$\begin{aligned} N_{t_0}(A) &= N((0, t_0] \times A), \\ N_{t_1}(A) &= N_{t_0}(A) = N((t_0, t_1] \times A), \ldots, \\ N_{t_n}(A) &= N_{t_{n-1}}(A) = N((t_{n-1}, t_n] \times A) \end{aligned}$$

are independent for part (1) of Definition 2.3.1, we have that $N_t(A)$ is a Poisson process with intensity $\nu(A)$.

$N_t(A)$ counts the number of points in A, but does not tell us where they are in A. Their distribution on A is the conditional distribution of ν:

Theorem 2.3.2 *For all measurable $A \subset D_0$ with $\nu(A) < +\infty$, denote the jump times of $N_t(A)$ by*

$$T_n(A) = \inf\{t \geq 0 : N_t(A) = n\} : \qquad n \geq 1$$

Then

$$Z_n(A) = \Delta_{T_n(A)} : \qquad n \geq 1$$

are independent of $N_t(A)$ and i.i.d. with common distribution $\nu(\cdot \cap A)/\nu(A)$.

2.3.3 Sums over Poisson point processes

It is now natural to aggregate the jumps of different ranges. For the sake of notation, we focus on positive jumps only (that is, ν concentrated on $(0, +\infty)$), since the same procedure can be exactly replicated for negative jumps. Thinking of Δ_s as a jump size at time s, we are going to study

$$X_t = \sum_{0 \leq s \leq t} \Delta_s$$

that is, the process performing all these jumps.

Finite activity

If $\nu(A) < +\infty$, by Theorem 2.3.2, the process $X_t^{(A)} = \sum_{0 \leq s \leq t} \Delta_s \mathbb{I}_{\{\Delta_s \in A\}}$, is a compound Poisson process with rate $\nu(A)$ and

$$X_t^{(A)} = \sum_{k=1}^{N_t(A)} \Delta_{T_n(A)}$$

In particular $X_t^{(A)}$ and $X_t^{(B)}$ are independent for disjoint Borel sets A and B (this is a consequence of (1) of Definition 2.3.1). Moreover, $X_t = X_t^{(D_0)}$ and, if $\nu(D_0) < +\infty$, X_t is a compound Poisson process.

Characteristic function of sums over positive Poisson point processes

Theorem 2.3.3 *Let $(\Delta_t)_{t\geq 0}$ be a Poisson point process with locally finite intensity measure ν concentrated on $(0, +\infty)$. Then for all $\gamma \in \mathbb{R}$*

$$\phi_{X_t}(\gamma) = \mathbb{E}\left[\exp\left(i\gamma \sum_{0\leq s\leq t} \Delta_s\right)\right] = \exp\left(t \int_0^{+\infty} \left(e^{i\gamma x} - 1\right) \nu(dx)\right)$$

Proof. Local finiteness of ν on $(0, +\infty)$ means in particular that if $I_n = \left(2^n, 2^{n+1}\right], n \in \mathbb{Z}$, $\nu(I_n) < +\infty$. This way $X_t^{(I_n)}$ is a compound Poisson process with rate $\nu(I_n)$. By (2.10) and Theorem 2.3.2

$$\begin{aligned}\phi_{X_t^{(I_n)}}(\gamma) &= \mathbb{E}\left[\exp\left(i\gamma X_t^{(I_n)}\right)\right] = \\ &= \exp\left(\nu(I_n)t \int_{I_n} \left(e^{i\gamma x} - 1\right) \frac{\nu(dx)}{\nu(I_n)}\right) \\ &= \exp\left(t \int_{I_n} \left(e^{i\gamma x} - 1\right) \nu(dx)\right)\end{aligned}$$

Now we have

$$Z_m = \sum_{n=-m}^{m} X_t^{(I_n)} = \sum_{n=-m}^{m} \sum_{0\leq s\leq t} \Delta_s \mathbb{I}_{\{\Delta_s \in I_n\}} \uparrow \sum_{0\leq s\leq t} \Delta_s \qquad \text{as } m \to +\infty$$

and the associated characteristic functions (products of individual characteristic functions being $X_t^{(I_n)}$ independent processes) converge as required:

$$\prod_{n=-m}^{m} \exp\left(t \int_{2^n}^{2^{n+1}} (e^{i\gamma x} - 1)\nu(dx)\right) \to \exp\left(t \int_0^{+\infty} (e^{i\gamma x} - 1)\nu(dx)\right)$$

□

Infinite activity

We now take care of cases in which ν is not integrable. We can prove by a technical argument that it cannot be so at infinity. In fact, if ν is not integrable at infinity, the process would have an infinite number of jumps greater than a given threshold: $\#\{0 \leq s \leq t : \Delta_s > 1\} \overset{d}{=} \text{Poi}\,(t\nu((1, +\infty))) \overset{d}{=} \text{Poi}(+\infty)$, and this is in contrast with the nature of a right-continuous function with left limits.

On the other hand, the case in which ν is not integrable at zero is possible. Intuitively, if intensity grows larger and larger as the interval A shrinks towards zero, one would expect that the dimension of jumps should become smaller and smaller in such a way as to dampen the increase in the number of jumps. In a sense, this leads to a condition that variance of changes in price in a bounded neighbourhood around 0 must be finite.

Below we report the technical steps of the proof for completeness. We first derive the first and second moments of jumps by a Taylor expansion of the characteristic function.

Proposition 2.3.4 *Let $(\Delta_t)_{t\geq 0}$ be a Poisson point process with intensity measure ν on* $(0, +\infty)$.

(1) *If* $\int_0^{+\infty} x\nu(\mathrm{d}x) < +\infty$, *then*

$$\mathbb{E}\left[\sum_{s\le t}\Delta_s\right] = t\int_0^{+\infty} x\nu(\mathrm{d}x)$$

(2) *If* $\int_0^{+\infty} x^2\nu(\mathrm{d}x) < +\infty$, *then*

$$\mathrm{Var}\left(\sum_{s\le t}\Delta_s\right) = t\int_0^{+\infty} x^2\nu(\mathrm{d}x)$$

Proof. These are the two leading terms in the expansion with respect to γ of the Fourier transform of Theorem 2.3.3. □

Thanks to the independence and stationarity of increments, a Lévy process is a martingale if and only if it has zero mean. In the case of a compound Poisson process $C_t = \sum_{j=1}^{N_t} Z_j$, $\mathbb{E}[C_t] = \lambda t\mathbb{E}[Z_1] = t\int_{-\infty}^{+\infty} x\nu(\mathrm{d}x)$, and $C_t - t\int_{-\infty}^{+\infty} x\nu(\mathrm{d}x)$ is again not only a Lévy process but is also a martingale.

Consider now the following compound Poisson processes with drifts that turn them into martingales.

$$Z_t^\epsilon = \sum_{s\le t}\Delta_s\mathbb{I}_{\{\epsilon<\Delta_s\le 1\}} - t\int_\epsilon^1 x\nu(\mathrm{d}x) \tag{2.11}$$

We have deliberately excluded jumps in $(1, +\infty)$ as they are easily handled separately. What integrability condition on ν do we need for Z_t^ϵ to converge as $\epsilon \downarrow 0$?

Lemma 2.3.5 *Let* $(\Delta_t)_{t\ge 0}$ *be a Poisson point process with intensity measure* ν *on* $(0, 1)$. *With* Z^ϵ *defined in (2.11),* Z_t^ϵ *converges in* L^2 *if* $\int_0^1 x^2\nu(\mathrm{d}x) < +\infty$.

Proof. Note that for $0 < \delta < \epsilon < 1$, by (2) of Proposition 2.3.4 applied to ν restricted on $[\delta, \epsilon)$,

$$\mathbb{E}\left[\left|Z_t^\delta - Z_t^\epsilon\right|^2\right] = t\int_\delta^\epsilon x^2\nu(\mathrm{d}x)$$

so that $\left(Z_t^\epsilon\right)_{0<\epsilon<1}$ is a Cauchy family as $\epsilon \downarrow 0$, for the L^2-distance $d(X, Y) = \sqrt{\mathbb{E}[(X-Y)^2]}$. By completeness of the L^2-space, there is a limiting random variable Z_t as required. □

Theorem 2.3.6 *There exists a Lévy process whose jumps form a Poisson point process with intensity measure* ν *on* $(0, +\infty)$ *if and only if* $\int_0^{+\infty}(1\wedge x^2)\nu(\mathrm{d}x) < +\infty$.

Proof. The "only if" statement is a consequence of the Lévy-Khintchine characterization of Lévy processes (Theorem 2.2.9)

Let us prove the "if" part. By part (1) of Proposition 2.3.4, $\mathbb{E}\left[Z_t^\epsilon - Z_t^\delta\right] = 0$. The process $Z_t^\epsilon - Z_t^\delta$ is a martingale and is square integrable thanks to part (2) of Proposition 2.3.4. The maximal inequality shows that

$$\mathbb{E}\left[\sup_{0\le s\le t}|Z_s^\epsilon - Z_s^\delta|\right] \le 4\mathbb{E}\left[Z_t^\epsilon - Z_t^\delta|^2\right] = 4t\int_\delta^\epsilon x^2\nu(\mathrm{d}x)$$

so that $(Z_s^\epsilon, 0 \le s \le t)_{0<\epsilon<1}$ is a Cauchy family as $\epsilon \downarrow 0$, for the uniform L^2-distance $d_{[0,t]}(X, Y) = \sqrt{\mathbb{E}\left[\sup_{0\le s\le t} |X_s - Y_s|^2\right]}$. By completeness of the L^2-space, there is a limiting process $\left(Z_s^{(1)}\right)_{0\le s\le t}$ which is the uniform limit (in L^2) of $\left(Z_s^\epsilon\right)_{0\le s\le t}$ right-continuous with left limits. Also consider the independent compound Poisson process

$$Z_t^{(2)} = \sum_{s\le t} \Delta_s \mathbb{I}_{\{\Delta_s>1\}}$$

and set

$$Z = Z^{(1)} + Z^{(2)}$$

It is not difficult to show that Z is a Lévy process that incorporates all jumps $(\Delta_s)_{0\le s\le t}$. □

2.3.4 The decomposition theorem

We are now in a position to gather all the possible different shocks that may reach the price and collect them into the price process. A Lévy process is made of a drift, a diffusion, a set of finite jumps of large size and a set of infinite jumps of infinitesimal size. It does not come as a surprise that all of this will end in the same formula, like the one we saw above as the characteristic function of Lévy processes.

Theorem 2.3.7 (***Lévy–Itô decomposition theorem***) *Let (a, σ^2, ν) be Lévy–Khintchine characteristics, $(B_t)_{t\ge0}$ a standard Brownian motion and $(\Delta_t)_{t\ge0}$ an independent Poisson point process of jumps with intensity measure ν. There is a Lévy process*

$$Z_t = at + \sigma B_t + M_t + C_t$$

where

$$C_t = \sum_{s\le t} \Delta_s \mathbb{I}_{\{|\Delta_s|>1\}}$$

is a compound Poisson process (of big jumps) and

$$M_t = \lim_{\epsilon\downarrow0}\left(\sum_{s\le t} \Delta_s \mathbb{I}_{\{\epsilon<|\Delta_s|\le1\}} - t\int_{\{x\in\mathbb{R}:\epsilon<|x|\le1\}} x\nu(dx)\right) \tag{2.12}$$

is a martingale (of small jumps compensated by a linear drift).

Proof (Outline). The construction of $M_t = P_t - N_t$ can be made from two independent processes P_t and N_t with no negative jumps, as in Theorem 2.3.6. N_t will be built from a Poisson point process with intensity measure $\overline{\nu}((c, d]) = \nu([-d, -c))$, $0 < c < d \le 1$.

We check that the characteristic function of $Z_t = at + \sigma B_t + P_t - N_t + C_t$ is of Lévy–Khintchine type with parameters (a, σ^2, ν). We have five independent components:

$$\mathbb{E}\left[e^{i\gamma at}\right] = e^{i\gamma at}$$

$$\mathbb{E}\left[e^{i\gamma\sigma B_t}\right] = e^{-\frac{1}{2}\gamma^2\sigma^2 t}$$

$$\mathbb{E}\left[e^{i\gamma P_t}\right] = \exp\left\{t\int_0^1 (e^{i\gamma x} - 1 - i\gamma x)\nu(dx)\right\}$$

$$\mathbb{E}\left[e^{-i\gamma N_t}\right] = \exp\left\{t\int_0^1 (e^{-i\gamma x} - 1 + i\gamma x)\overline{\nu}(dx)\right\} = \exp\left\{t\int_{-1}^0 (e^{i\gamma x} - 1 - i\gamma x)\nu(dx)\right\}$$

$$\mathbb{E}\left[e^{i\gamma C_t}\right] = \exp\left\{t\int_{|x|>1} (e^{i\gamma x} - 1)\nu(dx)\right\}$$

Now the characteristic function of Z_t is the product of the characteristic functions of the independent components, and this yields the expected formula. □

The decomposition theorem provides us with a constructive treatment of the Lévy process, so that we are now able to recognize the different kinds of shocks represented in this family of processes.

Example 2.3.1 *The following is a list of Lévy processes and their characteristics:*

1. ***Brownian motion:*** *The Brownian motion is parameterized by the characteristics*

$$a = 0, \quad \sigma^2 > 0, \quad \nu = 0$$

2. ***Poisson process:*** *The Poisson process with intensity λ is parameterized by the characteristics*

$$a = 0, \quad \sigma = 0, \quad \nu(dx) = \lambda\delta_1(x)$$

3. ***Compound Poisson process:*** *The compound Poisson process* (see (2.9)) *is parameterized by the characteristics*

$$a = \sigma^2 = 0, \quad \nu(dx) = \lambda\mu(dx)$$

where μ is the law of the i.i.d. jumps and λ is the intensity of the Poisson process counting the jumps.

4. ***Stable process:*** *The α-stable process for $0 < \alpha < 2$ is parameterized by the characteristics*

$$\sigma = 0, \text{ and } \nu(dx) = \begin{cases} c_1 x^{-1-\alpha}\,dx, & \text{for } x > 0 \\ c_2 |x|^{-1-\alpha}\,dx, & \text{for } x < 0 \end{cases}$$

with $c_1, c_2 > 0$.

5. ***Gamma process:*** *The Gamma process, where $X_t \stackrel{d}{=} \Gamma(\alpha t, \beta)$, is parameterized by the characteristics*

$$a = \sigma^2 = 0, \nu(dx) = \alpha x^{-1} e^{-\beta x}\,dx, \; x > 0$$

By a straightforward computation one gets

$$\phi_{X_t}(\gamma) = \exp\left\{t\int_0^{+\infty} (e^{i\gamma x} - 1)\alpha x^{-1} e^{-\beta x}\,dx\right\} = \left(\frac{\beta}{\beta - i\gamma}\right)^{\alpha t}$$

6. ***Variance Gamma process:*** *The Variance Gamma process is defined as the difference $X = G - H$ of two independent Gamma processes G and H. Let $G_t \stackrel{d}{=} \Gamma(\alpha_+ t, \beta_+)$ and*

$H_t \stackrel{d}{=} \Gamma(\alpha_- t, \beta_-)$. *The characteristic function of the Variance Gamma process is*

$$\phi_{X_t}(\gamma) = \mathbb{E}\left[e^{i\gamma G_t}\right]\mathbb{E}\left[e^{-i\gamma H_t}\right] = \left(\frac{\beta_+}{\beta_+ - i\gamma}\right)^{\alpha_+ t}\left(\frac{\beta_-}{\beta_- + i\gamma}\right)^{\alpha_- t} =$$

$$= \exp\left\{t\int_0^{+\infty}(e^{i\gamma x} - 1)\alpha_+ x^{-1}\,e^{-\beta_+ x}\,dx\right\}\exp\left\{t\int_0^{+\infty}(e^{-i\gamma x} - 1)\alpha_- x^{-1}\,e^{-\beta_- x}\,dx\right\} =$$

$$= \exp\left\{t\int_0^{+\infty}(e^{i\gamma x} - 1)\alpha_+ x^{-1}\,e^{-\beta_+ x}\,dx + t\int_0^{+\infty}(e^{-i\gamma x} - 1)\alpha_- x^{-1}\,e^{-\beta_- x}\,dx\right\}$$

and the Lévy–Khintchine characteristics are

$$a = \sigma = 0, \text{ and } \nu(dx) = \begin{cases} \alpha_+|x|^{-1}\,e^{-\beta_+|x|}\,dx, & x > 0 \\ \alpha_-|x|^{-1}\,e^{-\beta_-|x|}\,dx, & x < 0 \end{cases}$$

7. ***CGMY process:*** *As a natural generalization of the Variance Gamma process, Carr, Geman, Madan and Yor (CGMY) suggested the following Lévy measure for financial price processes:*

$$\nu(dx) = \begin{cases} C\,e^{-G|x|}\,|x|^{-Y-1}\,dx, & x > 0 \\ C\,e^{-M|x|}\,|x|^{-Y-1}\,dx, & x < 0 \end{cases}$$

for parameters $C > 0$, $G \geq 0$, $M \geq 0$, $Y < 2$). *The condition* $Y < 2$ *is induced by the requirement that Lévy densities integrate* x^2 *in the neighbourhood of* 0. *The characteristic exponent is, for* $Y \neq 0, 1$,

$$\psi(u) = -\Gamma(-Y)\left[C\left((G - iu)^Y - G^Y\right) + C\left((M + iu)^Y - M^Y\right)\right]$$

The CGMY model contains the Gamma model for $Y = 0$. *When this model is fitted to financial data, there is usually significant evidence against* $Y = 0$, *so the CGMY model seems more appropriate than the Variance Gamma model.*

The parameters play an important role in capturing various aspects of the stochastic process. The parameter C *may be viewed as a measure of the overall level of activity. Keeping the other parameters constant and integrating over all moves exceeding a small level, we see that the aggregate activity level may be calibrated through movements in* C. *In the special case when* $G = M$, *the Lévy measure is symmetric and, in this case, Madan et al. (1998) show that the parameter* C *provides control over the kurtosis of the distribution of the process.*

The parameters G *and* M, *respectively, control the rate of the exponential decay on the right and left of the Lévy density, leading to skewed distributions when they are different. For* $G < M$, *the left tail of the distribution for* X_t *is heavier than the right tail.*

The parameter Y *(see Figure* 2.1) *was studied in Vershik and Yor* (1995) *and arises in the process for the stable law. The parameter* Y *is particularly useful in characterizing the fine structure of the stochastic process; in fact for* $Y > 0$ *it generates a process of infinite activity. Moreover, as we shall see below,* Y *characterizes whether the jumps of process have finite or infinite variation (see section* 2.4.1) *or are endowed with a completely monotone density (see section* 2.4.2).

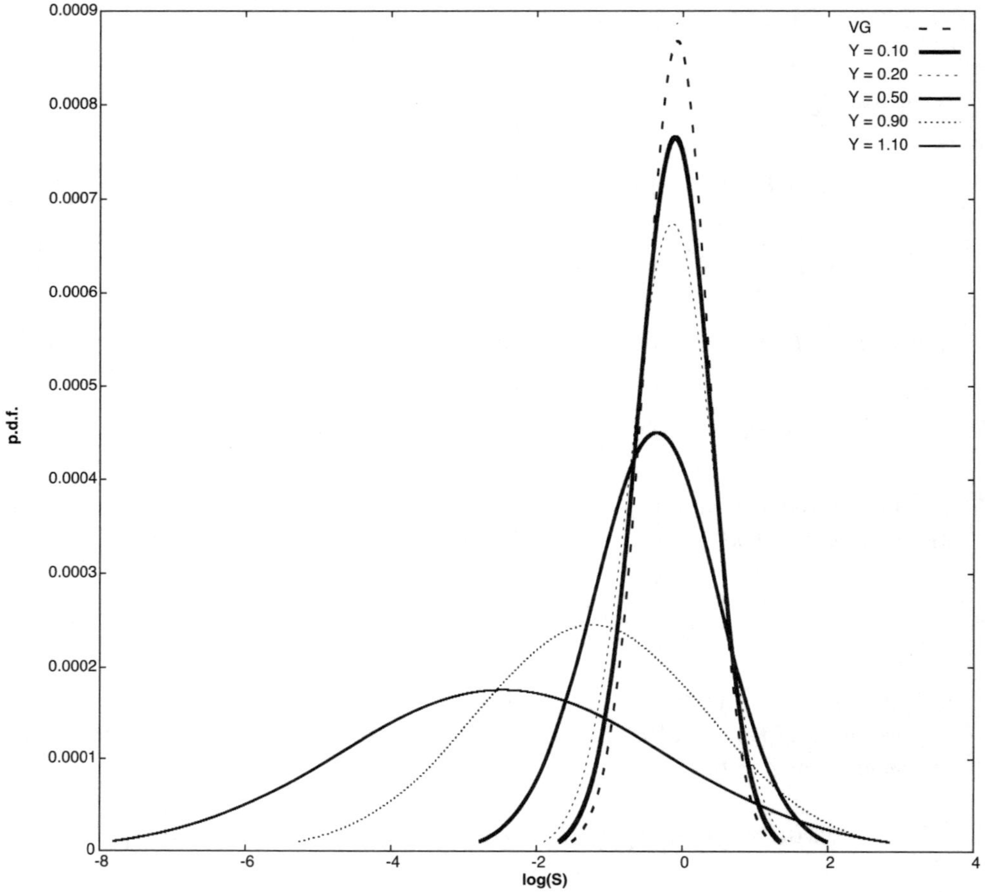

Figure 2.1 The p.d.f. for the CGMY model. Several values of Y

8. ***Meixner process:*** *The Meixner process is the Lévy process associated to the Meixner distribution. This is an infinitely divisible distribution with density function*

$$f_{\alpha,\beta,\delta,\mu}(x) = \frac{(2\cos(\beta/2))^{2\delta}}{2\alpha\pi\,\Gamma(2\delta)} \exp\left(\frac{\beta(x-\mu)}{\alpha}\right) \left|\Gamma\left(\delta + \frac{i(x-\mu)}{\alpha}\right)\right|^2$$

with $\alpha > 0$, $-\pi < \beta < \pi$, $\delta > 0$, $\mu \in \mathbb{R}$. The characteristics associated to the Meixner distribution are:

$$a = \alpha\delta\tan(\beta/2) - 2\delta \int_1^{+\infty} \frac{\sinh(\beta x/\alpha)}{\sinh(\pi x/\alpha)}\,\mathrm{d}x + \mu$$

and

$$\nu(\mathrm{d}x) = \delta \frac{\exp(\beta x/\alpha)}{x\sinh(\pi x/\alpha)}\,\mathrm{d}x$$

The characteristic exponent is

$$\psi(u) = -\log\left(\left(\frac{\cos(\beta/2)}{\cosh(\alpha u - i\beta)/2}\right)^{2\delta}\right) - i\mu u$$

9. ***The generalized hyperbolic process:*** *The generalized hyperbolic process is associated to the hyperbolic distribution whose density is*

$$f(x) = \frac{\sqrt{\alpha^2-\beta^2}}{2\alpha\delta K_1\left(\delta\sqrt{\alpha^2-\beta^2}\right)} e^{-\alpha\sqrt{\delta^2+(x-\mu)^2}+\beta(x-\mu)}$$

where $\mu \in \mathbb{R}$, $\delta > 0$, $0 \le |\beta| < \alpha$ and K_1 denotes the modified Bessel function with index 1. α and β determine the shape (β being responsible for skewness), δ and μ are respectively scale and location parameters. The hyperbolic distribution provides heavier tails.

The Lévy measure in the symmetric centered case ($\beta = \mu = 0$) has the following expression:

$$\nu(dx) = \frac{1}{|x|}\left[\int_0^{+\infty} \frac{e^{-\sqrt{2y+\alpha^2}|x|}\,dy}{\pi^2 y(J_1^2(\delta\sqrt{2y}) + Y_1^2(\delta\sqrt{2y}))} + e^{-\alpha|x|}\right] dx$$

where J_1 and Y_1 are Bessel functions.

By using asymptotes of the various Bessel functions, one can deduce that $\nu(dx) \sim 1/x^2\,dx$ for $x \to 0$; hence the Lévy measure is not integrable and the distribution defines an infinite activity setting.

The generalized hyperbolic distribution involves an extra parameter λ and has the following density:

$$f(x) = a(\lambda,\alpha,\beta,\delta)(\delta^2 + (x-\mu)^2)^{\frac{(\lambda-\frac{1}{2})}{2}} K_{\lambda-\frac{1}{2}}\left(\alpha\sqrt{\delta^2+(x-\mu)^2}\exp(\beta(x-\mu))\right)$$

where

$$a(\lambda,\alpha,\beta,\delta) = \frac{(\alpha^2-\beta^2)^{\lambda/2}}{\sqrt{2\pi}\alpha^{\lambda-\frac{1}{2}}\delta^\lambda K_\lambda\left(\delta\sqrt{\alpha^2-\beta^2}\right)}$$

and K_a denotes, as before, the Bessel function with index a. The extra parameter λ characterizes certain subclasses and has essentially an impact on the heaviness of the tails. For $\lambda = 1$, we recover the subclass of hyperbolic distributions, for $\lambda = -\frac{1}{2}$ the normal inverse Gaussian.

A Lévy process $(X_t)_{t\ge0}$, such that X_1 has the generalized hyperbolic distribution is called the generalized hyperbolic Lévy motion (this definition is due to Eberlein, 2001).

2.4 PROPERTIES OF LÉVY PROCESSES

In the following we refer to Cont and Tankov, 2004, and to Winkel, Lecture Notes, for details.

2.4.1 Pathwise properties of Lévy processes

Using the Lévy–Itô decomposition theorem, in this section we shall deduce some properties of the paths of Lévy processes.

Number of jumps

In subsection 2.3.3 we showed that, if $\nu(\mathbb{R} \setminus \{0\}) < +\infty$, the characteristic triplet $(0, 0, \nu)$ defines a compound Poisson process – that is, a process with piecewise constant paths. Since the number of jumps is given by a Poisson process with finite intensity, their number in each bounded time interval is finite.

If $\nu(\mathbb{R} \setminus \{0\}) = +\infty$ the set of jumps of every trajectory of the Lévy process associated to the characteristic triplet $(0, 0, \nu)$ is countably infinite and dense in $[0, +\infty)$. The countability follows directly from the fact that the paths are right-continuous with left limits.

To prove that the set of jump times is dense in $[0, +\infty)$, consider a time interval $(a, b]$. For every $n \in \mathbb{Z}$ let $I_n = (2^n, 2^{n+1}] \cup (-2^n, -2^{n+1}]$. Clearly $\cup_{n\in\mathbb{Z}} T_n = \mathbb{R} \setminus \{0\}$. Being I_n disjoint sets, $N_t(I_n)$ are independent Poisson processes with intensity $\nu(I_n)$, and the number of jumps in I_n, $N_b(I_n) - N_a(I_n)$ is a Poisson distributed random variable with parameter $(b-a)\nu(I_n)$. The total number of jumps in the time interval $(a, b]$ is

$$\sum_{n\in\mathbb{Z}} (N_b(I_n) - N_a(I_n)) .$$

But

$$Z_m = \sum_{n=-m}^{m} (N_b(I_n) - N_a(I_n)) \stackrel{d}{=} \text{Poi}\left((b-a) \sum_{n=-m}^{m} \nu(I_n)\right).$$

Being Z_m increasing,

$$Z_m \uparrow \sum_{n\in\mathbb{Z}} (N_b(I_n) - N_a(I_n))$$

and clearly $Z_m \stackrel{\text{d}}{\to} \text{Poi}(+\infty)$. This way $\mathbb{P}\left(\sum_{n\in\mathbb{Z}} (N_b(I_n) - N_a(I_n)) = +\infty\right) = 1$ for every time interval $(a, b]$. This means that the set of jump times is dense in $[0, +\infty)$.

The total variation of Lévy processes trajectories

Definition 2.4.1 *The total variation* (TV) *of a function $f : [0, t] \to \mathbb{R}$ is defined by*

$$\text{TV}_t(f) = \sup \sum_{i=1}^{n} |f(t_i) - f(t_{i-1})|$$

where the supremum is taken over all finite partitions $0 = t_0 < t_1 < \ldots < t_{n-1} < t_n = t$ of the interval $[0, t]$.

In particular every increasing or decreasing function is of finite variation and every function of finite variation is a difference of two increasing functions.

A Lévy process $(X_t)_{t\geq 0}$ is said to be of finite variation if

$$\mathbb{P}(\text{TV}_t(X) < +\infty) = 1$$

Proposition 2.4.1 *The Brownian motion is not of finite variation.*

Proof. The trajectories of a Brownian motion B_t have finite quadratic variation

$$\sum_{j=1}^{2^n} |B_{tj2^{-n}} - B_{t(j-1)2^{-n}}|^2 \to t, \text{ in the } L^2 \text{ sense}$$

since

$$\mathbb{E}\left(\sum_{j=1}^{2^n} |B_{tj2^{-n}} - B_{t(j-1)2^{-n}}|^2\right) = 2^n \mathbb{E}\left(B_{t2^{-n}}^2\right) = t$$

and

$$\mathbb{E}\left(\left(\sum_{j=1}^{2^n} |B_{tj2^{-n}} - B_{t(j-1)2^{-n}}|^2 - t\right)^2\right) = \text{Var}\left(\sum_{j=1}^{2^n} |B_{tj2^{-n}} - B_{t(j-1)2^{-n}}|^2\right)$$
$$\leq 2^n (2^{-n}t)^2 \text{Var}\left(B_1^2\right) \to 0$$

but then assuming finite total variation with positive probability, the uniform continuity of the Brownian paths implies

$$\sum_{j=1}^{2^n} |B_{tj2^{-n}} - B_{t(j-1)2^{-n}}|^2 \leq \left(\sup_{j=1,\dots,2^n} |B_{tj2^{-n}} - B_{t(j-1)2^{-n}}|\right)$$
$$\sum_{j=1}^{2^n} |B_{tj2^{-n}} - B_{t(j-1)2^{-n}}| \to 0$$

with positive probability, but this is inconsistent with convergence to t, so the assumption of finite total variation must have been wrong. □

On the other hand, a compound Poisson process is clearly of finite variation.

To understand how jumps influence total variation, let us focus for a moment on the case in which ν is concentrated in $(0, +\infty)$. The pure jumps process $\sum_{s\leq t} \Delta_s$ admits only positive jumps, thus it would be expected to be increasing – that is, of finite variation. This is true if small jumps are summable, that is if $\int_0^1 x\nu(\mathrm{d}x) < +\infty$. Nevertheless non-summable jumps are admissible: in this case $\sum_{\epsilon<s\leq 1} \Delta_s$ explodes to $+\infty$ as $\epsilon \downarrow 0$, but in order to let the limit (2.12) exist, the term $\int_\epsilon^1 x\nu(\mathrm{d}x)$ has to explode as well. The formalization of the above argument is the content of Proposition 2.4.3, but to prove it we need a preliminary lemma:

Lemma 2.4.2 *Let f be a right-continuous function with left limits and jumps $(\Delta f_s)_{0\leq s\leq t}$. Then*

$$\text{TV}_t(f) \geq \sum_{0\leq s\leq t} |\Delta f_s|$$

Proof. Enumerate the jumps in decreasing order of size by $\left(T_n, \Delta f_{T_n}\right)_{n\geq 0}$. Fix $N \in \mathbb{N}$ and $\delta > 0$. Choose $\epsilon > 0$ so small that $\cup[T_n - \epsilon, T_n]$ is a disjoint union and such that $|f(T_n - \epsilon) - f(T_n-)| < \delta/N$. Then for $\{T_n - \epsilon, T_n : n = 1, \dots, N\} = \{t_1, \dots, t_{2N}\}$ such that

$0 = t_0 < t_1 < \ldots < t_{2N+1} = t$, we have

$$\sum_{j=1}^{2N+1} |f(t_j) - f(t_{j-1})| \geq \sum_{n=1}^{N} |\Delta f(T_n)| - \delta$$

Since N and δ are arbitrary, this completes the proof, whether the right-hand side is finite or infinite. □

Proposition 2.4.3 *A Lévy process is of finite variation if and only if its characteristic triplet* (a, σ^2, ν) *satisfies*

$$\sigma^2 = 0 \text{ and } \int_{-\infty}^{+\infty} (|x| \wedge 1)\nu(\mathrm{d}x) < +\infty$$

Proof. The *if* part. Under the stated conditions, X_t can be represented in the following form

$$X_t = bt + \sum_{s \leq t} \Delta_s \mathbb{I}_{\{|\Delta_s| > 1\}} + \lim_{\epsilon \downarrow 0} \sum_{s \leq t} \Delta_s \mathbb{I}_{\{\epsilon < |\Delta_s| \leq 1\}} \tag{2.13}$$

where

$$b = a - \int_{\{x \in \mathbb{R}: |x| \leq 1\}} x\nu(\mathrm{d}x)$$

The first two terms of (2.13) are of finite variation, being a compound Poisson process plus a drift, therefore we only need to consider the third term. Its variation on the interval $[0, t]$ is

$$\mathrm{TV}_t\left(\sum_{s \leq t} \Delta_s \mathbb{I}_{\{\epsilon < |\Delta_s| \leq 1\}}\right) = \sum_{s \leq t} |\Delta_s| \mathbb{I}_{\{\epsilon < |\Delta_s| \leq 1\}}$$

Since the integrand in the right-hand side is positive, we obtain,

$$\mathbb{E}\left[\mathrm{TV}_t\left(\sum_{s \leq t} \Delta_s \mathbb{I}_{\{\epsilon < |\Delta_s| \leq 1\}}\right)\right] = t \int_{\{x \in \mathbb{R}: \epsilon < |x| \leq 1\}} |x|\nu(\mathrm{d}x)$$

which converges to a finite value when $\epsilon \to 0$. Therefore

$$\mathbb{E}\left[\mathrm{TV}_t\left(\lim_{\epsilon \downarrow 0} \sum_{s \leq t} \Delta_s \mathbb{I}_{\{\epsilon < |\Delta_s| \leq 1\}}\right)\right] < +\infty$$

which implies that the variation of X_t is almost certainly finite.

The *only if* part. Consider the Lévy–Itô decomposition of X_t. By Lemma 2.4.2, the variation of any right-continuous with left limits function is greater or equal to the sum of its jumps. We have for every $\epsilon > 0$

$$\begin{aligned}
\mathrm{TV}_t(X) &\geq \sum_{s \leq t} |\Delta_s| \mathbb{I}_{\{\epsilon < |\Delta_s| \leq 1\}} = \\
&= t \int_{\epsilon}^{1} |x|\nu(\mathrm{d}x) + \left(\sum_{s \leq t} |\Delta_s| \mathbb{I}_{\{\epsilon < \Delta_s \leq 1\}} - t \int_{\epsilon}^{1} |x|\nu(\mathrm{d}x)\right)
\end{aligned}$$

As shown in the proofs of Lemma 2.3.5 and Theorem 2.3.6, the second term converges to something finite. Therefore, if the condition $\int_{-\infty}^{+\infty}(|x| \wedge 1)\nu(\mathrm{d}x) < +\infty$ is not satisfied, the first term in the last line will diverge and the variation of X_t will be infinite.

Suppose now that this condition is satisfied. This means that X_t may be written as a term of finite variation plus a Brownian motion. Since the trajectories of a Brownian motion are almost surely of infinite variation (as we have shown at the beginning of the section), if σ^2 is non-zero, X_t will also have infinite variation. Therefore we must have $\sigma^2 = 0$. □

In this way a finite variation Lévy process can be expressed as the sum of its jumps and a linear drift term

$$X_t = bt + \sum_{s \in [0,t]} \Delta X_s \tag{2.14}$$

where

$$b = a - \int_{\{x \in \mathbb{R}:|x| \le 1\}} x\nu(\mathrm{d}x). \tag{2.15}$$

Example 2.4.1 *We refer to Example* 2.3.1 *for the notation.*

1. ***Stable process:*** *The stable process is of finite variation if and only if* $\alpha \in (0, 1)$. *In fact*

$$\int_0^1 x x^{-1-\alpha}\,\mathrm{d}x = \int_0^1 x^{-\alpha}\,\mathrm{d}x = \begin{cases} +\infty & \text{if } \alpha \ge 1 \\ \frac{1}{1-\alpha} & \text{if } \alpha < 1 \end{cases}$$

2. ***Gamma process:*** *The Gamma process is always of finite variation. In fact*

$$\alpha \int_0^1 x x^{-1}\,\mathrm{e}^{-\beta x}\,\mathrm{d}x = \alpha \int_0^1 \mathrm{e}^{-\beta x}\,\mathrm{d}x = \frac{\alpha}{\beta}\left(1 - \mathrm{e}^{-\beta}\right)$$

3. ***Variance Gamma process:*** *This is of finite variation, being the difference of two processes of finite variation.*
4. ***CGMY process:*** *The CGMY process is of finite variation if and only if* $Y < 1$. *In fact*

$$\int_0^1 x x^{-1-Y}\,\mathrm{e}^{-Gx}\,\mathrm{d}x = \int_0^1 x^{-Y}\,\mathrm{e}^{-Gx}\,\mathrm{d}x$$

and the last integral is finite if and only if $Y < 1$.

2.4.2 Completely monotone Lévy densities

Definition 2.4.2 *A function* $f(x)$ *on* $(0, +\infty)$ *is said to be completely monotone if it admits derivatives of all orders and if* $(-1)^n \frac{\mathrm{d}^n}{\mathrm{d}x^n} f(x) > 0$ *on* $(0, +\infty)$ *for* $n = 0, 1, \ldots$.

The Bernstein theorem tells us that f is completely monotone if and only if it can be expressed as a mixture of exponentials of type:

$$f(x) = \int_0^{+\infty} \mathrm{e}^{-xy}\rho(\mathrm{d}y)$$

for some measure ρ. The following result holds (see Sato, 1999, Theorem 51.6):

Theorem 2.4.4 *Consider a probability measure μ on $(0, +\infty)$, such that $\mu(\mathrm{d}x) = c\delta_{\{0\}} + f(x)\mathbb{I}_{(0,+\infty)}(x)\,\mathrm{d}x$ with $0 < c < 1$ and $f(x)$ being completely monotone. Then μ is infinitely divisible.*

Let us consider the family which is the union of $\{\delta_0\}$ and the class of exponential distributions. The class of mixtures of this family is called the *class ME* and it coincides with the class of laws μ considered in the above theorem.

For a variety of models on these lines the reader is referred to Geman *et al.* (2001). The characterizing feature of CM Lévy densities is that they structurally relate arrival rates of large jumps sizes to smaller jump sizes by requiring, among other things, that large jumps arrive less frequently than small jumps.

For example, the CGMY process has completely monotone Lévy density for $Y > -1$.

2.4.3 Moments of a Lévy process

Proposition 2.4.5 *Let X_t be a Lévy process with characteristic triplet (a, σ^2, ν). If $\sigma > 0$ or $\nu(\mathbb{R} \setminus \{0\}) = +\infty$, then X_t has a continuous density on $\mathbb{R}$.*

Proof. This is an immediate consequence of the Lévy–Khintchine representation and the properties of the Fourier transform (see Sato, 1999, Chapter 5). □

The tail behaviour of the distribution of a Lévy process and its moments are determined by the Lévy measure.

Proposition 2.4.6 *Let X_t be a Lévy process with characteristic triplet (a, σ^2, ν). The $n > 0$th absolute moment of X_t, $\mathbb{E}[|X_t|^n]$ is finite for some t, or equivalently for every $t > 0$, if and only if $\int_{|x|\geq 1} |x|^n \nu(\mathrm{d}x) < +\infty$. In this case, integer moments of X_t can be computed from its characteristic function by differentiation. In particular, the form of first moments of X_t is especially simple:*

$$\mu_1(X_t) = \mathbb{E}[X_t] = t\left(a + \int_{|x|\geq 1} x\nu(\mathrm{d}x)\right)$$

$$\mu_2(X_t) = \mathrm{Var}(X_t) = t\left(\sigma^2 + \int_{-\infty}^{+\infty} x^2\nu(\mathrm{d}x)\right)$$

$$\mu_3(X_t) = \mathbb{E}[(X_t - \mu_1(X_t))^3] = t\int_{-\infty}^{+\infty} x^3\nu(\mathrm{d}x)$$

Moreover, the skewness coefficient of X_t is

$$s(X_t) = \frac{\mu_3(X_t)}{\mu_2^{\frac{3}{2}}(X_t)}$$

and the kurtosis of X_t is

$$k(X_t) = \frac{t\int_{-\infty}^{+\infty} x^4\nu(\mathrm{d}x)}{\mu_2^2(X_t)}$$

Proof. See Corollary 25.8 in Sato (1999). □

The above proposition entails that all infinitely divisible distributions are leptokurtic since $k(X_t) > 0$. Moreover,

$$s(X_t) = \frac{s(X_1)}{\sqrt{t}}, \qquad k(X_t) = \frac{k(X_1)}{t}$$

Therefore the increments of a Lévy process or, equivalently, all infinitely divisible distributions are always leptokurtic, but the kurtosis and the skewness (if there is any) decrease with a different speed as the time interval increases.

In modelling assets dynamics we will be interested in exponentials of the Lévy process and, consequently, in exponential moments of the Lévy process.

Proposition 2.4.7 *Let X_t be a Lévy process with characteristic triplet (a, σ^2, ν) and let $u \in \mathbb{R}$. The exponential moment $\mathbb{E}[e^{uX_t}]$ is finite for some t or, equivalently, for all $t > 0$ if and only if $\int_{|x|\geq 1} e^{ux}\nu(dx) < +\infty$. In this case*

$$\mathbb{E}[e^{uX_t}] = e^{-t\psi(-iu)}$$

where ψ is the characteristic exponent of the Lévy process.

Proof. See Sato, 1999, Theorem 25.17.
This covers all (or almost all) we need to know about Lévy processes for our applications. As we casually observe, however, a feature of these processes is that stationarity of increments can be too restrictive. This assumption will be relaxed in the next chapter. Nevertheless we will see that what we learned of Lévy processes can still be useful in the study of non-stationary processes. □

3

Non-stationary Market Dynamics

It is a well-known empirical regularity of financial markets that the distribution of returns measured on the same period length is subject to change as time elapses, due to changes in general market conditions. A typical evidence is that huge price movements are concentrated in the same periods (*clustering of volatility*), separated by periods of relative calm. The stationarity of increments, which is a feature of Lévy processes, is at odds with this evidence. In this chapter we drop this stationarity assumption. We review two general approaches to the problem, with particular attention to the impact of such extension on the characteristic function that is to be used in the Fourier pricing machine. The first approach provides a generalization of the concept of infinite divisibility, to allow for non-stationary increments. This direction will lead to a general representation of the Lévy-Khintchine formula in which the diffusion parameter and the Lévy measure will change with time. The second approach will directly address the issue of modelling changes in volatility and intensity parameters by means of the so-called *time change* technique. The idea is that changes in degree activity in the market, as a result of changes in the process of information arrival and in the amount of liquidity, may be modelled by changing the clock measuring time, moving from the *calendar time* to what is called *business time*. Lévy processes and other processes can then be applied to represent the dynamics of market activity, and in this sense what we learned in the previous chapter will be found to be very useful, as promised. As for the characteristic function, it will become a composition of that representing the dynamics of the process of information arrival and that of price changes sampled at business time.

3.1 NON-STATIONARY PROCESSES

We begin by extending the analysis of the previous chapter to a wider class of processes for which the assumption of stationary increments is dropped. While doing that, we are willing to preserve weaker properties that may correspond to the empirical regularities that are observed in the dynamics of financial prices. For this reason, we will review the self-similarity and self-decomposability properties, and then we will extend them to a general setting of non-stationary increments.

3.1.1 Self-similar processes

As shown in Remark 2.2.4, if $\{X_t\}_{t\geq 0}$ is a stable process, then, for any $c > 0$, the process $\{X_{ct}\}_{t\geq 0}$ is identical in law to the process $\{c^{1/\alpha}X_t\}_{t\geq 0}$. This means that any change of time scale has the same effect as some change of spatial scale. This property is called the "self-similarity" of a stochastic process. Following the 1963 seminal work by Mandelbrot on fractal properties of cotton prices, a vast literature has developed on self-similarity properties of market prices. As shown in subsection 2.4.3 for Lévy processes, the term skewness falls at the rate of $1/\sqrt{t}$ while kurtosis decreases at $1/t$. Konikov and Madan (2002) empirically determined the term structures of these moments from market option prices and found that they may be slightly rising or constant, but they are not falling at all. Self-similar processes have the property that these higher moments are constant over the term by construction and hence they seem to be consistent with this empirical regularity.

Definition 3.1.1 *Let $\{X_t\}_{t\geq 0}$ be a stochastic process. It is called self-similar if for every $c > 0$ there exists a function $a(c)$ such that, for all t*

$$\{X_{ct}\}_{t\geq 0} \stackrel{d}{=} \{a(c)X_t\}_{t\geq 0}$$

It follows, for every $c, k > 0$

$$a(ck)X_t \stackrel{d}{=} X_{ckt} \stackrel{d}{=} a(c)X_{kt} \stackrel{d}{=} a(c)a(k)X_t$$

by which $a(ck) = a(c)a(k)$ and hence $a(c) = c^H$ for some exponent H. H is called the "self-similarity exponent" of the process $\{X_t\}_{t\geq 0}$.

In the α-stable case $c = 1/\alpha$ for $\alpha \in (0, 2]$. As a consequence of the above definition, we have that, for any $c > 0$ and any $t > 0$, $X_{ct} \stackrel{d}{=} c^H X_t$; choosing $c = 1/t$ yields $X_t \stackrel{d}{=} t^H X_1$ and the distribution of X_t is completely determined by the distribution of X_1.

A following question arises naturally: Are there any self-similar Lévy processes beyond the stable ones?

Let $\{X_t\}_{t\geq 0}$ be a Lévy process with characteristic exponent $\psi(\cdot)$. The self-similarity of $\{X_t\}_{t\geq 0}$ implies the following scaling relation for ψ:

$$\forall t > 0,\, X_t \stackrel{d}{=} t^H X_1 \iff \forall t > 0, \forall u \in \mathbb{R}, \psi(t^H u) = t\psi(u)$$

As noticed in Remark 2.2.2, the above relation characterizes normalized α-stable distributions. Therefore the only self-similar Lévy processes are the centred *alpha*-stable Lévy processes with self-similarity exponent $H = 1/\alpha$.

3.1.2 Self-decomposable distributions

Definition 3.1.2 *Consider a sequence $\{Z_k\}_{k\geq 1}$ of independent random variables and let $S_n = \sum_{k=1}^n Z_k$. Suppose that there are centring constants c_n and scaling constants b_n such that there exists a random variable X for which*

$$b_n S_n + c_n \stackrel{d}{\to} X$$

Then the random variable X is said to have the class L property.

These laws were studied by Lévy 1937 and Khintchine 1938 who coined the term "class L". This definition extends that of α-stable distributions (see Theorem 2.2.3). However, class L laws represent an important generalization of the stable ones, since they describe limit laws with more general scaling constants than $n^{-1/\alpha}$ (see Remark 2.2.2). In a financial context, this higher flexibility may be required if the independent influences being summed are of different orders of magnitude.

Definition 3.1.3 *The distribution of a random variable X is said to be self-decomposable* (Sato, 1999, Definition 15.1) *if for any constant c, $0 < c < 1$, there exists an independent random variable $X^{(c)}$ such that*

$$X \stackrel{d}{=} cX + X^{(c)}$$

In other words, a random variable is self-decomposable if it has the same distribution as the sum of cX (a scaled-down, or *shaved*, version of itself) and an independent residual random variable $X^{(c)}$. Self-decomposable laws have the property that the associated densities are unimodal (see Sato, 1999, page 404).

The following result shows that a random variable has a distribution of class L if and only if the law of the random variable is self-decomposable.

Theorem 3.1.1 (i) *Let $\{Z_n\}_{n\geq 1}$ be independent random variables and $S_n = \sum_{k=1}^{n} Z_k$. Let X be a random variable and suppose that there are $b_n > 0$ and $c_n \in \mathbb{R}$ for $n \geq 1$ such that*

$$b_n S_n + c_n \xrightarrow{d} X \tag{3.1}$$

and that

$$\{b_n Z_k : k = 1, \ldots, n; n = 1, \ldots\} \text{ is a null array} \tag{3.2}$$

Then X has a self-decomposable distribution.

(ii) *For any random variable X with a self-decomposable distribution we can find $\{Z_n\}$ independent, $b_n > 0$ and $c_n \in \mathbb{R}$ satisfying* (3.1) *and* (3.2).

Proof. See Sato (1999), Theorem 15.3.

As a consequence of Theorem 3.1.1 and Theorem 2.2.5 self-decomposable laws are an important subclass of the class of infinitely divisible laws (see Proposition 15.5 in Sato, 1999): self-decomposable laws are between α-stable distributions and infinitely divisible distributions.

Specifically, the characteristic function of these laws has the form

$$\phi(u) = \exp(iau - \frac{1}{2}\sigma^2 u^2 + \int_{-\infty}^{+\infty} \left(e^{iux} - 1 - iux\mathbb{I}_{|x|\leq 1}\right) \frac{h(x)}{|x|} dx)$$

where a is a real constant,

$$\sigma^2 \geq 0, h(x) \geq 0, \int_{-\infty}^{+\infty} (|x|^2 \wedge 1)\frac{h(x)}{|x|} dx < +\infty$$

and $h(x)$ is increasing on $(-\infty, 0)$ and decreasing in $(0, +\infty)$. (This is the content of Corollary 15.11 in Sato, 1999.) Since the function $h(x)$ characterizes every self-decomposable distribution, we call it the *self-decomposable characteristic* (SDC) of the random variable X.

Note that if X_t is a Lévy process, then X_1 is self-decomposable if and only if X_t is self-decomposable for every t. Moreover, the SDC representation holds for both processes of bounded and unbounded variation (see Theorem 2.4.3).

Since it is desirable that a return distribution could be motivated as a limit law and that it be infinitely divisible, a considerable stream of literature has led to consider self-decomposable laws as candidates for the unit period distribution of financial returns. This is a huge innovation with respect to the older jump-diffusion option pricing models with Gaussian or exponential jump sizes. In fact, the Lévy measures of these compound Poisson processes do not assume the necessary shape to allow for the self-decomposability property. In contrast, α-stable processes, Variance Gamma process, CGMY processes and Meixner proceses (for an adequate choice of parameters) enjoy the self-decomposability property (see Carr *et al.*, 2007, for details and other examples).

3.1.3 Additive processes

Additive processes are obtained from Lévy processes by relaxing the condition of stationarity of increments. We refer to Cont and Tankov (2004) for details.

Definition 3.1.4 *A real-valued stochastic processes $X = (X_t)_{t\geq 0}$ is called an additive process if*

(1) *the random variables $X_{t_0}, X_{t_1} - X_{t_0}, \ldots, X_{t_n} - X_{t_{n-1}}$ are independent for all $n \geq 1$ and $0 \leq t_0 < t_1 < \ldots < t_n$;*
(2) $\mathbb{P}(X_0 = 0) = 1$;
(3) *it is stochastically continuous: for every $t \geq 0$ and $\epsilon > 0$*

$$\lim_{s\to t}\mathbb{P}\left[|X_s - X_t| > \epsilon\right] = 0$$

(4) *the paths $t \to X_t$ are right-continuous with left limits with probability* 1.

It is an immediate consequence of (1) that an additive process is a Markov process.

Theorem 3.1.2 *If $\{X_t\}_{t\geq 0}$ is an additive process then for every t, the distribution of X_t is infinitely divisible.*

Proof. First notice that, for every ϵ and for every η there is a δ such that, if $s, r \in [0, t]$ and $|s - r| < \delta$, then $\mathbb{P}(|X_s - X_r| > \epsilon) < \eta$. In fact, by the stochastic continuity, for every $s \in [0, t]$, there is $\delta_s > 0$ such that $\mathbb{P}(|X_r - X_s| > \epsilon/2) < \eta/2$ for $|r - s| < \delta_s$. Let $I_s = (s - \delta_s/2, s + \delta_s/2)$. Then $\{I_s : s \in [0, t]\}$ covers the interval $[0, t]$. hence there is a finite subcovering $\{I_{s_j} : j = 1, \ldots, N\}$ of $[0, t]$. Let δ be the minimum of δ_{t_j} for $j = 1, \ldots, N$. If $|s - r| < \delta$ and $s, r \in [0, t]$ then $r \in I_{t_j}$ for some j, hence $|s - t_j| < \delta_{t_j}$ and

$$\mathbb{P}(|X_s - X_r| > \epsilon) \leq \mathbb{P}\left(|X_s - X_{t_j}| > \frac{\epsilon}{2}\right) + \mathbb{P}\left(|X_r - X_{t_j}| > \frac{\epsilon}{2}\right) < \eta$$

Fix $t > 0$ and let $t_{nk} = kt/n$ for $n = 1, 2, \ldots$ and $k = 0, 1, \ldots, n$. Let $r_n = n$ and $Z_{nk} = X_{t_{nk}} - X_{t_{n,k-1}}$ for $k = 1, \ldots, n$. By previous arguments, it follows that $\{Z_{nk}\}$ is a null array. Hence the thesis follows from Theorem 2.2.5, assuming that $b_n = 0$ and having that S_n equals X_t. □

By the above theorem, for every t, the characteristic function of X_t has a Lévy–Khintchine representation;

$$\phi_{X_t}(\lambda) = \exp(-\psi_t(\lambda))$$

where

$$\psi_t(\lambda) = -ia_t\lambda + \frac{1}{2}\sigma_t^2\lambda^2 - \int_{-\infty}^{+\infty}\left(e^{i\lambda x} - 1 - i\lambda x\mathbb{I}_{\{|x|\leq 1\}}\right)\nu_t(dx), \qquad \lambda \in \mathbb{R}$$

Notice that, unlike the case of Lévy processes, $\psi_t(\lambda)$ is no longer linear in t. The independence of increments implies that

$$\phi_{X_t - X_s}(\lambda) = \frac{\phi_{X_t}(\lambda)}{\phi_{X_s}(\lambda)}$$

hence,

$$\phi_{X_t - X_s}(\lambda) = \exp(-\psi_{s,t}(\lambda))$$

where

$$\psi_{s,t}(\lambda) = -i(a_t - a_s)\lambda + \frac{1}{2}\left(\sigma_t^2 - \sigma_s^2\right)\lambda^2 - \int_{-\infty}^{+\infty}\left(e^{i\lambda x} - 1 - i\lambda x \mathbb{I}_{\{|x|\leq 1\}}\right)(\nu_t(dx) - \nu_s(dx))$$

For any $t > s$, $X_t - X_s$ is again infinitely divisible. If

$$\sigma_t^2 - \sigma_s^2 > 0 \quad \text{and} \quad \nu_t - \nu_s \quad \text{is a Lévy measure,} \tag{3.3}$$

the above equation is the Lévy–Khintchine representation of $X_t - X_s$. But (3.3) implies that the volatility σ_t^2 and the Lévy measure ν_t should increase with t. Theorem 3.1.3, drawn from Sato (1999), Theorem 9.8, shows that the above conditions are also sufficient to specify an additive process.

Theorem 3.1.3 *Let $\{X_t\}_{t\geq 0}$ be an additive process. The law of X_t is uniquely determined by its spot characteristics $(a_t, \sigma_t^2, \nu_t)_{t\geq 0}$:*

$$\phi_{X_t}(\lambda) = \exp(-\psi_t(\lambda))$$

where

$$\psi_t(\lambda) = -ia_t\lambda + \frac{1}{2}\sigma_t^2\lambda^2 - \int_{-\infty}^{+\infty}\left(e^{i\lambda x} - 1 - i\lambda x \mathbb{I}_{\{|x|\leq 1\}}\right)\nu_t(dx), \qquad \lambda \in \mathbb{R}$$

The spot characteristic triplets $(a_t, \sigma_t^2, \nu_t)_{t\geq 0}$ satisfy the following conditions:

1. *For all t, $\sigma_t^2 \geq 0$ and ν_t is a non-negative measure on $\mathbb{R}$ satisfying $\nu_t(\{0\}) = 0$ and $\int_{-\infty}^{+\infty}(|x|^2 \wedge 1)\nu_t(dx) < +\infty$.*
2. *$\sigma_0 = 0$, $a_0 = 0$, $\nu_0 = 0$, and for all s, t with $s \leq t$ $\sigma_t^2 - \sigma_s^2 \geq 0$ and $\nu_t(B) - \nu_s(B) \geq 0$ for all real Borel sets B.*
3. *Continuity: $\lim_{s\to t} a_s = a_t$, $\lim_{s\to t}\sigma_s^2 = \sigma_t^2$ and $\lim_{s\to t}\nu_s(B) = \nu_t(B)$ for every real Borel set B.*

Conversely, for families of triplets $(a_t, \sigma_t^2, \nu_t)_{t\geq 0}$ satisfying all the above conditions there exists an additive $\{X_t\}_{t\geq 0}$ with $(a_t, \sigma_t^2, \nu_t)_{t\geq 0}$ as spot characteristic triplets.

Additive processes satisfy a generalized version of the decomposition theorem. For this purpose we need to extend the definitions in the context of Lévy processes to the more general additive processes setting.

Definition 3.1.5 *Let ν_t be a locally finite measure on $D_0 \subset \mathbb{R}\setminus\{0\}$. Assume that ν_t is increasing with respect to t. A process $(\Delta_t)_{t\geq 0}$ in $D_0 \cup \{0\}$ such that*

$$N((a,b] \times A) = \#\{s \in (a,b] : \Delta_s \in A\}, \qquad 0 \leq a < b,\ A \subset D_0\ (\textit{measurable})$$

satisfies

(1) *for all $n \geq 1$ and disjoint $A_1, A_2, \ldots A_n \subset [0, +\infty) \times D_0$, the random variables $N(A_1), \ldots, N(A_n)$ are independent, and*
(2) *$N((a,b] \times A)$ is a Poisson random variable with parameter $\int_a^b \nu_t(A)dt$*

is called "Poisson point process with time inhomogeneous intensity measure $\nu_t(\mathrm{d}x)\mathrm{d}t$*".*

By a similar argument as in the Lévy case, it is possible to prove that the additive process $\{X_t\}_{t\geq 0}$ satisfies the following generalized version of Lévy–Itô decomposition:

$$X_t = a_t + \int_0^t \sigma_s \mathrm{d}B_s + M_t + C_t$$

where

$$C_t = \sum_{s\leq t} \Delta_s \mathbb{I}_{\{|\Delta_s|>1\}}$$

and

$$M_t = \lim_{\epsilon\downarrow 0}\left(\sum_{s\leq t} \Delta_s \mathbb{I}_{\{\epsilon<|\Delta_s|\leq 1\}} - \int_{s\in(0,t],\{x\in\mathbb{R}:\epsilon<|x|\leq 1\}} x\nu_s(\mathrm{d}x)\mathrm{d}s\right)$$

Example 3.1.1 (***The time-dependent volatility case***) *Let* $\{W_t\}_{t\geq 0}$ *be a standard Brownian motion and let* $\sigma : \mathbb{R}^+ \to \mathbb{R}^+$ *be a measurable function such that* $\int_0^t \sigma^2(s)\mathrm{d}s < +\infty$ *for all* $t > 0$ *and* $b : \mathbb{R}^+ \to \mathbb{R}$ *be a continuous function. Then the process*

$$X_t = b(t) + \int_0^t \sigma(s)\mathrm{d}W_s$$

is an additive process. Its characteristic triplet is $(b(t), \sigma^2(t), 0)$.

Example 3.1.2 (***Cox process with deterministic intensity***) *Let* $\lambda : \mathbb{R}^+ \to \mathbb{R}^+$ *be a measurable function such that* $\Lambda(t) = \int_0^t \lambda(s)\mathrm{d}s < +\infty$ *for all* t*. If* $\{N_t\}_{t\geq 0}$ *is a standard Poisson process, then the process* $\{X_t\}_{t\geq 0}$ *defined by*

$$X_t = N_{\Lambda(t)}$$

is an additive process. The independent increments property follows from the properties of Poisson processes, while the regularity of trajectories is a consequence of the continuity of the time change $\Lambda(t)$*. This process is a Poisson process with time dependent intensity* $\lambda(t)$*: the probability of having a jump between* t *and* $t + \delta$ *is given by* $\lambda(t)\delta + o(\delta)$*. It is an example of a Cox process, which is a generalization of the Poisson process allowing for stochastic intensity* (see Kingman, 1993). *Its characteristic triplet is* $(0, 0, \Lambda(t)\delta_1)$.

Example 3.1.3 (***Time inhomogeneous jump-diffusion***) *Given positive functions* σ *and* λ *as in previous examples, and a sequence of i.i.d. random variables* $\{Y_k\}$ *the process*

$$X_t = \int_0^t \sigma(s)dW_s + \sum_{k=1}^{N_{\Lambda(t)}} Y_k$$

is an additive process. Its characteristic triplet is $(0, \sigma^2(t), \Lambda(t)\mu)$*, where* μ *is the law of* Y_1.

Example 3.1.4 (***Lévy processes with deterministic volatility***) *Extending Example* 3.1.1, *we can consider Lévy processes with time-dependent volatility. Consider a continuous function* $\sigma : \mathbb{R}^+ \to \mathbb{R}^+$*. Let* $\{L_t\}_{t\geq 0}$ *be a Lévy process. Then*

$$X_t = \int_0^t \sigma(s)\mathrm{d}L_s$$

is an additive process. If (a, σ^2, ν) is the characteristic triplet of L, the characteristic triplet of X_t is $(a\int_0^t \sigma(s)\mathrm{d}s, \sigma^2\sigma^2(t), t\sigma(t)\nu)$.

Example 3.1.5 (***Time-changed Lévy processes***) *Along the same lines as Example 3.1.2, we can provide a similar extension to the general Lévy processes. Let $\{L_t\}_{t\geq 0}$ be a Lévy process and let $T : \mathbb{R}^+ \rightarrow \mathbb{R}^+$ be a continuous increasing function such that $T(0) = 0$. Then*

$$X_t = L_{T(t)}$$

is an additive process. This follows from the independent increment property of L and the continuity of the time change. If (a, σ^2, ν) is the characteristic triplet of L, the characteristic triplet of X_t is $(aT(t), T(t)\sigma^2, T(t)\nu)$.

3.1.4 Sato processes

Definition 3.1.6 *A self-similar additive process $\{X_t\}_{t\geq 0}$ such that the law of X_1 is self-decomposable is called a Sato process.*

This definition is based on an important result proved in Sato (1991).

Theorem 3.1.4 *A law is self-decomposable if and only if it is the law at unit time of an additive process that is also a self-similar process.*

Proof. See Sato (1991 and 1999), Theorem 16.1.

Moreover, given a self-decomposable distribution with SDC h, then there exists a self-similar process X_t defined through the scaling exponent H with characteristic function

$$\phi_{X_t}(u) = \exp\left(\int_0^{+\infty}\int_{-\infty}^{+\infty} (\mathrm{e}^{iuy} - 1)g(y, s)\mathrm{d}y\mathrm{d}s\right)$$

where

$$g(y, t) = \begin{cases} -\frac{h'(y/t^H)H}{t^{1+H}} & \text{if } y > 0 \\ \frac{h'(y/t^H)H}{t^{1+H}} & \text{if } y < 0 \end{cases}$$

(See Carr *et al.*, 2007, Theorem 1.)

3.2 TIME CHANGES

Lévy and additive processes can be used to address the shortcomings of the Black–Scholes model. Another possibility, linked to it, is to modify the Black–Scholes model, including a separate specification of the speed of the market. The rationale behind this goes back to the work by Clark (1973) and is to capture periods with increased activity (and hence larger price movements) distinguishing between *business* and *calendar* time. In business time, the price follows the Black–Scholes model. In calendar time, a busy day may instead correspond to several business time days, while a quiet day corresponds to a fraction of a day. And when the market is closed, time may not elapse at all.

This passage from calendar to business time is naturally modelled by a time change $t \rightarrow T_t$, which can be represented by an increasing stochastic process. If $(S_t^B)_{t\geq 0}$ is the price process in the Black–Scholes model, the time-changed price process is $(S_{T_t}^B)_{t\geq 0}$. The process T_t cannot be observed directly in practice, but it can be approximated by estimating the quadratic variation of the price process and by market quotes of realized variance swaps (see Carr *et al.*, 2005, for more details on the latter possibility).

3.2.1 Stochastic clocks

We first lay out the basic technique for the change of time. We begin by discussing the link between motion of a process under different clocks.

Let $(X_t)_{t\geq 0}$ be a stochastic process. Assume that the trajectory $T_t(\omega)$ is a continuous strictly increasing function $T_t(\omega) : [0, +\infty) \to [0, +\infty)$ with $T_0(\omega) = 0$ and $T_{+\infty}(\omega) = +\infty$. In this case, the time-changed process $Z_t = X_{T_t}$ visits the same states as X, in the same order and performing the same jumps as X, but at a different speed. Specifically, if $T_t(\omega) < t$, then at time t, the process X will have gone to $X_t(\omega)$, but Z only to $Z_t(\omega) = X_{T_t(\omega)}(\omega)$. We say that Z has evolved more slowly than X. It would move faster in the opposite case $T_t(\omega) > t$.

If the trajectory $T_t(\omega)$ is not strictly increasing, that is, if there exists an interval $[t_1, t_2)$ on which it is constant, then $Z_t(\omega) = X_{T_t(\omega)}(\omega) = X_{T_{t_1}(\omega)}(\omega)$ for all $t \in [t_1, t_2)$. For a financial market model this can be interpreted as a time interval with no market activity, when the price will not change.

If $T_t(\omega)$ admits (upward) jumps, then $Z_t(\omega) = X_{T_t(\omega)}(\omega)$ does not evaluate X everywhere. Specifically, if $\Delta T_t(\omega) > 0$ is the first jump of $T_t(\omega)$, then $Z_t(\omega)$ will visit the same points as $X_t(\omega)$ and in the same order until $X_{T_t^-(\omega)}(\omega)$ and then skip over $\left(X_{T_t^-(\omega)+s}(\omega)\right)_{0\leq s<\Delta T_t(\omega)}$ to directly jump to $X_{T_t(\omega)}(\omega)$. In general, this is the behaviour at every jump of $T_t(\omega)$.

We now introduce the basic result that allows us to represent general processes by a suitable change of time. In its most general form it is due to Monroe (1978).

Theorem 3.2.1 (***Monroe***) *Every semi-martingale is equivalent to a time change of Brownian motion.*

This powerful result provides a generalization of earlier findings by Dambis (1965) and Dubins and Schwartz (1965), that were limited to martingales. In the latter case the time change is performed by using the quadratic variation of the process.

Below we will show how to apply Theorem 3.2.1 using Lévy processes and more general clocks to perform the time change.

3.2.2 Subordinators

We now discuss a class of processes, known as *subordinators*, that are extensively used as stochastic clocks to apply the time-change technique. These processes are simply increasing Lévy processes (that is, for $t \geq s$ we have almost surely (a.s.) that $X_t \geq X_s$).

As for their properties, by having almost surely increasing trajectories, subordinators are of finite variation.

Theorem 3.2.2 *A Lévy process is a subordinator if and only if admits a representation of finite variation of the type*

$$X_t = bt + \sum_{s\in[0,t]} \Delta X_s$$

with $b \geq 0$ *and with intensity measure* ν *such that* $\nu((-\infty, 0]) = 0$ *and* $\int_0^{+\infty}(x \wedge 1)\nu(\mathrm{d}x) < +\infty$.

Proof. The *if* part is trivial.

The *only if* part. Being the trajectories of finite variation, $\sigma^2 = 0$ and $\int_0^{+\infty}(x \wedge 1)\nu(\mathrm{d}x) < +\infty$. For the trajectories to be increasing, there must be no negative jumps, hence $\nu((-\infty, 0]) = 0$. If a function is increasing, then after removing some of its jumps, we obtain another increasing function. When we remove all jumps from a trajectory of X_t, we obtain a deterministic function bt which must therefore be increasing. This allows us to conclude that $b \geq 0$. □

Remark 3.2.1 A non-negative Lévy process, i.e. one where $X_t \geq 0$ a.s., for all $t \geq 0$, is automatically increasing, since every increment $X_{s+t} - X_s$ has the same distribution as X_t and therefore it is also non-negative.

Moreover, if $X_t \geq 0$ for some $t > 0$, then it is automatically a subordinator. In fact, for every n, X_t is the sum of n i.i.d. random variables $X_{\frac{t}{n}}, X_{2\frac{t}{n}} - X_{\frac{t}{n}}, \ldots, X_t - X_{(n-1)\frac{t}{n}}$. This means that all these variables are non-negative almost surely. With the same logic we can prove that for any two rational numbers p and q such that $0 < p < q$, $X_{qt} - X_{pt} \geq 0$ a.s. Since the trajectories are right-continuous, this indicates that they are increasing.

From the previous remark the following result holds trivially:

Corollary 3.2.3 *Given a non-negative random variable Y with infinitely divisible distribution, there exists a subordinator* $(X_t)_{t\geq 0}$ *with* $X_1 \stackrel{d}{=} Y$.

Remark 3.2.2 Remember that there exist Lévy processes without diffusion component, having no negative jumps, but satisfying $\int_0^{+\infty}(x \wedge 1)\nu(\mathrm{d}x) = +\infty$. In this case the above result entails that these processes cannot have increasing trajectories, whatever drift coefficient they may have. In this case, in fact, the sum of jumps is compensated by a term with an infinitely negative drift.

Example 3.2.1 *From the processes described in Chapter 2 section 2.3.4 the following examples can be used to build a subordinator:*

1. ***Gamma process***: *The Gamma process is an increasing Lévy process by definition since the support of the marginal distributions is the positive real line.*
2. ***Poisson process***: *The Poisson process is an increasing Lévy process by definition since the support of the marginal distributions is the set of natural numbers.*
3. ***Increasing compound process***: *The compound Poisson process* $C_t = Z_1 + \ldots + Z_{N_t}$, *for a Poisson process* N_t *and identically distributed non-negative* $Z_1, Z_2, \ldots$ *with probability density function concentrated in* $(0, +\infty)$. *We can add a drift and consider* $\tilde{C}_t = at - C_t$ *for some* $a > 0$ *to get a compound Poisson process with drift.*
4. ***Stable subordinator***: *The stable subordinator is best defined in terms of its Lévy–Khintchine characteristics* $a = 0$ *and* $\nu(\mathrm{d}x) = x^{-\alpha-1}\mathrm{d}x$, *for* $x > 0$ *and* $\alpha < 1$. *This gives*

$$\mathbb{E}(\exp(i\gamma X_t)) = \exp\left\{t\int_0^{+\infty}(\mathrm{e}^{i\gamma x} - 1)x^{-\alpha-1}\mathrm{d}x\right\}$$

   More generally we can also consider tempered stable processes with $\nu(\mathrm{d}x) = x^{-\alpha-1}\mathrm{e}^{-\rho x}$, $\rho > 0$, $x > 0$, $\alpha < 1$.
5. ***Inverse Gaussian process***: *Let* X_t *be the first time that a Brownian motion with drift* v *reaches the positive level* t. *As shown in the previous chapter, the distribution of* X_t *is of the Inverse Gaussian type and it is an* α*-stable processes with* $\alpha = \frac{1}{2}$ *and skewness parameter*

$\beta = 1$. *Its characteristic function is*

$$\phi(\gamma) = \exp it(\sqrt{2\gamma + v^2} - v)$$

3.2.3 Stochastic volatility

Another possible model for the random time T_t can be made in terms of its local intensity $v(t)$,

$$T_t = \int_0^t v(s_-)\mathrm{d}s \tag{3.4}$$

where $v(t)$ is the *instantaneous (business) activity rate*. A more active business day, captured by a higher activity rate, generates higher volatility for the economy. Randomness in business activity generates randomness in volatility. In particular, changes in the business activity rate can be correlated with innovations in X_t, due, for example, to the so-called *leverage effect*.

Note that although T_t has been assumed to be continuous, the instantaneous activity rate process $v(t)$ can jump. However, it needs to be non-negative in order for T_t not to decrease. In this sense, we intend the term "volatility" not as the simple standard deviation of returns, but as a more general representation of the uncertainty in the economy. In fact, when the driving process X_t is the Brownian motion, the activity rate is proportional to the rate of the instantaneous variance rate of the Brownian motion. When X_t is a pure jump Lévy process, $v(t)$ is proportional to the Lévy density of the jumps.

Example 3.2.2 (***CIR Stochastic Clock***) *The activity rate is the Cox–Ingersoll–Ross* (*CIR*) *process that solves the SDE:*

$$\mathrm{d}v(t) = k(\eta - v(t))\mathrm{d}t + \lambda v^{1/2}(t)\mathrm{d}W_t$$

where W_t is a standard Brownian motion. The characteristic function of T_t (given $V(0)$) is explicitly known (see Cox, *et al.*, 1985):

$$\begin{aligned}\phi_{T_t}(u) &= \mathbb{E}\left[\exp(iuT_t)|v(0)\right] \\ &= \frac{\exp(k^2\eta t/\lambda^2)\exp(2v(0)iu/(k+\gamma\coth(\gamma t/2)))}{(\cosh(\gamma t/2) + k\sinh(\gamma t/2)/\gamma)^{2k\eta/\lambda^2}}\end{aligned}$$

where $\gamma = \sqrt{k^2 - 2\lambda^2 iu}$.

Example 3.2.3 (***Gamma–OU Stochastic Clock***) *The activity rate is the solution of the SDE*

$$\mathrm{d}v(t) = -\lambda v(t)\mathrm{d}t + \mathrm{d}z_t$$

where the process z_t is a compound Poisson process

$$z_t = \sum_{n=1}^{N_t} Y_n$$

and N_t is a Poisson process with intensity a and each Y_n follows an exponential law with mean $1/b$. One can show that $v(t)$ is a stationary process with marginal law that follows a Gamma distribution with mean a and variance a/b. In this case the characteristic function of T_t (given

$v(0)$) *can be given explicitly*

$$\begin{aligned}\phi_{T_t}(u) &= \mathbb{E}\left[\exp(iuT_t)|v(0)\right] \\ &= \exp\left(iuv(0)\lambda^{-1}(1-e^{-\lambda t}) + \frac{\lambda a}{iu-\lambda b}\left(b\log\left(\frac{b}{b-iu\lambda^{-1}(1-e^{-\lambda t})}\right) - iut\right)\right)\end{aligned}$$

3.2.4 The time-change technique

From now on, we shall concentrate on a special case of time-changed stochastic process: we assume $(X_t)_{t\geq 0}$ to be a Lévy process and $(T_t)_{t\geq 0}$ to be a subordinator. Moreover, we assume that the two processes involved are independent. We now introduce a technique, called *subordination*, to construct a time-changed Lévy process (see Winkel, Lecture notes, for more details).

Theorem 3.2.4 (***Bochner***) *Let $(X_t)_{t\geq 0}$ be a Lévy process and $(T_t)_{t\geq 0}$ an independent increasing process with $T_0 = 0$. Then the process $Z_t = X_{T_t}$ has characteristic function*

$$\phi_{Z_t}(\lambda) = e^{-t\Phi(\psi(\lambda))}$$

where

$$\phi_{X_t}(\lambda) = e^{-t\psi(\lambda)} \quad \text{and} \quad \mathbb{E}\left[e^{-\lambda T_t}\right] = e^{-t\Phi(\lambda)}$$

In particular, if $(T_t)_{t\geq 0}$ is a subordinator, then $(Z_t)_{t\geq 0}$ is a Lévy process.

Proof. Let μ_{T_t} be the law of T_t. By independence

$$\begin{aligned}\phi_{Z_t}(\lambda) = \mathbb{E}\left[e^{i\lambda X_{T_t}}\right] &= \int_0^{+\infty} \mathbb{E}\left[e^{i\lambda X_s}\right]\mu_{T_t}(ds) \\ &= \int_0^{+\infty} \phi_{X_s}(\lambda)\mu_{T_t}(ds) = \int_0^{+\infty} e^{-s\psi(\lambda)}\mu_{T_t}(ds) \\ &= \mathbb{E}\left[e^{-\psi(\lambda)T_t}\right] = e^{-t\Phi(\psi(\lambda))}\end{aligned}$$

Now, if $(T_t)_{t\geq 0}$ is a subordinator, for $r, s \geq 0$,

$$\begin{aligned}&\mathbb{E}\left[\exp(i\lambda Z_t + i\mu(Z_{t+s} - Z_t))\right] \\ &= \int_0^{+\infty}\int_0^{+\infty} \mathbb{E}\left[\exp(i\lambda X_v + i\mu(X_{v+u} - Z_v))\right]\mu_{T_t, T_{t+s}-T_t}(dv, du) \\ &= \int_0^{+\infty}\int_0^{+\infty} e^{-v\psi(\lambda)}e^{-u\psi(\mu)}\mu_{T_t}(dv)\mu_{T_s}(du) \\ &= e^{-t\Phi(\psi(\lambda))}e^{-s\Phi(\psi(\lambda))},\end{aligned}$$

so that Z_t and $Z_{t+s} - Z_t$ are independent and $Z_{t+s} - Z_t \stackrel{d}{=} Z_s$.

For the right-continuity of the paths, notice that

$$\lim_{\epsilon\downarrow 0} Z_{t+\epsilon} = \lim_{\epsilon\downarrow 0} X_{T_{t+\epsilon}} = X_{T_t} = Z_t$$

since $T_{t+\epsilon} = T_t + \delta \downarrow T_t$ and therefore $X_{T_t+\delta} \to X_{T_t}$. For left limits, the same argument applies. □

Being X_t and T_t independent, they cannot jump simultaneously apart from a set of measure zero. More formally, the countable set of times $\{T_{t^-}, T_t : t \geq 0 \text{ and } \Delta T_t \neq 0\}$ is almost surely disjoint from $\{t \geq 0 : \Delta X \neq 0\}$. Then, $\Delta Z_t = Z_t - Z_{t^-} = X_{T_t} - X_{T_{t^-}}$ can be non-zero if either $\Delta T_t \neq 0$ or $\Delta X_{T_t} \neq 0$, so Z inherits jumps from T_t and from X_t. We have, with probability 1 for all $t \geq 0$ that

$$\Delta Z_t = X_{T_t} - X_{T_{t^-}} = \begin{cases} (\Delta X)_{T_t} & \text{if } (\Delta X)_{T_t} \neq 0 \\ X_{T_t} - X_{T_{t^-}} & \text{if } \Delta T_t \neq 0 \end{cases}$$

Put in plain words, the jumps of the time-changed process can be due only to jumps in the clock or jumps in the price process separately, and no other possibility is allowed.

From all these arguments, we can expect that, if X_t has law μ_t and T_t has Lévy measure ν, then Z will have Lévy measure

$$\tilde{\nu}(\mathrm{d}z) = \int_0^{+\infty} \mu_t(\mathrm{d}z)\nu(\mathrm{d}t), \quad z \in \mathbb{R}$$

since every jump of T of size $\Delta T_t = s$ leads to a jump $X_{T_t} - X_{T_{t^-}} \stackrel{d}{=} X_s$, and the total intensity of jumps of size z receives contributions from T-jumps of all sizes $s \in (0, +\infty)$.

We make this precise as follows:

Theorem 3.2.5 *Let X be a Lévy process with probability distribution μ_t of X_t, for all $t \geq 0$, T a subordinator with a Lévy–Khintchine characteristic triplet $(0, 0, \nu)$, then $Z_t = X_{T_t}$ has a Lévy–Khintchine characteristic triplet $(0, 0, \tilde{\nu})$, where*

$$\tilde{\nu}(\mathrm{d}z) = \int_0^{+\infty} \mu_t(\mathrm{d}z)\nu(\mathrm{d}t), \quad z \in \mathbb{R}$$

Example 3.2.4

1. ***The Variance Gamma process***: *The Variance Gamma process can be defined as the difference of two independent Gamma processes, as was shown in Chapter 2, but it can also be defined by time changing a Brownian motion with drift a and volatility σ by an independent Gamma process with unit mean rate and variance rate v. If T_t is the Gamma process, then the Variance Gamma process may be written as*

$$X_t = aT_t + \sigma B_{T_t}$$

where B is an independent Brownian motion. Applying Theorem 3.2.4 and using Proposition 2.4.7 to compute the function ϕ in the statement, it can be proved that

$$\phi_{X_t}(\gamma) = \left(\frac{1}{1 + \frac{v}{2}\gamma^2\sigma^2 - i\gamma v a}\right)^{\frac{t}{v}}$$

The relationship between the parameters of the Lévy measure associated to the Gamma process and those of the time-changed model is (see Chapter 2 for notation):

$$\alpha_+ = \alpha_- = \frac{1}{v},$$

$$\beta_- = \left(\sqrt{\frac{a^2v^2}{4} + \frac{\sigma^2v^2}{2}} - \frac{av}{2}\right)^{-1}$$

$$\beta_+ = \left(\sqrt{\frac{a^2v^2}{4} + \frac{\sigma^2v^2}{2}} + \frac{av}{2}\right)^{-1}$$

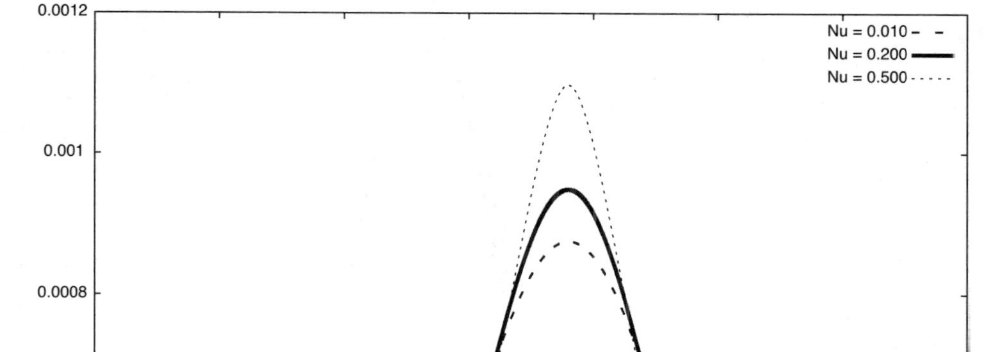

Figure 3.1 The p.d.f for the VG model. Several values of ν

In Figures 3.1 and 3.2 we report the sensitivity of the p.d.f. of the process to changes in the relevant parameters.

2. ***The NIG process***: *The normal Inverse Gaussian model is defined by time changing a Brownian motion with drift through an Inverse Gaussian process. More precisely, let us consider a Brownian motion B with a drift a and volatility σ and an Inverse Gaussian process T_t defined as the first time that an independent Brownian motion with drift v reaches a positive level t (see Example* 3.2.1)*. The process*

$$X_t = aT_t + \sigma_{T_t}$$

is called the NIG process. Since

$$\mathbb{E}\left[e^{-\lambda T_t}\right] = \exp(-t(\sqrt{2\lambda + v^2} - v))$$

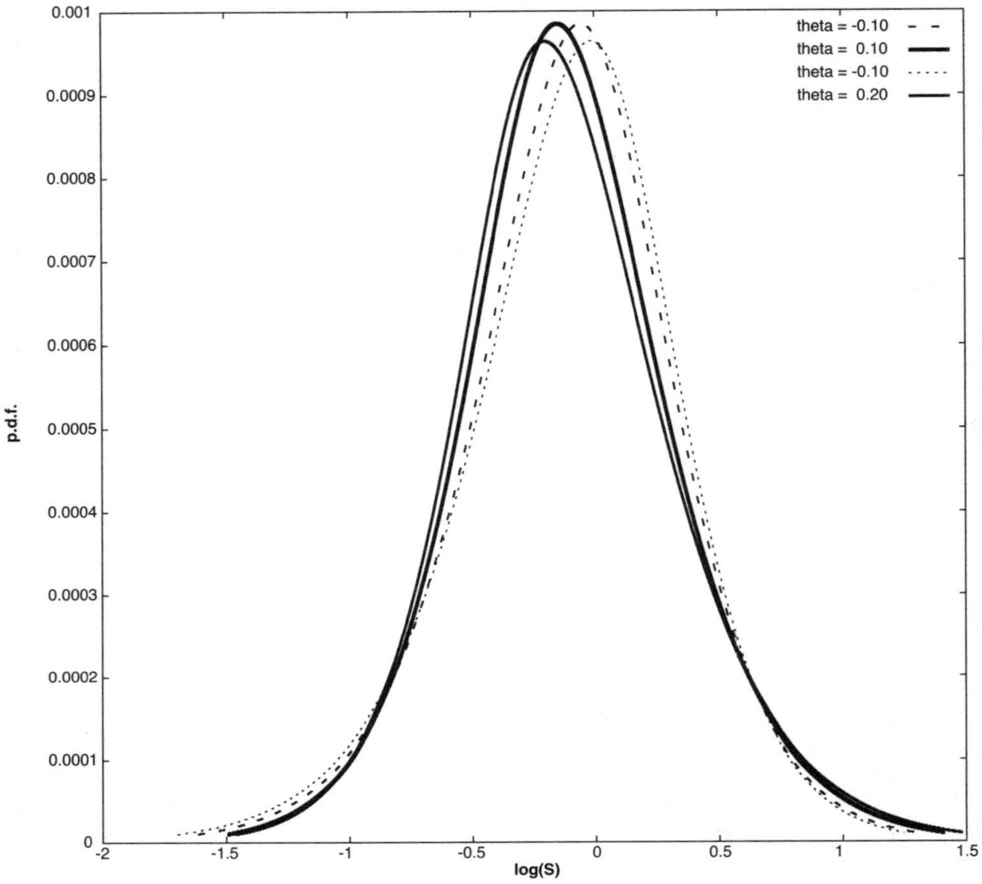

Figure 3.2 The p.d.f for the VG model. Several values of θ

by applying Theorem 3.2.4 *we get the characteristic function of* X_t

$$\phi(u) = \exp\left(-t\sigma\left(\sqrt{\frac{v^2}{\sigma^2}+\frac{a^2}{\sigma^4}-\left(\frac{a}{\sigma^2}+iu\right)^2}-\frac{v}{\sigma^2}\right)\right)$$

If

$$\beta = \frac{a}{\sigma^2}, \quad \alpha^2 = \frac{v^2}{\sigma^2}+\frac{a^2}{\sigma^4}, \quad \delta = \sigma$$

the NIG process can be written as

$$X_t = \beta\delta^2 T_t^{\delta\sqrt{\alpha^2-\beta^2}} + \delta B_{T_t^{\delta\sqrt{\alpha^2-\beta^2}}}$$

The Lévy measure for the NIG process is

$$\nu(\mathrm{d}x) = \sqrt{\frac{2}{\pi}}\delta\alpha^2\frac{\mathrm{e}^{\beta x}K_1(|x|)}{|x|}\mathrm{d}x$$

where $K_a(x)$ is the Bessel function

$$K_a(x) = \frac{1}{2}\left(\frac{x}{2}\right)^a \int_0^{+\infty} \exp\left(-\left(t + \frac{x^2}{4t}\right)\right) t^{-a-1} dt$$

3. ***The CGMY process***: *The CGMY model can also be written as a time-changed Brownian motion, that is, in the form*

$$X_t = \theta T_t + B_{T_t}$$

for an independent subordinator T_t. It may be proved that

$$\mathbb{E}\left[e^{-\lambda T_t}\right] = \exp\left(tC\Gamma(-Y)\left[2r^Y \cos(\eta Y) - M^Y - G^Y\right]\right)$$

with

$$r = \sqrt{2\lambda + GM}$$

$$\eta = \arctan\left(\frac{\sqrt{2\lambda - \left(\frac{G-M}{2}\right)^2}}{\left(\frac{G+M}{2}\right)}\right)$$

The Lévy measure of the subordinator T_t is

$$\nu(dx) = \frac{KBe^{-\frac{x}{2}(B^2-A^2)}}{x^{\frac{Y+1}{2}}} \int_0^{+\infty} e^{-\frac{xB^2h}{2}} \frac{h^{\frac{Y-1}{2}}}{(1+h)^{\frac{Y}{2}}} \frac{dh}{\sqrt{2\pi}} dx$$

where

$$A = \frac{G-M}{2}, \; B = \frac{G+M}{2}, \; K = \left[\frac{C\Gamma(\frac{Y}{4})\Gamma(1-\frac{Y}{4})}{2\Gamma(1+\frac{Y}{2})}\right]$$

4. ***The Meixner process***: *The Meixner process can be written as a time-changed Brownian motion as*

$$X_t = \theta T_t + B_{T_t}$$

for an independent subordinator T_t. The Lévy density of the subordinator is

$$\nu(dx) = \frac{\delta\alpha}{\sqrt{2\pi u^3}} \exp\left(-\frac{A^2 u}{2}\right) \sum_{n=-\infty}^{+\infty} (-1)^n e^{-\frac{n^2\pi^2}{2C^2 u}} dx$$

where

$$A = \frac{\beta}{\alpha}, \quad C = \frac{\pi}{\alpha}$$

5. ***Heston model***: *In the Heston stochastic volatility model the log-returns follow the SDE in which the volatility is behaving stochastically over time. Formally,*

$$dX_t = (r - q)dt + \sigma_t dW_t$$

with the squared volatility following the classical CIR process

$$d\sigma_t^2 = k(\eta - \sigma_t^2)dt + \theta\sigma_t d\tilde{W}_t$$

where W_t and $\tilde{W}_t$ are two correlated standard Brownian motions such that $\mathrm{Cov}(\mathrm{d}W_t \mathrm{d}\tilde{W}_t) = \rho \mathrm{d}t$. *The characteristic function given by X_0 and σ_0 is*

$$\phi_{X_t}(u) = \exp(iu(X_0 + (r-q)t)) \exp\left(\eta k \theta^{-2}\left((k - \rho\theta ui - d)t - 2\log\left(\frac{1 - g\mathrm{e}^{-\mathrm{d}t}}{1-g}\right)\right)\right)$$
$$\times \exp\left(\frac{\sigma_0^2 \theta^{-2}(k - \rho\theta iu - d)\left(1 - \mathrm{e}^{-\mathrm{d}t}\right)}{1 - g\mathrm{e}^{-\mathrm{d}t}}\right)$$

where

$$d = \left((\rho\theta ui - k)^2 - \theta^2\left(-iu - u^2\right)\right)^{1/2}$$
$$g = \frac{k - \rho\theta ui - d}{k - \rho\theta ui + d}$$

This model can clearly be obtained by time changing a Brownian motion with drift through the CIR stochastic clock (see Example 3.2.2).

6. ***The Barndorff-Nielsen–Shephard model***: *This class of models were introduced in Barndorff–Nielsen and Shephard* (2001) *and have a comparable structure to Heston model. The volatility is now modelled by a Gamma–OU process. Volatility can only jump upward and then it will decay exponentially. A co-movement effect between upward jumps in volatility and downward jumps in the process is also incorporated. The process will be more likely to jump downwards when an up-jump in volatility takes place. In the absence of a jump, the process moves continuously and the volatility also decays continuously. The squared volatility now follows a SDE of the form*

$$\mathrm{d}\sigma^2(t) = -\lambda\sigma^2(t)\mathrm{d}t + \mathrm{d}z_{\lambda t}$$

where the process z_t is a compound Poisson process

$$z_t = \sum_{n=1}^{N_t} Y_n$$

where N_t is a Poisson process with intensity a and each Y_n follows an exponential law with mean $1/b$. *One can show that $v(t)$ is a stationary process with a marginal law that follows a Gamma distribution with mean a and variance a/b. We consider the process satisfying the SDE*

$$\mathrm{d}X_t = (r - q - \lambda k(-\rho) - \sigma_t^2/2)\mathrm{d}t + \sigma_t \mathrm{d}W_t + \rho \mathrm{d}z_{\lambda t}$$

where W_t is a Brownian motion independent of z_t. Note that the parameter ρ is introducing a co-movement effect between the volatility and the process.

In this case the characteristic function is

$$\phi_{X_t}(u) = \exp\left(iu\left(X_0 + \left(r - q - a\lambda\rho(b-\rho)^{-1}\right)t\right)\right)$$
$$\times \exp\left(-\lambda^{-1}(u^2 + iu)(1 - \exp(-\lambda t))\frac{\sigma_0^2}{2}\right)$$
$$\times \exp\left(a(b - f_2)^{-1}\left(b\log\left(\frac{b - f_1}{b - iu\rho}\right) + f_2\lambda t\right)\right)$$

where

$$f_1 = iu\rho - \lambda^{-1}(u^2 + iu)\frac{(1 - \exp(-\lambda t))}{2}$$

$$f_2 = iu\rho - \lambda^{-1}\frac{(u^2 + iu)}{2}$$

This model can clearly be obtained by time changing a Brownian motion with drift through the Gamma–OU stochastic clock (see Example 3.2.3).

3.3 SIMULATION OF LÉVY PROCESSES

In many instances, once a model has been chosen and the parameters are calibrated to market data, one would need techniques to simulate scenarios from the process that was specified in order to carry out risk analysis or to price exotic options.

For this reason, in this section we provide a bird's-eye view of the tools available to accomplish this task – that is the main techniques by which to simulate trajectories of Lévy processes. We refer to Cont and Tankov (2004) and Winkel, Lecture Notes, for details. As an example, in Figure 3.3 we report the dynamics of the geometric Brownian motion underlying the Black–Scholes model.

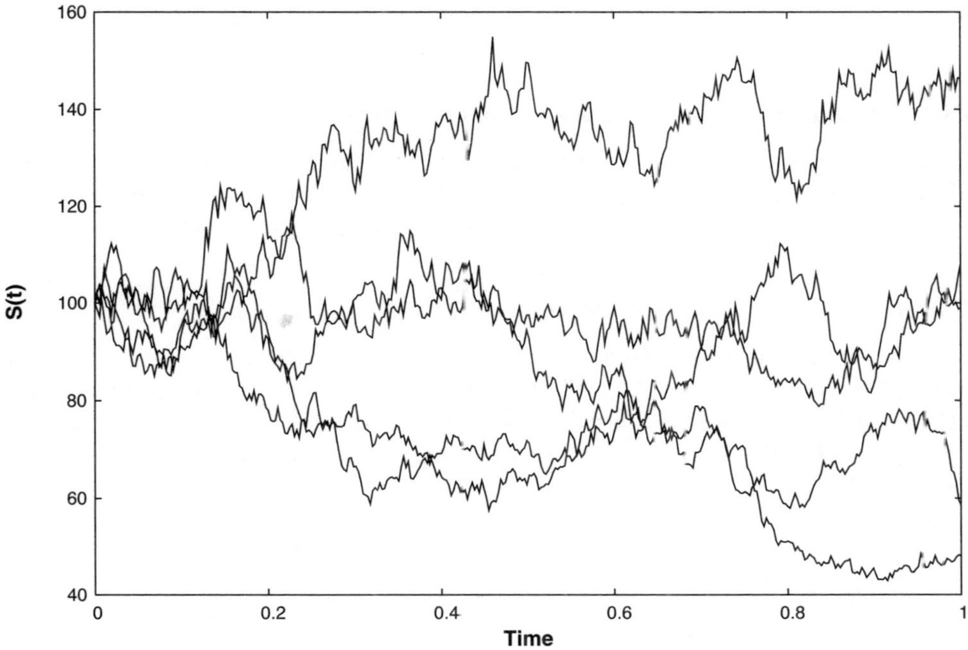

Figure 3.3 Sample trajectories for the Black–Scholes diffusion model

3.3.1 Simulation via embedded random walks

Assume a sequence $(U_k)_{k\geq 1}$ of i.i.d. random variables with U_k uniformly distributed on the interval (0, 1). If the increments distribution is explicitly known, we can simulate the process *via* time discretization.

Let $(X_t)_{t\geq 0}$ be a Lévy process so that X_t has cumulative distribution function F_t. Fix a time lag $\delta > 0$ and let $F_t^{-1}(u) = \inf\{x \in \mathbb{R} : F_t(x) > u\}$. Then the process

$$X_t^{(1,\delta)} = S_{[\frac{t}{\delta}]}, \quad \text{where} \quad S_n = \sum_{k=1}^{n} Y_k \quad \text{and} \quad Y_k = F_\delta^{-1}(U_k)$$

is called the *time discretization* of X with time lag δ.

Proposition 3.3.1 *As $\delta \downarrow 0$, we have $X_t^{(1,\delta)} \xrightarrow{d} X_t$.*

Proof. Notice that $X_t^{(1,\delta)} \stackrel{d}{=} X_{[\frac{t}{\delta}]\delta}$. By stochastic continuity we have $X_{[\frac{t}{\delta}]\delta} \to X_t$ in all probability. This way, $X_t^{(1,\delta)} \xrightarrow{d} X_t$. □

This simulation method requires, almost always, the numerical approximation of F_t^{-1}.

Example 3.3.1 (*Gamma processes*) *For Gamma processes, F_t is a Gamma function, which has no closed-form expression, and F_t^{-1} is also not explicit, but numerical evaluations have been implemented in many statistical packages. There are also Gamma generators based on several uniform random variables. Once a process X with distribution $\Gamma(1, 1)$ has been generated, the process $\left(\beta^{-1}X_{\alpha t}\right)_{t\geq 1}$ is $\Gamma(\alpha, \beta)$.*

Example 3.3.2 (*Variance Gamma processes*) *We simulate the Variance Gamma process as the difference of two independent Gamma processes. An example is reported in Figure* 3.4.

3.3.2 Simulation via truncated Poisson point processes

Since, in practice, the Lévy characteristics are known, we can simulate the pure jump component of a Lévy process by throwing away the small jumps and analysing the error incurred.

We start by simulating a compound Poisson process (an example is reported in Figure 3.5).

Simulation of a compound Poisson process

Let $(X_t)_{t\geq 0}$ be a compound Poisson process with Lévy measure $\nu(\mathrm{d}x) = \lambda\mu(\mathrm{d}x)$, where μ is a distribution, and let H be the associated cumulated distribution function. Denote by $H^{-1}(u) = \inf\{x \in \mathbb{R} : H(x) > u\}$ the generalized inverse. Let $Y_k = H^{-1}(U_{2k})$ and $Z_k = -\lambda^{-1}\ln(U_{2k-1})$, $k \geq 1$ (remember that time intervals between jumps are exponentially distributed random variables). Then the process

$$X_t^{(2)} = S_{N_t}$$

where $\quad S_n = \sum_{k=1}^{n} Y_k$ and if $T_n = \sum_{k=1}^{n} Z_k$

$$N_t = \#\{n \geq 1 : T_n \leq t\}$$

has the same distribution as X.

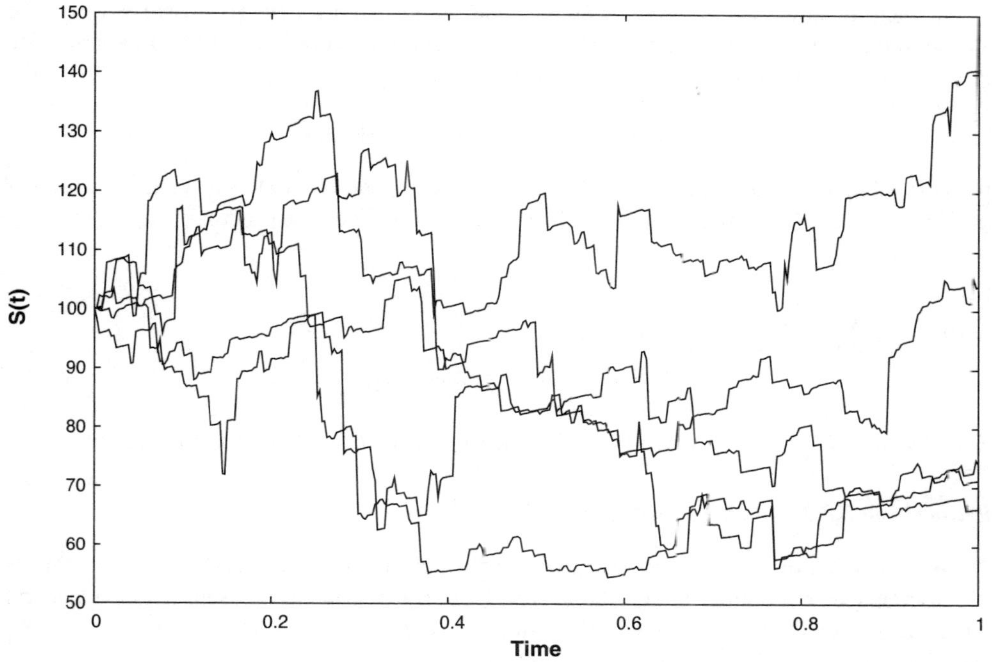

Figure 3.4 Sample trajectories for the Variance Gamma model

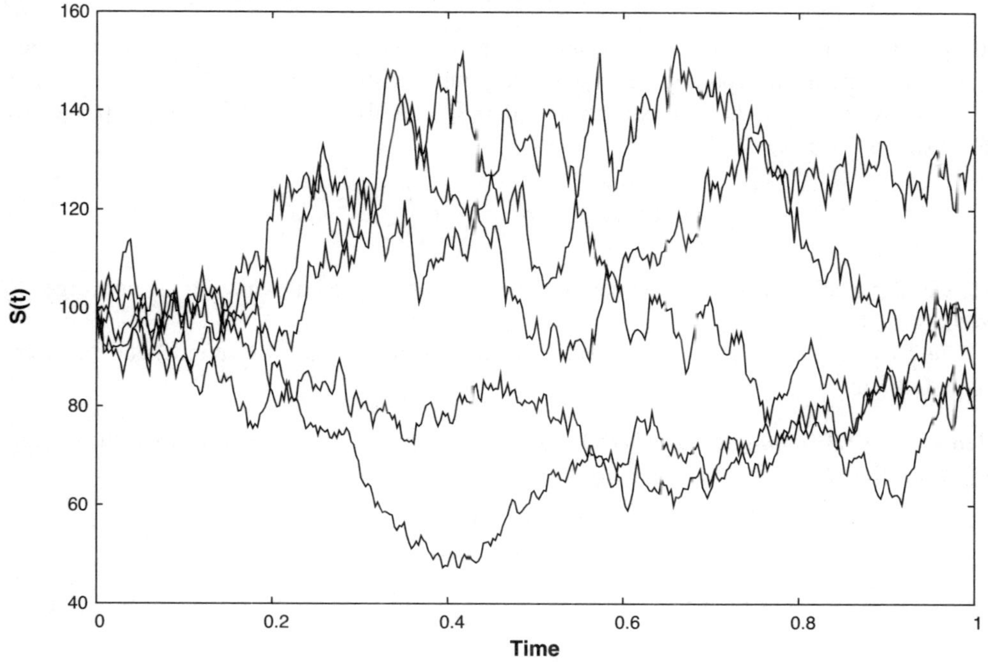

Figure 3.5 Sample trajectories for the Merton jump diffusion model: log-normal jumps

We want to show how to throw away the small jumps from a Lévy process. Let $(X_t)_{t\geq}$ be a Lévy process with characteristic triplet $(a, 0, \nu)$, where ν is not integrable. Fix a jump size threshold $\epsilon > 0$ so that $\lambda_\epsilon = \int_{\{x\in\mathbb{R}:|x|>\epsilon\}} \nu(\mathrm{d}x) > 0$, and write

$$\nu(\mathrm{d}x) = \lambda_\epsilon \mu_\epsilon(\mathrm{d}x), \qquad |x| > \epsilon, \qquad \mu_\epsilon([-\epsilon, \epsilon]) = 0$$

for a probability measure μ_ϵ. Denote $H(x) = \mu_\epsilon(-\infty, x)$ and $H_t^{-1}(u) = \inf\{x \in \mathbb{R} : H_t(x) > u\}$. Let $Y_k = H^{-1}(U_{2k})$ and $Z_k = -\lambda_\epsilon^{-1} \ln(U_{2k-1})$, $k \geq 1$. Then the process

$$X_t^{(2,\epsilon)} = S_{N_t} - b_\epsilon t$$

where $S_n = \sum_{k=1}^n Y_k$ and $T_n = \sum_{k=1}^n Z_k$

$$N_t = \#\{n \geq 1 : T_n \leq t\}$$

and $b_\epsilon = a - \int_{\{x\in\mathbb{R}:\epsilon<|x|\leq 1\}} x\nu(\mathrm{d}x)$ is called the *process with small jumps thrown away.*

Proposition 3.3.2 *As $\epsilon \downarrow 0$, we have $X_t^{(2,\epsilon)} \xrightarrow{d} X_t$.*

Proof. For a process with no negative jumps and characteristic triplet $(0, 0, \nu)$, this is a consequence of Lemma 2.3.5, which gives convergence in the L^2 sense. For a general Lévy process with characteristics $(a, 0, \nu)$ we can write $X_t = at + P_t - N_t$ with P_t and N_t independent with no negative jumps and deduce

$$\begin{aligned}\mathbb{E}\left[\exp\left(i\lambda X_t^{(2,\epsilon)}\right)\right] &= \mathrm{e}^{iat}\mathbb{E}\left[\exp\left(i\lambda P_t^{(2,\epsilon)}\right)\right]\mathbb{E}\left[\exp\left(-i\lambda N_t^{(2,\epsilon)}\right)\right]\\ &\to \mathrm{e}^{iat}\mathbb{E}\left[\exp(i\lambda P_t)\right]\mathbb{E}\left[\exp(-i\lambda N_t)\right] = \mathbb{E}\left[\exp(i\lambda X_t)\right]\end{aligned}$$ □

We now show how to recover error bounds for the simulation. By the decomposition theorem, the residual term (incorporating compensated jumps smaller than ϵ) $R_t^{(2,\epsilon)} = X_t - X_t^{(2,\epsilon)}$ is a Lévy process with characteristic triplet $(0, 0, \mathbb{I}_{[-\epsilon,\epsilon]}\nu(\mathrm{d}x))$ with $\mathbb{E}[R_t(2, \epsilon)] = 0$ and, by Proposition 2.4.6,

$$\mathrm{Var}(R_t(2, \epsilon)) = t\sigma^2(\epsilon) = t\int_{|x|\leq\epsilon} x^2\nu(\mathrm{d}x)$$

Hence, the quality of the approximation depends on the speed at which $\sigma^2(\epsilon)$ converges to zero as $\epsilon \downarrow 0$.

The following result justifies the approximations of small jumps by an independent Brownian motion.

Theorem 3.3.3 **(*Asmussen–Rosinski*)** *Let $(X_t)_{t\geq 0}$ be a Lévy process with characteristics $(a, 0, \nu)$. Denote*

$$\sigma^2(\epsilon) = \int_{[-\epsilon,\epsilon]} x^2\nu(\mathrm{d}x)$$

If

$$\frac{\sigma(\epsilon)}{\epsilon} \to +\infty$$

as $\epsilon \downarrow 0$, then

$$\frac{X_t - X_t^{(2,\epsilon)}}{\sigma(\epsilon)} \xrightarrow{d} B_t \text{ as } \epsilon \downarrow 0$$

for an independent Brownian motion $(B_t)_{t\geq 0}$.

Proof. See Asmussen and Rosinski (2001). □

Hence, if $\sigma(\epsilon)/\epsilon \to +\infty$, it is well justified to adjust the method setting

$$X_t^{(2^{++},\epsilon)} = X_t^{(2,\epsilon)} + \sigma(\epsilon)B_t$$

for an independent Brownian motion.

Example 3.3.3 (***Symmetric stable processes***) *Symmetric stable processes $(X_t)_{t\geq 0}$ are Lévy processes with characteristic triplet $(0, 0, \nu)$ where $\nu(\mathrm{d}x) = c|x|^{-\alpha-1}\mathrm{d}x$, $x \in \mathbb{R}\setminus\{0\}$ for some $\alpha \in (0, 2)$. We decompose $X_t = P_t - N_t$ for two independent processes with no negative jumps and simulate P_t and N_t. By doing this we have*

$$\lambda_\epsilon = \int_\epsilon^{+\infty} cx^{-\alpha-1}\mathrm{d}x = \frac{c}{\alpha}\epsilon^{-\alpha}, \quad H_\epsilon(x) = 1 - \left(\frac{\epsilon}{x}\right)^\alpha \quad \text{and} \quad H_\epsilon^{-1}(u) = \epsilon(1-u)^{-\frac{1}{\alpha}}$$

Example 3.3.4 (***Stable processes***) *For symmetric stable processes $\sigma(\epsilon) \equiv \epsilon^{1-\frac{\alpha}{2}}$, the condition in Example 3.3.3 is satisfied and the normal approximation holds. It is easy to check that it also holds for general stable processes and for all Lévy processes with a law of type $\frac{1}{|x|^{1+\alpha}}$ (with $\alpha > 0$) behaviour near the origin, for example normal inverse Gaussian, truncated stable, etc. The normal approximation does not hold for compound Poisson processes ($\sigma(\epsilon) \equiv o(\epsilon)$), nor for the Gamma process ($\sigma(\epsilon) \equiv \epsilon$).*

Example 3.3.5 (***CGMY process***) *In this case*

$$\sigma^2(\epsilon) = \int_{[-\epsilon,\epsilon]} x^2\nu(\mathrm{d}x) \leq C\int_{[-\epsilon,\epsilon]} |x|^{1-Y}\mathrm{d}x = \frac{2C}{2-Y}\epsilon^{2-Y}$$

and for a given $\delta > 0$ and all $\epsilon > 0$ small enough, the same quantity with C replaced by $C - \delta$ is a lower bound so that

$$\lim_{\epsilon\downarrow o}\frac{\sigma(\epsilon)}{\epsilon} = \lim_{\epsilon\downarrow o}\sqrt{\frac{2C}{2-Y}}\epsilon^{-\frac{Y}{2}} = +\infty \iff Y > 0$$

Hence an approximation of the small jumps of size $(-\epsilon, \epsilon)$ thrown away by a Brownian motion $\sigma(\epsilon)B_t$ is appropriate if and only if $Y > 0$. In fact, for $Y < 0$, the process has finite jump intensity, so all jumps can be simulated. Therefore only the case $Y = 0$ is problematic, but this is the Variance Gamma process.

4
Arbitrage-Free Pricing

4.1 INTRODUCTION

In this chapter we address the issue of using the dynamics described in Chapters 2 and 3 to price contingent claims, that is *derivative contracts*. In dealing rooms, one happens to hear statements that may not seem sound at first judgement. Pricers say that they do not care about the expected returns of an asset, because they only have to price a derivative contract written on that underlying. This would sound strange and staggering to those who are not accustomed to the world of finance, and pricing in particular. Both physicists and economists would complain that this would only give the right price if one does not take into account the risk premium. Economists would also insist that the price would be right only in a world of *rational expectations*. They are wrong, as we are going to see in this chapter. The reason is that, in finance, probability is used in a way that is unique to every other field of application. Probability in finance does not have much to do with beliefs or experiments, but much more with the concepts of arbitrage and of replicating portfolios. In an economy in which the risk is priced into the assets, the assumption that one does not care about such premium in order to price derivatives means that one has accounted for risk in another way, that is by changing the probability assigned to the scenarios. This *change of measure* technique is the main instrument of work for pricers. How to perform this change of measure is of interest in this book because we want to realize how this will impact on characteristic functions if we want to use them for pricing. Since some of our readers may be somewhat new to economics and finance, we also take the opportunity to give a review of the main concepts of the theory of option pricing, with particular attention to those that are used in Fourier pricing.

4.2 EQUILIBRIUM AND ARBITRAGE

In Chapters 2 and 3 we exploited the main principles of the Efficient Market Hypothesis to discuss plausible dynamics for the prices of financial assets. The hypothesis in the background is a general equilibrium model in which financial prices immediately adjust the supply and demand of assets in response to new information flowing to market. It is intuitive that such an equilibrium relationship will impose a constraint across the drift of the prices of assets, i.e. their expected return. Investors will be ready to absorb or get rid of infinite amounts of assets if the tradeoff between expected return and risk is favourable, and will move in and out from assets until the risk premium on every investment is proportional to the amount of risk. In equilibrium, the *Sharpe ratio* – that is, the ratio of expected excess return and volatility – will have to be the same across all assets and markets. In equilibrium, riskier assets will have higher expected returns than the less risky ones. How much higher will depend on the risk aversion of the marginal investor, that is the least risk averse: he will be ready to buy and sell until the risk premium is consistent with his degree of risk aversion. In case of the marginal investor being risk-neutral, all assets will share the same drift, equal to the risk-free return.

The general equilibrium condition imposes a very hard structure on the model. However, as far as the cross-section restriction on the movement of asset prices is concerned, this condition is not necessary. A much weaker assumption, requiring absence of arbitrage opportunities, is sufficient to yield much the same cross-section restriction. The existence of arbitrage opportunities means that one could make money for sure, that is without taking any risk at all. If two portfolios will for sure yield the same value in the future, they must have the same value today, otherwise anyone could exploit infinite profits by going long the cheaper portfolio and being short the dearer one. This would move up the price of the underpriced portfolio and would push down that of the overvalued one, until they were equal. This would result in the same restriction as that of the general equilibrium model, i.e. that portfolios with the same risk must have the same price, and then on average must yield the same return. The literature on arbitrage pricing has gone even further to derive a very strong restriction on the dynamics of prices. Namely, in an arbitrage-free setting, prices can be assumed to grow at the same rate as that of the risk-free asset, once that a suitable change of measure has been performed. More precisely, if the dynamics of asset prices is consistent with absence of arbitrage, it may be proved that, by a change of measure, this dynamics can be reduced to that of the risk-less asset, and this must obtain for all the assets in the economy. The new measure is called *risk-neutral* because the dynamics of the assets would be the same as that of a general equilibrium model in which the marginal investor is risk-neutral.

From a technical point of view, the restriction above can be stated as the requirement that the prices of all risky assets in the economy deflated by the risk-less asset should be martingale processes – that is, future expected values must be equal to the current value. For this reason, the stochastic processes described in the previous chapters should be changed to martingale if they are to be used for pricing financial products or strategies whose payoffs are linked to the dynamics of a risky asset – *derivatives*. Among them, non-linear contracts, namely options, are highly developed and traded in very liquid markets across the world. The prices of these contracts provide precious information on the probability distributions of the assets on which the contracts are written (*underlying assets*). So, the current development of markets has made the task of specifying the dynamics of assets even harder. Not only should they be martingale processes, but they must also be such as to deliver consistent prices for the option contracts traded in the market. This is the so-called *calibration* problem. In this chapter we formalize the problem with particular reference to the general class of additive processes discussed in Chapter 3.

4.3 ARBITRAGE-FREE PRICING

We first lay out the basic concept of arbitrage and the results that obtain for linear products, namely the risky assets traded in the economy. Derivative contracts will be included later.

4.3.1 Arbitrage pricing theory

If we go back to the seminal paper by Ross (1976), the *arbitrage pricing theory* (APT) is built upon very weak assumptions that describe the working of the financial market. In its simplest form, the basic assumption is that the prices of all traded assets are linear functions of a limited set of risk factors. The other assumption is that there exists a risk-less asset yielding a risk-free rate. So, assuming a single risk factor, we have that the return on asset i over a given holding period is given by the following *data generating process* (DGP)

$$r_i = a_i + b_i f$$

where f denotes the risk factor, and a_i and b_i are constants (the latter is known as *factor loading*). The risk factor is assumed to be scaled in such a way as to have zero mean, and unit variance (if it exists). So, we have $E(r_i) = a_i$. Denote further by r the risk-less asset return over the same holding period. If we now combine whatever couple of assets in a way that forms a portfolio whose factor loading is equal to zero, we can easily prove that this implies a restriction on the expected holding period returns of all assets i. More precisely, we have

$$E(r_i) = r + \lambda b_i$$

where it is crucial that λ has to be the same across all assets. This is called the *market price of risk* and is an attribute of the market, while the factor loading parameter b_i is an attribute of the asset. The model can be easily extended to the case of k risk factors. The only extension would be that now a portfolio of $k + 1$ assets should be constructed to yield a position with zero factor loadings. The result concerning expected returns will end up simply modified to

$$E(r_i) = r + \sum_{i=1}^{k} \lambda_k b_{ik}$$

and once again the market prices of risk must be the same across all assets. The principle is then that in order to avoid arbitrage the expected returns of all assets must include a risk premium, and for each risk factor the risk premium must be proportional to the corresponding factor loadings for all assets.

4.3.2 Martingale pricing theory

An important technical fallout of the previous theory is that one can actually enforce a much stronger relationship among the dynamics of assets. While we postpone the details of the technique to the rest of the chapter, here we give an intuitive illustration of the point. Arbitrage pricing theory requires that the realized return on every asset can be decomposed into the risk-free rate, a risk premium component and an idiosyncratic zero mean disturbance. Let us consider absorbing the risk premium on the asset in the stochastic disturbance:

$$r_i = r + \epsilon_i$$

Of course, the disturbance has mean equal to the risk premium. Assume that under a suitable change of measure we could change the mean of the disturbance to zero. Formally we denote a new measure Q such that $E^Q(\epsilon_i) = 0$. Then, there must exist a Radon–Nikodym derivative μ such that

$$\int \epsilon_i \mu \, \mathrm{d}P = 0$$

A fundamental theorem of finance (Harrison and Kreps, 1979) holds that this change of measure exists if and only if there are no arbitrage opportunities left unexploited in the market. Notice that, in general, the functional μ does not need to be unique. If this is instead the case, the market is also said to be *complete* (Harrison and Pliska, 1981).

Ruling out arbitrage then is equivalent to assuming that the price of any asset yields a risk free rate when computed under measure Q, and for this reason this measure is called *risk-neutral*. From a technical point of view, the same result could be spelled out in a different way. Assume that each and every asset is measured using the risk-free asset as the numeraire. Denote by $B(0, T)$ the risk-free assets, with dynamics

$$\mathrm{d}B(t, T) = rB(t, T)\,\mathrm{d}t, \quad r > 0$$

It is easy to check that the value of risky assets $S_i(t) = \mathrm{e}^{X_i(t)}$ computed using the risk-free rate as numeraire is

$$Z_{i,t} \equiv \frac{S_{i,t}}{B(t, T)} = S_{i,t}\,\mathrm{e}^{-r(T-t)} = \mathrm{e}^{X_{i,t}-r(T-t)}$$

and

$$E_Q(\mathrm{d}\ln Z_{i,t}) = E_Q(\mathrm{d}X_{i,t}) - r\,\mathrm{d}t = (E_Q(r_i) - r)\,\mathrm{d}t = 0$$

where we have used the no-arbitrage restriction on the log-return of assets. This means that once the price of any risky asset is deflated by the risk-free asset it must be expected to yield a zero return, and this must be true for all $T \geq t$ under the measure Q. This property is called a *martingale*. So, the result can be restated by saying that in the economy there are no arbitrage opportunities if and only if there exists a measure Q under which the price of each and every asset computed using the risk-less asset as the numeraire is a martingale. A further technical point requires that the change of measure does not change the set of assets to which the original measure was giving zero weight. This property is denoted *equivalence* between measures. For this reason the new measure is also called *equivalent martingale measure* (EMM).

4.3.3 Radon–Nikodym derivative

Formally, any cadlag process can clearly be considered as a random variable on the space $\Omega = D([0, T])$ of right-continuous with left limits paths, equipped with its σ-algebra $\mathcal{F}$ telling us which events are measurable or, in other words, which statements can be made about these paths. The probability distribution of $(X_t)_{t\geq 0}$ then defines a probability measure $\mathbb{P}^X$ on this space of paths. Now, if $(Y_t)_{t\geq 0}$ is another cadlag process and $\mathbb{P}^Y$ is its distribution on the paths space Ω, then $\mathbb{P}^X$ and $\mathbb{P}^Y$ are equivalent probability measures if they define the same set of possible scenarios:

$$\mathbb{P}^X(A) = 1 \iff \mathbb{P}^Y(A) = 1$$

If $\mathbb{P}^X$ and $\mathbb{P}^Y$ are equivalent, then the stochastic models X and Y define the same set of possible evolutions. The construction of a new process on the same set of paths by assigning new probabilities to events is called a *change of measure*.

Given a probability measure $\mathbb{P}$ on the path space Ω, equivalent measures may be generated in many ways: given any random variable $Z > 0$ on Ω with $\mathbb{E}^{\mathbb{P}}[Z] = 1$, the new probability measure $\mathbb{Q}$, defined by

$$\frac{\mathrm{d}\mathbb{Q}}{\mathrm{d}\mathbb{P}} = Z, \text{ i.e. } \forall A \in \mathcal{F}, \quad \mathbb{Q}(A) = \mathbb{E}^{\mathbb{P}}[Z\mathbb{I}_A]$$

is equivalent to $\mathbb{P}$.

If we restrict our attention to events occurring between 0 and t, then each path is weighted by $Z_t(\omega) = \mathbb{E}[Z|\mathcal{F}_t]$, i.e.

$$\forall A \in \mathcal{F}_t, \quad \mathbb{Q}(A) = \mathbb{E}^{\mathbb{P}}[Z_t \mathbb{I}_A] \tag{4.1}$$

By construction, $(Z_t)_{t\geq 0}$ is a strictly positive martingale verifying $\mathbb{E}^{\mathbb{P}}[Z_t] = 1$. Conversely, any strictly positive martingale $(Z_t)_{t\geq 0}$ with $\mathbb{E}[Z_t] = 1$ defines a new measure described by (4.1).

4.4 DERIVATIVES

We now apply the principles stated above to the pricing of derivatives. Since in a derivative contract the payoff is determined as a specific function of some risky asset, ruling out arbitrage opportunities means that we select a portfolio or a strategy that would yield the same payoff as the derivative contract and evaluate that strategy at market prices. If this portfolio or strategy exists, the derivative contract is called *attainable*. If all assets are attainable, the market is said to be *complete*, corresponding to the existence of a unique martingale measure.

4.4.1 The replicating portfolio

To understand the basics of the replicating portfolio technique, take a linear derivative contract, giving a payoff $S_T - F$ at time T (a *forward* contract). It is immediate to check that the same payoff at time T can be replicated by buying S_t on the spot market and issuing $B(t, T)F$ debt. This is the replicating portfolio of this *forward* contract. Even this simple structure of the portfolio reminds us of the general feature of derivative contracts: the replicating portfolio of all derivative contracts include positions of different sign, and underwriting derivative contracts means investing in some asset by issuing debt in some other. The simple example also allows us to comment on the financial meaning of market completeness: this would mean that the value of the forward contract would be equal to the value of the replicating portfolio in all possible states of nature. It would not be so if, for example, one would allow for the possibility that the counterparty of the contract could default before the payoff becomes due. So, if this is the case, either more products are included in the replicating portfolio to allow for the possibility of default, or a perfect replication of the contract is not possible, and the market is incomplete. A third comment that is in order about forward contracts is that the linear shape of the payoff function allows as to perform a *static replication* of the derivative: the replicating portfolio can be set up once and for all and held until maturity of the contract. In cases in which the payoff is not linear, static replication is not effective, unless it uses an infinite set of special securities, as will be discussed below. Consider plain vanilla options. It is intuitive to grasp that a European call option, promising the payoff $\max[S_T - K, 0]$ at time T (K the *strike* price) must correspond to a long position in the underlying asset for some physical quantity Δ. Comparing the replicating portfolio with the value obtained using the equivalent martingale measure we have

$$E_t^Q[\max(S_T - K, 0)] = \Delta S_t - B(t, T)W$$

It is also clear that the position in the risk-free asset must be short, that is, a debt position. In fact, if we assume that Q is such that the option is exercised with probability 1 (it is *in-the-money*), it is immediate to see that the product is actually a forward contract, and we have $\Delta = 1$ and $W = K$. On the other hand, if the option is not exercised, the price would be zero independently of the value of the underlying asset, and so we would have $\Delta = W = 0$. So, a call option implies an amount of debt ranging from zero to the *strike*. In order to emphasize this limit we may write

$$E_Q[\max(S_T - K, 0)] = \Delta S_t - B(t, T)\alpha K$$

where the parameter α ranges from 0 to 1 and so does Δ. The replicating portfolio of put options is derived immediately from the *put–call parity* relationship, which we recall reads

$$\mathcal{C} + B(t, T)K = \mathcal{P} + S_t$$

where $\mathcal{C}$ and $\mathcal{P}$ denote European call and put options with same strike and exercise date. From this we have

$$E_t^Q[\max(K - S_T, 0)] = -(1 - \Delta)S_t + B(t, T)(1 - \alpha)K$$

and the put option corresponds to a short position in the underlying asset associated with credit in the risk-free asset. A point that is worth noting for the future development of this chapter, and this book, is that both Δ and α range between 0 and 1, like a probability.

4.4.2 Options and pricing kernels

Actually, from a mathematical point of view plain vanilla options do not look like the most straightforward bet that one could conceive. The most natural contract would have been one paying a fixed sum, say 1 dollar, if some underlying asset at date T is above a given threshold (call) or below it (put). In the financial options world, this natural product is instead exotic and called *digital*. However, it is easy to verify that this is directly linked to plain vanilla options, as was first pointed out by Breeden and Litzenberger (1978). The idea is to observe a set of options for the same exercise date, but for a large range of strike prices – ideally a continuum of them. Then take a spread strategy, say, using put options. To be more precise, assume the strategy

$$\frac{\mathcal{P}(K + h) - \mathcal{P}(K)}{h}$$

where $\mathcal{P}(x)$ denotes the put option value at time t with strike price x. So, the spread strategy is the purchase of $1/h$ units of put options with strike $K + h$ and the sale of the same number of put options with strike K. In the limit in which h approaches zero, the payoff of the spread at the exercise converges to the *Heavyside* step function assigning 1 to the set $S_T \leq K$ and zero to the complement. Throughout this book, we denote this function with the notation $\theta(-\ln(S_T/K))$. From the martingale pricing theory above, we then know that the price of a *digital put option* has to be

$$\mathcal{P}_{\text{CoN}}(K) = B(t, T)E_t^Q(\theta(-\ln(S_T/K)) = B(t, T)Q(S_T \leq K)$$

But taking the limit of the spread (see Figure 4.1), we also obtain

$$\lim_{h \to 0} \frac{\mathcal{P}(K + h) - \mathcal{P}(K)}{h} = \frac{\partial \mathcal{P}(K)}{\partial K}$$

Putting the two results together we have that

$$\mathcal{P}_{\text{CoN}}(K) = \frac{\partial \mathcal{P}(K)}{\partial K} = B(t, T)Q(S_T \leq K)$$

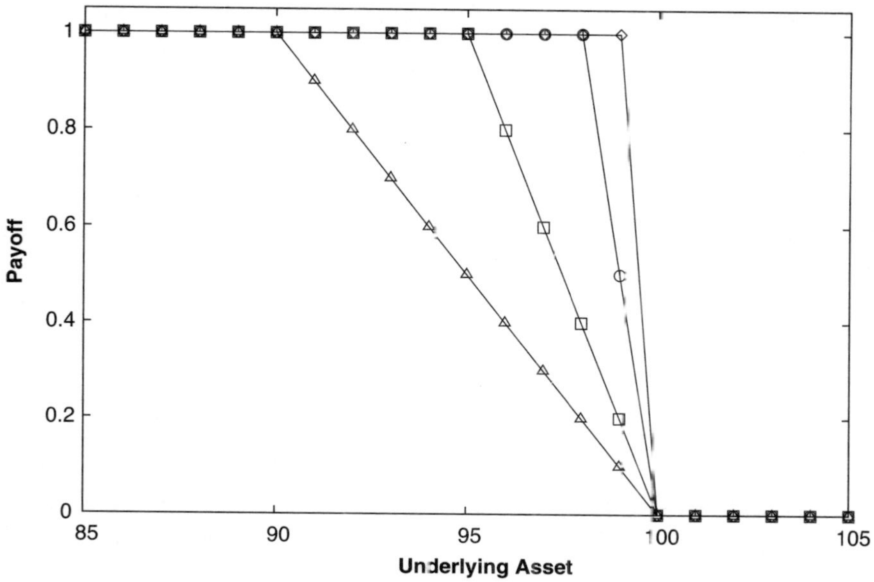

Figure 4.1 A digital option is the limit of spreads of options

By the same token, a *digital call option* paying 1 dollar in the event $S_T > K$ will be worth

$$\mathcal{P}_{\text{CoN}}(K) = -\frac{\partial C(K)}{\partial K} = B(t,T)(1 - Q(S_T \leq K))$$

Taking the argument one step further we have that the option strategy known as *butterfly spread*, defined as

$$\frac{\mathcal{P}(K+h) - 2\mathcal{P}(K) + \mathcal{P}(K-h)}{h^2}$$

is actually a spread over two spreads. Therefore, it does not come as a surprise that if we take the limit for h that tends to 0 we obtain

$$\lim_{h \to 0} \frac{\mathcal{P}(K+h) - 2\mathcal{P}(K) + \mathcal{P}(K-h)}{h^2} = \frac{\partial^2 \mathcal{P}(K)}{\partial K^2}$$

Also, recalling that the first derivative of the put is the discounted value of the *cumulative distribution function* (c.d.f.) Q, the derivative of it will be the discounted probability density function (p.d.f.), if it exists. As for the payoff of this product, it can be seen that while reducing h more and more, the triangle-shaped payoff becomes more and more concentrated around K (see Figure 4.2). In the limit the payoff will become a spike of infinite height at K, and 0 for all other values. This function is called the *Dirac delta* function. In financial terms, it plays the same role as *Arrow–Debreu securities* in a continuous variable setting. The property of the *Dirac delta* function confirms that the price of the Arrow–Debreu security is a probability density function.

The system of financial prices is uniquely determined by a set of cumulative distribution functions or the corresponding density functions, which corresponds to the set of Arrow–Debreu prices in a discrete sample space setting. Equivalently, in Fourier space any set of

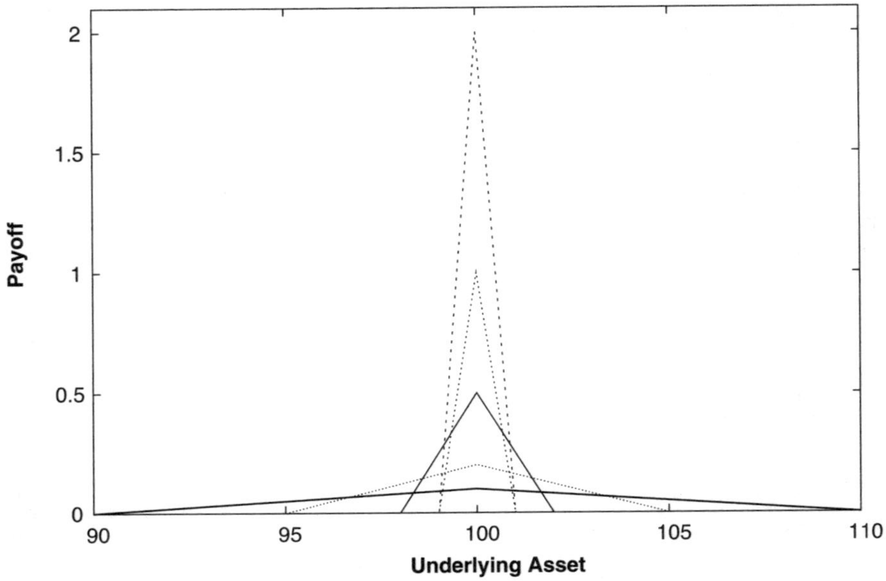

Figure 4.2 Dirac delta function is the limit of butterfly spreads

prices is uniquely defined by a *characteristic function*. The distribution driving the prices is also called the *pricing kernel* of the economy. The relationships between option strategies, digital and Arrow–Debreu securities, and risk-neutral probabilities are reported in Table 4.1.

4.4.3 Plain vanilla options and digital options

We now elicit another relationship between digital products and plain vanilla options. In the previous section we have focused on digital options that pay cash at exercise. These are called *cash-or-nothing* digital options, and we recall that the price is

$$\mathcal{O}_{\text{CoN}} = B(t,T)E_t^Q(\theta(\omega(\ln(S_T) - \ln(K)))) = \begin{cases} B(t,T)(1 - Q(K)) & \text{if } \omega = 1 \\ B(t,T)Q(K) & \text{if } \omega = -1 \end{cases}$$

where ω is equal to 1 for calls and -1 for puts. One may actually conceive many another digital contracts paying contingent payoffs instead of cash. The most straightforward case is a contract paying one unit of the underlying asset if the same is above (call) or below (put) the strike. This contract is called an *asset-or-nothing* digital option, and its value is given by

$$\mathcal{O}_{\text{AoN}} = B(t,T)E_t^Q(S_T\theta(\omega(\ln(S_T) - \ln(K))))$$

Table 4.1 The pricing kernel

Product	Payoff function	Approximation	Price
Digital	Heavyside step	Call/put spread	Discounted c.d.f.
Arrow–Debreu	Dirac delta	Butterfly spread	Discounted p.d.f.

We recall the following general result linking conditional expectations with respect to different equivalent probabilities. If Q and Q^* are equivalent probability measures, we have that

$$E_t^Q[Y] = E_t^{Q^*}\left[Y\frac{\mathrm{d}Q}{\mathrm{d}Q^*}\right]E_t^Q\left[\frac{\mathrm{d}Q^*}{\mathrm{d}Q}\right], \quad \text{for all random variables } Y$$

This way, choosing

$$\frac{\mathrm{d}Q^*}{\mathrm{d}Q} = \frac{B(0,T)S_T}{E_0^Q[B(0,T)S_T]} = \frac{B(0,T)S_T}{S_0}$$

we get

$$\begin{aligned}
&B(t,T)E_t^Q[S_T\theta(\omega(\ln(S_T) - \ln(K)))] = \\
&= B(t,T)E_t^{Q^*}\left[\frac{S_0}{B(0,T)}\theta(\omega(\ln(S_T) - \ln(K)))\right]E_t^Q\left[\frac{B(0,T)S_T}{S_0}\right] \\
&= B(t,T)E_t^{Q^*}\left[\theta(\omega(\ln(S_T) - \ln(K)))\right]E_t^Q\left[S_T\right]
\end{aligned}$$

Then, remembering that by the property of the risk-neutral measure we have $S_t = B(t,T)E_t^Q(S_T)$ we obtain

$$\mathcal{O}_{\text{AoN}} = S_t E_t^{Q^*}[\theta(\omega(\ln(S_T) - \ln(K)))] = \begin{cases} S_t(1 - Q^*(K)) & \text{if } \omega = 1 \\ S_t Q^*(K) & \text{if } \omega = -1 \end{cases}$$

and the price of the *asset-or-nothing* digital can be factorized into the spot price of the asset and the value of a cash-or-nothing option under the new measure Q^*.

Now consider the following portfolio. Buy an *asset-or-nothing* digital call and short K *cash-or-nothing* calls with same strike K and exercise date T. It is easy to check that the payoff at time T will be $\max(S_T - K, 0)$. So, we found another replicating portfolio for a call option

$$\mathcal{C} = \mathcal{C}_{\text{AoN}} - K\mathcal{C}_{\text{AoN}}$$

This arbitrage relationship holds for all option pricing models. If we now compare the replicating portfolio with that based on the underlying asset and debt:

$$\mathcal{C} = \Delta S_t - B(t,T)\alpha K$$

we see immediately $\mathcal{C}_{\text{AoN}} = \Delta * S_t$ and $\mathcal{C}_{\text{CoN}} = B(t,T) * \alpha$. If we now introduce the pricing formulas for the digital options we have

$$\mathcal{C} = (1 - Q^*(K))S_t - B(t,T)(1 - Q(K))K$$

Using put–call parity we recover the price of put options as

$$\mathcal{P} = -Q^*(K)S_t + B(t,T)Q(K)K$$

For a general dynamics of the underlying asset, the prices of put and call options for every strike and exercise date can be obtained by simply computing the conditional distributions Q and Q^*.

4.4.4 The Black–Scholes model

To give an example of the relationship between the dynamic model chosen for the underlying asset and the price of options we apply the above arguments to the world famous Black–Scholes model. Here the main assumption is that the underlying asset follows a geometric Brownian motion (GBM)

$$dS_t = \mu S_t \, dt + \sigma S_t \, dW_t$$

where μ and σ are the *drift* and *diffusion* parameter respectively. The no-arbitrage pricing model requires that

$$dS_t = (r + \lambda\sigma) S_t \, dt + \sigma S_t \, dW_t$$

where the *market price of risk* λ must be the same for all assets. We may change the measure by using the Girsanov theorem. Define $dW^* = dW + \lambda \, dt$, a Wiener process under measure Q. The new measure is risk-neutral, because under it we have

$$dS_t = r S_t \, dt + \sigma S_t \, dW_t^*$$

Now, assume we have a derivative contract written on S_t, say a European option with value $\mathcal{O}(t)$ at time t. No-arbitrage requires that under the same measure Q

$$E_t^Q(d\mathcal{O}(t)) = r\mathcal{O}(t) \, dt$$

Furthermore, since the call option is a function of $S(t)$ and t, by Itô's lemma we obtain

$$E_t^Q(d\mathcal{O}(t)) = (\mathcal{O}_t + rS\mathcal{O}_S + {}^1\!/\!{}_2\sigma^2\mathcal{O}_{SS}) \, dt = r\mathcal{O}(t) \, dt$$

and the so-called *fundamental PDE* of the Black–Scholes model

$$\mathcal{O}_t + rS\mathcal{O}_S + {}^1\!/\!{}_2\sigma^2\mathcal{O}_{SS} - r\mathcal{O}(t) = 0$$

where the underscores to $\mathcal{O}$ denote partial derivatives. No-arbitrage then implies that the prices of derivatives must be solutions of this equation with appropriate boundary conditions. Alternatively, the solutions can be recovered by computing the expectations of final payoffs for plain vanilla options this is easily done considering that the distribution of the underlying asset is log-normal. The result is the famous Black–Scholes formula

$$\mathcal{O}(S, t) = \omega \, [S_t N(\omega \, d_1) - B(t, T) K N(\omega \, d_2)]$$

where $N(x)$ is the standard normal c.d.f. and

$$d_1 = \frac{\ln(S_t/K) + (r + 1/2\sigma^2)(T - t)}{\sigma\sqrt{T - t}}$$

$$d_2 = d_1 - \sigma\sqrt{T - t}$$

It is easy to see that

$$Q^*(K) = N(-d_1) \quad Q(K) = N(-d_2)$$

and that, by the symmetry property of the standard Normal distribution, we have $1 - Q^*(K) = N(d_1)$ and $1 - Q(K) = N(d_2)$.

This model, which was the market standard until the crash of 19 October 1987, is a specific case of the more general class of Lévy models. The extension to the general class of these models requires the definition of a relationship between the characteristic function and the cumulative distribution function.

4.5 LÉVY MARTINGALE PROCESSES

From the analysis above we saw that imposing the no-arbitrage condition requires us to select a martingale process. So, once a stochastic process for the dynamics of the discounted price has been chosen, we have to make sure that it is a martingale, or to define a suitable change of measure to transform it into a martingale. While it is well known how to do that for diffusion processes, it is not straightforward to apply the change of measure technique to Lévy processes in general. We will address that topic in this section.

4.5.1 Construction of martingales through Lévy processes

In the previous chapter we reviewed processes with independent increments. Thanks to this property, an additive process or a Lévy process is a martingale if and only if the conditional expectations of all increments are null. So, different martingales can be constructed from Lévy processes by modelling independent increments. In the proposition below we give some hint at how to construct martingale processes starting from independent increments processes.

Proposition 4.5.1 *Let X_t be a process with independent increments. Then:*

1. *If for some $u \in \mathbb{R}$, $\mathbb{E}[e^{uX_t}] < +\infty\ \forall t \geq 0$, then $\left(\frac{e^{uX_t}}{\mathbb{E}[e^{uX_t}]}\right)_{t\geq 0}$ is a martingale.*
2. *If $\mathbb{E}[|X_t|] < +\infty\ \forall t \geq 0$, then $M_t = X_t - \mathbb{E}[X_t]$ is a martingale (and also a process with independent increments).*
3. *If $\mathrm{Var}\,[X_t] < +\infty\ \forall t \geq 0$, then $M_t^2 - \mathbb{E}[M_t^2]$ is a martingale, where M_t is the martingale defined above.*

If X_t is a Lévy process, for all the processes of this proposition to be martingales it suffices that the corresponding moments be finite for one value of t (see Theorems 25.17 and 25.3 in Sato, 1999).

Proof. This follows directly from the independent increments property. □

Sometimes, particularly in financial applications, it is important to check whether a given Lévy process or its exponential is a martingale. It is then paramount to derive the conditions to be satisfied by the characteristic triplet:

Proposition 4.5.2 *Let X_t be a Lévy process with characteristic triplet (a, σ^2, ν).*

1. *X_t is a martingale if and only if $\int_{|x|>1} |x|\nu(dx) < +\infty$ and*

$$a + \int_{|x|>1} x\nu(dx) = 0$$

2. *e^{X_t} is a martingale if and only if $\int_{|x|>1} e^x \nu(dx) < +\infty$ and*

$$\frac{\sigma^2}{2} + a + \int_{-\infty}^{+\infty} (e^x - 1 - x\mathbb{I}_{|x|\leq 1})\nu(dx) = 0 \tag{4.2}$$

Proof. This is an immediate consequence of Proposition 4.5.1, Proposition 2.4.7 and the Lévy–Khintchine formula. □

4.5.2 Change of equivalent measures for Lévy processes

We now illustrate how to apply the change of measure technique to the family of Lévy processes. Remember that $\mathbb{P}$ and $\mathbb{Q}$ must be equivalent measures on $\Omega = D([0, T])$ provided by the σ-algebra $\mathcal{F}$. Though the processes defined by $\mathbb{P}$ and $\mathbb{Q}$ share the same paths, they can have quite different analytical and statistical properties. For example, if $\mathbb{P}$ defines a Lévy process X, the process Y defined by Q is not necessarily a Lévy process with increments that are neither independent nor stationary.

Clearly, if both X and Y are Lévy processes, the equivalence of their probability distributions $\mathbb{P}^X$ and $\mathbb{P}^Y$ implies relationships between their parameters. As an example, take a Poisson process with jump size equal to 1 and intensity λ. Then, the paths of X are piecewise constant with jumps equal to 1. Let Y be another Poisson process on the same paths space with intensity λ and jump size equal to 2. The probability measures $\mathbb{P}^X$ and $\mathbb{P}^Y$ are clearly not equivalent since all the trajectories of Y that have jumps have zero probability of being trajectories of X, and vice versa. However, if Y has the same jump size as X but a different intensity $\hat{\lambda}$, then every trajectory of X on $[0, T]$ can also be a possible trajectory of Y and vice versa, so the two measures have a chance of being equivalent. The following general results of equivalence of probability measures for Lévy processes hold (we refer to Cont and Tankov (2004) for more details):

Theorem 4.5.3 *Let X_t and Y_t be two Lévy processes with characteristic triplets (a, σ^2, ν) and $(\hat{a}, \hat{\sigma}^2, \hat{\nu})$. Then $\mathbb{P}^X_{|\mathcal{F}_t}$ and $\mathbb{P}^Y_{|\mathcal{F}_t}$ are equivalent for all t (or equivalently for one $t > 0$) if and only if the following conditions are satisfied:*

1. $\sigma = \hat{\sigma}$;
2. *The Lévy measures are equivalent with*

$$\int_{-\infty}^{+\infty} (\mathrm{e}^{\frac{\phi(x)}{2}} - 1)^2 \nu(\mathrm{d}x) < +\infty$$

where $\phi(x) = \ln\left(\frac{\mathrm{d}\hat{\nu}}{\mathrm{d}\nu}\right)$.

3. *If $\sigma = 0$ then we must in addition have*

$$\hat{a} - a = \int_{|x|\leq 1} x(\hat{\nu} - \nu)(\mathrm{d}x)$$

When $\mathbb{P}^X$ and $\mathbb{P}^Y$ are equivalent, the Radon–Nikodym derivative is

$$\frac{\mathrm{d}\mathbb{P}^Y_{|\mathcal{F}_t}}{\mathrm{d}\mathbb{P}^X_{|\mathcal{F}_t}} = \mathrm{e}^{U_t}$$

with

$$U_t = \eta X^c_t - \frac{\eta^2\sigma^2 t}{2} - \eta a t + \lim_{\epsilon\downarrow 0}\left(\sum_{s\leq t}\phi(\Delta X_s)\mathbb{I}_{|\Delta_s|>\epsilon} - t\int_{|x|>\epsilon}(\mathrm{e}^{\phi(x)} - 1)\nu(\mathrm{d}x)\right)$$

Here X_t^c is the continuous part of X_t and η is such that

$$\hat{a} - a - \int_{|x|\leq 1} x(\hat{\nu} - \nu)\nu(\mathrm{d}x) = \sigma^2 \eta$$

if $\sigma > 0$ and zero if $\sigma = 0$. U_t is a Lévy process with characteristic triplet (a_U, σ_U^2, ν_U) given by:

$$a_U = -\frac{1}{2}a\eta^2 - \int_{-\infty}^{+\infty} (e^y - 1 - y\mathbb{I}_{|y|\leq 1})(\nu\phi^{-1})(dy)$$
$$\sigma_U^2 = \sigma^2\eta^2$$
$$\nu_U = \nu\phi^{-1}$$

Proof. For the proof see Sato (1999), Theorems 33.1 and 33.2. □

The following corollaries are particular cases included in Theorem 4.5.3.

Corollary 4.5.4 *Let $N^{(1)}$ and $N^{(2)}$ be two Poisson processes with intensities λ_1 and λ_2 and jump size a_1 and a_2 respectively.*

1. *If $a_1 = a_2$, then $\mathbb{P}^{N^{(1)}}$ and $\mathbb{P}^{N^{(2)}}$ are equivalent, with Radon–Nikodym density*

$$\frac{\mathrm{d}\mathbb{P}^{N^{(1)}}}{\mathrm{d}\mathbb{P}^{N^{(2)}}} = \mathrm{e}^{\left[(\lambda_2-\lambda_1)T - N_T^{(1)} \ln\left(\frac{\lambda_2}{\lambda_1}\right)\right]}$$

2. *If $a_1 \neq a_2$, then $\mathbb{P}^{N^{(1)}}$ and $\mathbb{P}^{N^{(2)}}$ are not equivalent.*

Corollary 4.5.5 *Let X and Y be two compound Poisson processes with Lévy measures ν^X and ν^Y. $\mathbb{P}^X$ and $\mathbb{P}^Y$ are equivalent if and only if ν^X and ν^Y are equivalent. In this case the Radon–Nikodym density is*

$$\frac{\mathrm{d}\mathbb{P}^X}{\mathrm{d}\mathbb{P}^Y} = \mathrm{e}^{\left[(\lambda^Y-\lambda^X)T + \sum_{s\leq T}\phi(\Delta X_s)\right]}$$

where $\lambda^X = \nu^X(\mathbb{R})$ and $\lambda^Y = \nu^Y(\mathbb{R})$ are the jump intensities of the two processes and $\phi =$ $\ln\left(\mathrm{d}\nu^X/\mathrm{d}\nu^Y\right)$.

Corollary 4.5.6 *Let Z and W be two Brownian motions with volatilities $\sigma_Z > 0$ and $\sigma_W > 0$ and drifts μ_Z and μ_W, respectively. $\mathbb{P}^Z$ and $\mathbb{P}^W$ are equivalent if and only if $\sigma_Z = \sigma_W = \sigma$. In this case the Radon–Nikodym derivative is*

$$\frac{\mathrm{d}\mathbb{P}^Z}{\mathrm{d}\mathbb{P}^W} = \mathrm{e}^{\frac{\mu_Z-\mu_W}{c^2}W_T - \frac{1}{2}\frac{\mu_Z^2-\mu_W^2}{\sigma^2}T}$$

Theorem 4.5.3 shows that, contrary to what happens with diffusion models, there is considerable freedom in changing the Lévy measure while preserving the equivalence of measures, but, unless a diffusion component is present, we cannot freely change the drift.

4.5.3 The Esscher transform

The Esscher transform is a particular change of measure according to Theorem 4.5.3. Let $(X_t)_{t\geq 0}$ be a Lévy process with characteristic triplet (a, σ^2, ν) such that $\int_{|x|>1} \mathrm{e}^{\theta x}\nu(\mathrm{d}x) < +\infty$.

For $\theta \in \mathbb{R}$, let $\phi(x) = \theta x$. Thanks to Theorem 4.5.3 we get an equivalent probability under which $(X_t)_{t\geq 0}$ is a Lévy process with zero Gaussian component, Lévy measure $\tilde{\nu}(\mathrm{d}x) = \mathrm{e}^{\theta x}\nu(\mathrm{d}x)$ and drift $\tilde{a} = a + \int_{|x|\leq 1} x(\mathrm{e}^{\theta x} - 1)\nu(\mathrm{d}x)$. The Radon–Nikodym derivative corresponding to this measure change is

$$\frac{\mathrm{d}\mathbb{Q}_{|\mathcal{F}_t}}{\mathrm{d}\mathbb{P}_{|\mathcal{F}_t}} = \frac{\mathrm{e}^{\theta X_t}}{\mathbb{E}[\mathrm{e}^{\theta X_t}]} = \exp\left(\theta X_t + \gamma(\theta)t\right)$$

where $\gamma(\theta) = -\ln \mathbb{E}[\exp(\theta X_1)]$ is the log of the moment generating function of X_1 which, up to the change of variable $\theta \to -i\theta$, is given by the characteristic exponent of the Lévy process $(X_t)_{t\geq 0}$.

The Esscher transform can be used to construct equivalent martingale measures in exponential Lévy market models, as we shall see below (see Cont and Tankov (2004) for more details).

4.6 LÉVY MARKETS

We use the above analysis to extend the market model from the Black–Scholes setting to the case in which the dynamics of the underlying asset is described by general Lévy processes. We call this a Lévy market. We recall that we assume a model with two assets. The first is a deterministic risk-free bank account process, $B(0, t) = \mathrm{e}^{rt}$, $r \geq 0$, $t \geq 0$. The second is now a risky asset $S_t = \mathrm{e}^{X_t}$ for a Lévy process $(X_t)_{t\geq 0}$. We exclude deterministic drift $X_t = \mu t$ in the sequel. The same basic principle as in the discussion about the Black–Scholes model holds true: no-arbitrage is closely related to the existence of martingale probabilities. Formally, an equivalent martingale measure $\mathbb{Q}$ is a probability measure which has the same sets of zero probability as $\mathbb{P}$, i.e. under which the same things are possible or impossible as under $\mathbb{P}$, and under which the process $\left(\mathrm{e}^{-rt}S_t\right)_{t\geq 0}$ is a martingale. Since we are working with logarithms of prices, it is convenient to state the no-arbitrage theorem accordingly.

Theorem 4.6.1 *Let $(X_t)_{t\geq 0}$ be a Lévy process. The Lévy market is arbitrage free if and only if $X_t - rt$ is not an increasing process and $rt - X_t$ is not an increasing process.*

Proof. The *only if* part. If $X_t - rt$ is an increasing process, i.e. a subordinator, then the portfolio $V_t = -B(0, t) + S_t = -\mathrm{e}^{rt} + \mathrm{e}^{X_t} = \mathrm{e}^{rt}\left(\mathrm{e}^{X_t - rt} - 1\right)$ is an arbitrage portfolio.

The *if* part. Let $(X_t)_{t\geq 0}$ have characteristic triplet (a, σ^2, ν). As a consequence, $X_t - rt$ is a Lévy process with characteristic triplet $(a - r, \sigma^2, \nu)$.

If $\sigma > 0$, an equivalent martingale measure can be obtained by changing the drift without changing the Lévy measure: condition 2 of Theorem 4.5.3 is automatically satisfied and the drift can be chosen in order to satisfy equation (4.2).

Let us focus on the case $\sigma = 0$. First, let us apply, according to Theorem 4.5.3, a measure transformation with $\phi(x) = -x^2$: we obtain an equivalent probability under which $X_t - rt$ is a Lévy process with zero Gaussian component, the same location coefficient $a - r$ and Lévy measure $\hat{\nu}(\mathrm{d}x) = \mathrm{e}^{-x^2}\nu(\mathrm{d}x)$, which satisfies $\int_{|x|\geq 1} \mathrm{e}^{\theta x}\hat{\nu}(\mathrm{d}x) < +\infty$. Let $(a - r, 0, \hat{\nu})$ be the new characteristic triplet.

We are now in the position to apply the Esscher transform in order to construct a martingale measure. Once we have performed such transformation with parameter θ, the characteristic triplet of $X_t - rt$ becomes $(\tilde{a}, 0, \tilde{\nu})$ with $\tilde{\nu}(\mathrm{d}x) = \mathrm{e}^{\theta x}\hat{\nu}(\mathrm{d}x)$ and $\tilde{a} = a - r + \int_{|x|\leq 1} x(\mathrm{e}^{\theta x} - 1)$

$\hat{\nu}(dx)$. For $e^{X_t - rt}$ to be a martingale under the new probability, the new triplet must satisfy

$$\tilde{a} + \int_{-\infty}^{+\infty} (e^x - 1 - x\mathbb{I}_{|x|\leq 1})\tilde{\nu}(dx) = 0.$$

To prove the theorem we must now show that there exists a θ solving the equation $f(\theta) = -a + r$ where

$$f(\theta) = \int_{-\infty}^{+\infty} (e^x - 1 - x\mathbb{I}_{|x|\leq 1})\, e^{\theta x}\,\hat{\nu}(dx) + \int_{|x|\leq 1} x(e^{\theta x} - 1)\hat{\nu}(dx).$$

By dominated convergence we have that f is continuous and that $f'(\theta) = \int_{-\infty}^{+\infty} x(e^x - 1)\, e^{\theta x}\,\hat{\nu}(dx) \geq 0$, therefore f is an increasing function. Moreover, if $\hat{\nu}((0, +\infty)) > 0$ and $\hat{\nu}((-\infty, 0)) > 0$ then f' is everywhere bounded from below by a positive number. Therefore in this case $f(+\infty) = +\infty$, $f(-\infty) = -\infty$ and we have a solution.

It remains to consider the case when ν is concentrated on one of the half line. We start assuming $\hat{\nu}((-\infty, 0)) = 0$. By similar arguments, we have $f(+\infty) = +\infty$ but $f(-\infty)$ need not be equal to $-\infty$. When $\theta \to -\infty$, the first term in the definition of $f(\theta)$ always converges to 0. As for the second term, if $\int_{0\leq x\leq 1} x\hat{\nu}(dx) = +\infty$, then it goes to $-\infty$ as $\theta \to -\infty$, and in this case also we have a solution. Let $-\int_{0\leq x\leq 1} x\hat{\nu}(dx)$ be a negative number. By Proposition 2.4.3, the Lévy process is of finite variation type and, by equation (2.15), $-\int_{0\leq x\leq 1} x\hat{\nu}(dx)$ is equal to $b - a + r$, where b is the drift of the process. If the drift is negative $b - a + r < -a + r$ and a solution also exists. To sum up, we have proved that a solution exists unless $\nu((-\infty, 0)) = 0$, $\int_{0\leq x\leq 1} x\hat{\nu}(dx) < +\infty$ and the drift is positive, that is, unless the Lévy process $X_t - rt$ is a subordinator by Theorem 3.2.2 (notice that upon the change of measure a subordinator remains a subordinator). By symmetry, the case $\nu((0, +\infty)) = 0$ is proved analogously. □

Once the above proposition has proved the condition under which the Lévy process used for pricing satisfies the martingale condition, a natural question arises: How many martingale processes can be found for the same price? Actually, in general Lévy processes the presence of different sources of shocks suggests that we may construct martingale processes in many ways. Technically, this would imply that the market is not *complete*.

Theorem 4.6.2 (***Completeness***) *A Lévy market is complete if and only if $(X_t)_{t\geq 0}$ is either a multiple of Brownian motion with drift, $X_t = \mu t + \sigma B_t$ or a multiple of the Poisson process with drift, $X_t = at + bN_t$ (with $(a - r)b < 0$ to get no arbitrage).*

In an incomplete market there are infinitely many of these martingale probabilities from which to choose. This raises the question of how to make the right choice. As a result, while we can determine an arbitrage-free system of prices for all contingent claims, we are not allowed to perfectly hedge all of them, and so a residual amount of risk, called *hedging error*, is unavoidable.

5 Generalized Functions

5.1 INTRODUCTION

One of the main achievements of nineteenth century mathematics was to carefully analyse concepts such as the continuity and differentiability of functions. While it was always clear that not every continuous function is differentiable, (e.g. the function $f : R \to R$ given by $f(x) = |x|$ is not differentiable at 0), it was not until the work by Bolzano and Weierstrass that the full extent of the problem became clear: there exist continuous functions that are nowhere differentiable. However even in these pathological cases one can make sense of f', and even the nth order derivative of f, for any continuous f if one relaxes the requirement that f' be a function. In particular, the theory of distributions frees differential calculus from the difficulties that are brought about by the existence of non-differentiable functions. This is done by providing an extension to a class of objects which is much larger than the class of differentiable functions to which calculus applies in its original form. These objects are called *distributions* or *generalized functions*, but we will adopt the latter definition to avoid confusion with the term "distribution" used in probability. We will see that the introduction of distributions allows us to extend the concept of derivatives to all integrable functions and beyond.

The basic idea is to identify functions with abstract linear functionals on a space of unproblematic *test functions* (conventional and well-behaved functions). Operators on generalized functions can be understood by moving them to the test function. A prerequisite to fully understand the concepts addressed in this chapter is the theory of vector spaces. To save space, a bird's-eye review of the main concepts is collected in Appendix D. Here instead we focus on the theory of distributions, which is the core technical concept in this book.

5.2 THE VECTOR SPACE OF TEST FUNCTIONS

We start by introducing a vector space that is fundamental for our approach: the space of **test functions**. First of all recall some useful definitions:

Definition 5.2.1 *When a function has continuous derivatives of all orders on some set of points, we shall say that the function is infinitely smooth on that set. If this is true for all points of the set, we shall say that the function is simply infinitely smooth*

In particular, let us consider a subspace of the vector space of complex-valued functions defined on $\mathbf{R}^n$.

Definition 5.2.2 *A function $\varphi : \Omega \subset \mathbf{R}^n \to \mathbf{C}$ is said to have compact support if there exists a compact subset K of Ω such that $\varphi(x) = 0$ for all x in $\Omega - K$.*

The space of testing functions, which we shall denote by $\mathcal{D}$, is defined as that function vector space which consists of all complex-valued functions $\varphi(x)$ that are infinitely smooth and have compact support. Obviously K is not the same for all functions $\varphi \in \mathcal{D}$.

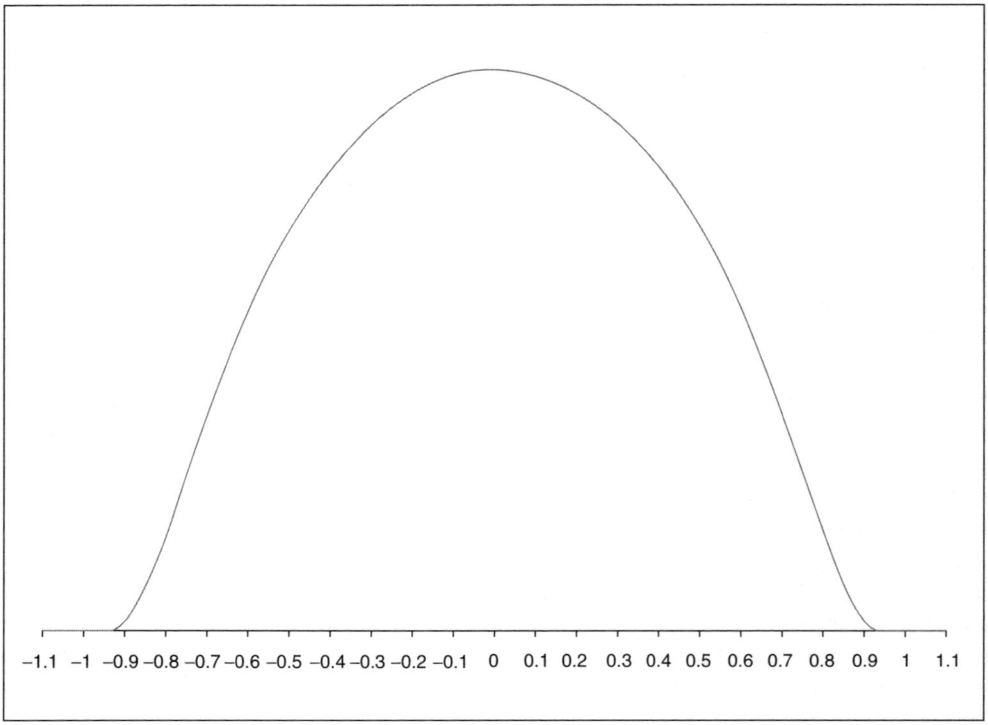

Figure 5.1 The function $\zeta(x)$

A common example of a testing function in $\mathcal{D}$ is

$$\zeta(x) = \begin{cases} 0 & |x| \geq 1 \\ \exp\left(\frac{1}{x^2-1}\right) & |x| < 1 \end{cases} \tag{5.1}$$

This function is infinitely differentiable for $|x| > 1$ since then it is identically zero as well as for $|x| < 1$ since it is then the exponential of an infinitely differentiable function. It is easily shown to be infinitely differentiable everywhere since its derivatives of all order are zero at $x \pm 1$ (Figure 5.1).

It is possible to prove (Schwartz, 1961, Chapter 2, Theorem 1, p. 72) that any complex-valued function $f(t)$ that is continuous for all t and zero outside a finite interval can be approximated uniformly by a sequence of testing functions.

It is possible to turn this space into a *topological vector space* by defining the concept of convergence in $\mathcal{D}$.

We will say that a sequence of testing functions $\{\varphi_\nu(t)\}_{\nu=1}^{\infty}$ converges to 0 in $\mathcal{D}$ if and only if there exists a compact subset K of Ω such that all $\{\varphi_\nu(t)\}_{\nu=1}^{\infty}$ are identically zero outside K, and if for every $\epsilon > 0$ and natural number $d \geq 0$ there exists a natural number k_0 such that for all $k \geq k_0$ the absolute value of all dth derivatives of $\{\varphi_k\}$ is smaller than ϵ. This is equivalent to requiring that the convergence of the sequence $\{\varphi_\nu(t)\}_{\nu=1}^{\infty}$ and of its derivatives be uniform.

As an example, the sequence $\zeta(x)/n$, where $\zeta(x)$ is given by equation (5.1) converges in $\mathcal{D}$ to zero as $n \to \infty$. On the other hand, the sequence $\zeta(x/n)/n$ does not converge in $\mathcal{D}$, even though it and all its derivatives converge uniformly to zero, since there does not exist a fixed finite interval outside which all the $\varphi_\nu(x)$ are zero.

A sequence of testing functions $\{\varphi_\nu(t)\}_{\nu=1}^{\infty}$ is said to converge in $\mathcal{D}$ if the $\varphi_\nu(t)$ are all in $\mathcal{D}$, if they are all zero outside some fixed finite interval I, and if for every fixed non-negative integer k the sequence $\{\varphi_\nu(t)\}_{\nu=1}^{\infty}$ converges uniformly for $-\infty < t < +\infty$. Let $\varphi(t)$ be the limit function of the sequence $\{\varphi_\nu(t)\}_{\nu=1}^{\infty}$. Uniformity of convergence ensures that, for each k, $\varphi^{(k)}(t)$ is continuous and is the limit of $\{\varphi_\nu^{(k)}(t)\}_{\nu=1}^{\infty}$.

It is possible to demonstrate that the limit of every sequence that converges in $\mathcal{D}$ is also in $\mathcal{D}$. We shall refer to this property of $\mathcal{D}$ by saying that $\mathcal{D}$ is *closed under convergence*. With this definition, $\mathcal{D}(\Omega)$ becomes a complete topological vector space.

Another concept that we shall use is that of the support of a testing function. The support is the closure of the set E of all points where $\varphi(t)$ is different from zero. Thus a testing function in $\mathcal{D}$ is simply an infinitely smooth function whose support is a closed bounded set.

5.3 DISTRIBUTIONS

As we have seen above, a *functional* is a rule that assigns a number to every member of a given set of functions. Actually the idea of specifying a function not by its values but by its behaviour as a functional on some space of testing functions, is a concept that is quite familiar to mathematicians and scientists, mainly because they are well acquainted with the classical Fourier and Laplace transformations. In fact, when specifying a function $f(x)$ by its Fourier transform

$$\tilde{f}(k) = \int_{-\infty}^{+\infty} f(x)\, \mathrm{e}^{2\pi ikx}\, \mathrm{d}x$$

the function is being considered as a functional on the set of testing functions consisting of all exponential functions $\mathrm{e}^{2\pi ikx}$ having imaginary exponents.

For our purposes, the set of functions will be taken to be the space $\mathcal{D}$ and we shall consider functionals that assign a complex number to every member of $\mathcal{D}$. Denoting a functional by the symbol f we designate the number that f assigns to a particular testing function by $\langle f, \varphi\rangle$ *Distributions* are particular functionals on the space $\mathcal{D}$ that possess two essential properties: *linearity* and *continuity*.

A functional f on $\mathcal{D}$ is said to be continuous if, for any sequence of testing functions $\{\varphi_k\}$ that converges in $\mathcal{D}$ to φ, the sequence of numbers $\langle f, \varphi_k\rangle$ converges to the number $\langle f, \varphi\rangle$ in the ordinary sense. If f is known to be linear, the definition of continuity may be somewhat simplified. In this case, f will be continuous if the numerical sequence $\langle f, \varphi_k\rangle$ converges to zero whenever the sequence $\{\varphi_k\}$ converges in $\mathcal{D}$ to zero.

So we can state the following

Definition 5.3.1 *A continuous linear functional on the space $\mathcal{D}$ is a distribution. The space of all such distributions is denoted by $\mathcal{D}'$. $\mathcal{D}'$ is called the dual space of $\mathcal{D}$*

In the following example we shall see a possible way to generate a distribution.

Example 5.3.1 *Let $f : R \rightarrow R$ be a locally integrable function (i.e. a function that is integrable in the Lebesgue sense over every finite interval), and let $\varphi : R \rightarrow R$ be a smooth (that is, infinitely differentiable) function with compact support (i.e. identically zero outside of some* *bounded set). The function φ is the test function. We then set*

$$\langle f, \varphi \rangle = \int_{\mathbf{R}} f(x)\varphi(x)\, \mathrm{d}x$$

This is a real number which linearly and continuously depends on φ (Zemanian, 1987, p. 7). One can therefore think of the function f as a continuous linear functional on the space which consists of all the "test functions" φ. Actually the limits on this integral can be altered to finite values since $\varphi(x)$ has a bounded support.

Similarly, if P is a probability distribution on the reals and φ is a test function, then

$$\langle P, \varphi \rangle = \int_{\mathbf{R}} \varphi \quad \mathrm{d}P$$

is a real number that continuously and linearly depends on φ: probability distributions can thus also be viewed as continuous linear functionals on the space of test functions.

Distributions that can be generated, as in the above example, from locally integrable functions are called *regular distributions*. For these distributions a remarkable result holds: *two continuous functions that produce the same regular distribution are identical*. From this it follows that each testing function in $\mathcal{D}$ uniquely determines a regular distribution in $\mathcal{D}'$ and is, in turn, uniquely determined by this regular distribution. This important result can be extended to functions that are merely locally integrable, relaxing the assumption of continuity. In fact, since our integrals are Lebesgue integrals, we can alter the values of $f(x)$ on a set of measure zero without altering the corresponding regular distribution. We can then state a more general result:

Definition 5.3.2 *If $f(x)$ and $g(x)$ are locally integrable and if their corresponding regular distributions agree (i.e. $\langle f, \varphi \rangle = \langle g, \varphi \rangle \quad \forall \varphi \in \mathcal{D}$), then $f(x)$ and $g(x)$ differ at most on a set of measure zero.*

The relevance of the class of distributions stems from the fact that not only does it include representations of locally integrable functions (i.e. regular distributions) but, in addition, it contains many other entities. Moreover, many operations, such as integration, differentiation and other limiting processes that were originally developed for functions, can be extended to these new entities. It should be mentioned, however, that other operations such as the multiplication of functions $f(x)g(x)$ or the formation of composite functions $f(g(x))$ cannot be extended in general to all distributions.

5.3.1 Dirac delta and other singular distributions

A distribution that is not a regular distribution is called a *singular distribution*. One of the most famous singular distributions is the so-called *Dirac delta* (after the name of the famous British theoretical physicist Paul Dirac). Informally, it is a function representing an infinitely sharp peak bounding a unit area: a function $\delta(x)$ that has the value zero everywhere except at $x = 0$ where its value is infinitely large in such a way that its total integral is 1. It is a continuous analogue of the discrete *Kronecker delta*. In the context of signal processing it is often referred to as the unit impulse function. In finance, we saw in Chapter 4 that it is the limit of a sequence of butterfly spreads with the same payoff but strike prices closer and closer. Notice that the Dirac delta is not strictly a function, while for many purposes it can be manipulated as such;

formally it can be correctly defined as a distribution as follows:

$$\langle \delta, \varphi \rangle = \varphi(0)$$

A helpful identity is the scaling property (taking α non-zero),

$$\int_{-\infty}^{\infty} \delta(\alpha x)\,\mathrm{d}x = \int_{-\infty}^{\infty} \delta(\alpha x)\,|\alpha| \frac{\mathrm{d}x}{|\alpha|} = \int_{-\infty}^{\infty} \delta(u)\,\frac{\mathrm{d}u}{|\alpha|} = \frac{1}{|\alpha|}$$

where in the third step we have put $u = |\alpha|x$, so:

$$\delta(\alpha x) = \frac{\delta(x)}{|\alpha|}$$

The scaling property may be generalized to:

$$\delta(g(x)) = \sum_i \frac{\delta(x - x_i)}{|g'(x_i)|}$$

where x_i are the real roots of $g(x)$ (assumed simple roots) and,

$$\delta(\alpha g(x)) = \frac{1}{|\alpha|}\delta(g(x))$$

Thus, for example,

$$\delta(x^2 - \alpha^2) = \frac{1}{2|\alpha|}[\delta(x + \alpha) + \delta(x - \alpha)]$$

In the integral form the generalized scaling property may be written as

$$\int_{-\infty}^{\infty} f(x)\,\delta(g(x))\,\mathrm{d}x = \sum_i \frac{f(x_i)}{|g'(x_i)|}$$

In an n-dimensional space with position vector $\mathbf{r}$, this is generalized to:

$$\int_V f(\mathbf{r})\,\delta(g(\mathbf{r}))\,\mathrm{d}^n r = \int_{\partial V} \frac{f(\mathbf{r})}{|\nabla g|}\,\mathrm{d}^{n-1} r$$

where the integral on the right is over ∂V, the $n - 1$ dimensional surface defined by $g(\mathbf{r}) = 0$. The integral of the *time-delayed* Dirac delta is given by:

$$\int_{-\infty}^{\infty} f(t)\delta(t - T)\,\mathrm{d}t = f(T)$$

(the *shifting* property). The delta function is said to "shift out" the value at $t = T$.

5.4 THE CALCULUS OF DISTRIBUTIONS

The power of distributional analysis in large part rests on the facts that every distribution possesses derivatives of all orders and that differentiation is a continuous operation in this theory. As a consequence, distributional differentiation commutes with various limiting processes such as infinite summation and integration. This is in contrast to classical analysis wherein either such operations cannot be interchanged or the inversion of order must be justified by additional arguments.

5.4.1 Distribution derivative

To define the derivative of a distribution, we first consider the case of a differentiable and integrable function $f : R \to R$. If φ is a test function, then we have

$$\int_{\mathbf{R}} f'\varphi \quad \mathrm{d}x = -\int_{\mathbf{R}} f\varphi' \quad \mathrm{d}x$$

using integration by parts (note that φ is zero outside of a bounded set and that therefore no boundary values have to be taken into account). This suggests that if S is a distribution, we should define its derivative S' by

$$\langle S', \varphi \rangle = -\langle S, \varphi' \rangle$$

It turns out that this is the proper definition; it extends the ordinary definition of derivative, every distribution becomes infinitely differentiable and the usual properties of derivatives hold.

Example 5.4.1 *The Dirac delta, defined by*

$$\langle \delta, \varphi \rangle = \varphi(0)$$

is the derivative of the Heaviside step function. Notice that this is the same function that is denoted $\theta(x)$ in a setting with unbounded support.

In fact for any test function φ,

$$\begin{aligned}\langle H', \varphi \rangle = -\langle H, \varphi' \rangle &= -\int_{-\infty}^{\infty} H(x)\varphi'(x)\,\mathrm{d}x \\ &= -\int_{0}^{\infty} \varphi'(x)\,\mathrm{d}x = \varphi(0) - \varphi(\infty) = \varphi(0) = \langle \delta, \varphi \rangle\end{aligned}$$

so $H' = \delta$. $\varphi(\infty) = 0$ because of compact support. Similarly, the derivative of the Dirac delta is the distribution δ' such that:

$$\langle \delta', \varphi \rangle = -\varphi'(0)$$

5.4.2 Special examples of distributions

We shall now present the computation of some particular examples of distributions arising in the pricing formulas introduced in Chapter 1.

Example 5.4.2 *As an example of a singular distribution we shall take up the Cauchy principal value of the divergent integral*

$$\int_{-\infty}^{+\infty} \frac{\varphi(x)}{x} \mathrm{d}x$$

by definition this is the finite quantity

$$\left\langle \text{p.v.}\,\frac{1}{x}, \varphi(x) \right\rangle = \lim_{\epsilon\to 0} \int_{|x|>\epsilon} \mathrm{d}x \frac{\varphi(x)}{x}$$
$$= \lim_{\epsilon\to 0} \int_{|x|>\epsilon} \mathrm{d}x \frac{\varphi(x) - \varphi(0) + \varphi(0)}{x}$$
$$= \int_{-\infty}^{+\infty} \mathrm{d}x \frac{\varphi(x) - \varphi(0)}{x} - \varphi(0) \lim_{\epsilon\to 0} \int_{|x|>\epsilon} \mathrm{d}x \frac{1}{x}$$

where we use the following abbreviation

$$\int_{|x|>\epsilon} = \int_{-\infty}^{-\epsilon} + \int_{+\epsilon}^{+\infty}$$

It is worth noting that the expression

$$\frac{\varphi(x) - \varphi(0)}{x} \to \varphi'(x)$$

is well defined everywhere due to the differentiability of $\varphi(x)$, which in turn is due to the fact that $\varphi(x)$ is a testing function. Furthermore, $1/x$ is an odd function, so the second term is zero and we conclude:

$$\left\langle \text{p.v.}\,\frac{1}{x}, \varphi(x) \right\rangle = \int_{-\infty}^{+\infty} \mathrm{d}x \frac{\varphi(x) - \varphi(0)}{x}$$

Example 5.4.3 *Compute the distributional value of*

$$g^{+}(x) = \lim_{\epsilon\to 0^+} \frac{1}{x + i\epsilon}$$

The computation is as follows:

$$\langle g^{+}(x), \varphi(x) \rangle = \lim_{\epsilon\to 0^+} \int \mathrm{d}x \frac{\varphi(x)}{x + i\epsilon}$$
$$= \lim_{\epsilon\to 0^+} \int \mathrm{d}x \frac{\varphi(x) - \varphi(0)}{x + i\epsilon} + \varphi(0) \lim_{\epsilon\to 0^-} \int \mathrm{d}x \frac{1}{x + i\epsilon}$$
$$= \left\langle \text{p.v.}\,\frac{1}{x}, \varphi(x) \right\rangle + \varphi(0) \lim_{\epsilon\to 0^+} \int \mathrm{d}x \frac{1}{x + i\epsilon}$$

Let us concentrate on the following integral

$$\lim_{\epsilon\to 0^+} \int \mathrm{d}x \frac{1}{x + i\epsilon}$$

If we rewrite it as a complex integral and consider that the integrand has a pole in $z = -i\epsilon$, we can conclude that with reference to the contour in Figure 5.2

$$\oint_C \mathrm{d}z \frac{1}{z + i\epsilon} = 0 = \int_{-R}^{R} \mathrm{d}x \frac{1}{x + i\epsilon} + i \int_0^{\pi} \frac{\rho\, \mathrm{e}^{i\theta}}{\rho\, \mathrm{e}^{i\theta} + i\epsilon} \mathrm{d}\theta$$

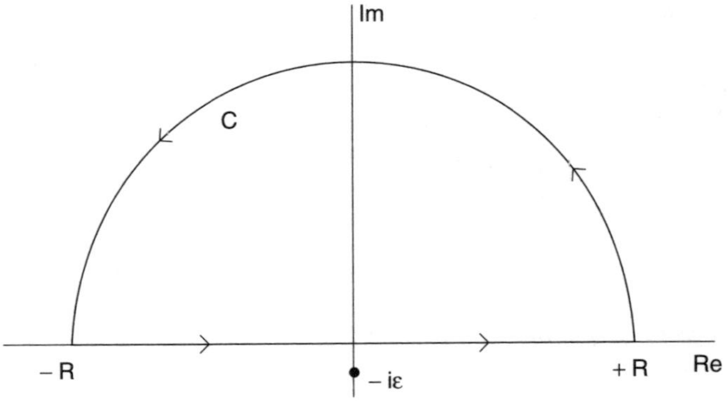

Figure 5.2 Complex integral contour for Example 5.4.3

where we have set $z = \rho\, e^{i\theta}$ *along the arc and* $\rho = |R|$. *So, taking the limit* $\epsilon \to 0$ *and* $\rho \to \infty$ *we get*

$$\lim_{\epsilon \to 0} \int_{-\infty}^{+\infty} dx \frac{1}{x + i\epsilon} = -i \int_0^{\pi} d\theta = -i\pi$$

and we conclude that:

$$\begin{aligned}\langle g^+(x), \varphi(x) \rangle &= \left\langle P\frac{1}{x}, \varphi(x) \right\rangle - i\pi\varphi(0) \\ &= \left\langle \text{p.v.}\,\frac{1}{x}, \varphi(x) \right\rangle - i\pi \langle \delta(x), \varphi(x) \rangle \\ &= \left\langle \text{p.v.}\,\frac{1}{x} - i\pi\delta(x), \varphi(x) \right\rangle\end{aligned}$$

so

$$g^+(x) = \lim_{\epsilon \to 0^+} \frac{1}{x + i\epsilon} = \text{p.v.}\,\frac{1}{x} - i\pi\delta(x)$$

Example 5.4.4 *Compute the distribution value of*

$$g^-(x) = \lim_{\epsilon \to 0^+} \frac{1}{x - i\epsilon}$$

Applying the same techniques as before, we can write:

$$\begin{aligned}\langle g^-(x), \varphi(x) \rangle &= \lim_{\epsilon \to 0^+} \int dx \frac{\varphi(x)}{x - i\epsilon} \\ &= \left\langle \text{p.v.}\,\frac{1}{x}, \varphi(x) \right\rangle + i\varphi(0) \int_{\pi}^{2\pi} d\omega\end{aligned}$$

and

$$g^-(x) = \text{p.v.}\,\frac{1}{x} + i\pi\delta(x)$$

5.5 SLOW GROWTH DISTRIBUTIONS

An important type of distributions, namely the *distributions of slow growth*, arise quite naturally in the development of the Fourier transform in the framework of distributions. The distributions of slow growth comprise a proper subspace of $\mathcal{D}'$ but, on the other hand, they can be defined as continuous linear functionals on a class of testing functions that is wider than $\mathcal{D}$. This extended class of testing functions are known as *rapid descent* functions.

Let $t \equiv \{t_1, \ldots, t_n\}$ be the n-dimensional real variable and let $|t|$ denote $\sqrt{t_1^2 + t_2^2 + \cdots + t_n^2}$; $\mathcal{S}$ is the space of all complex-valued functions $\varphi(t)$ that are infinitely smooth and such that, as $|t| \to \infty$, they and all their partial derivatives decrease to zero faster than every power of $1/|t|$. In other words, for every set of non-negative integers $m, k_1, k_2, \ldots, k_n$,

$$|t|^m \left| \frac{\partial^{k_1+\cdots+k_n}}{\partial t_1^{k_1} \partial t_2^{k_2} \cdots \partial t_n^{k_n}} \varphi(t_1, t_2, \ldots, t_n) \right| \le C_{mk_1k_2\ldots k_n}$$

over all of $\mathbb{R}^n$, where the quantity on the right-hand side is a constant with respect to t but depends upon the choices of $m, k_1, k_2, \ldots, k_n$.

The elements of $\mathcal{S}$ are called *testing functions of rapid descent*. $\mathcal{S}$ is a linear space, and if φ is in $\mathcal{S}$ every one of its partial derivatives is again in $\mathcal{S}$. Furthermore, all testing functions in $\mathcal{D}$ are also in $\mathcal{S}$. However, there are testing functions in $\mathcal{S}$ that are not in $\mathcal{D}$ such as, for example:

$$\exp\left(-t_1^2 - t_2^2 - \ldots - t_n^2\right)$$

Thus $\mathcal{D}$ is a proper subspace of $\mathcal{S}$.

A distribution f is said to be *of slow growth* if it is a continuous linear functional on the space $\mathcal{S}$ of testing functions of rapid descent. Such distributions are also called *tempered* distributions. The space of all tempered distributions is denoted by $\mathcal{S}'$.

In order for a locally integrable function $f(t)$ to assign a finite number $\langle f, \varphi\rangle$ to every testing function $\varphi \in \mathcal{S}$ through the expression

$$\langle f, \varphi\rangle \equiv \int_{-\infty}^{+\infty} f(t)\varphi(t)\,\mathrm{d}t \tag{5.2}$$

the behaviour of $f(t)$ as $|t| \to \infty$ must be restricted in such a way that the integral converges for all $\varphi \in \mathcal{S}$. This is certainly assured if $f(t)$ satisfies the condition

$$\lim_{t\to\infty} |t|^{-N} f(t) = 0 \tag{5.3}$$

for some integer N. Functions that satisfy equation (5.3) are said to be *functions of slow growth*. Every locally integrable function of slow growth defines a regular distribution of slow growth through equation (5.2).

Since each testing function in $\mathcal{S}$ certainly satisfies (5.3), it generates a regular distribution of slow growth. Another fact that can be readily proved is that every distribution in $\mathcal{D}'$ with a bounded support is of slow growth. Thus the delta functional and its derivatives are distributions of slow growth.

Since $\mathcal{S}'$ is a subspace of $\mathcal{D}'$, it follows that all the operations that were defined for distributions in $\mathcal{D}'$ also apply to distributions in $\mathcal{S}'$. However, the application of some operations to a distribution in $\mathcal{S}'$ need not result in a distribution that is also in $\mathcal{S}'$. When a given operation does produce distributions of slow growth from distributions of slow growth, the space $\mathcal{S}'$ is said to be *closed under that operation*. The following is a list of such operations:

- Addition of distributions
- Multiplication of a distribution by a constant
- Shifting of a distribution
- Transposition of a distribution
- Multiplication of the independent variable by a positive constant
- Differentiation of a distribution

5.6 FUNCTION CONVOLUTION

5.6.1 Definitions

A convolution between two functions is an integral that expresses the amount of overlap of one function g as it is shifted over another function f.

Convolution of two functions f and g over a finite range $[0, t]$ is given by

$$[f \star g](t) = \int_0^t f(\tau)g(t-\tau)\,\mathrm{d}\tau \tag{5.4}$$

where the symbol $[f \star g](t)$ denotes the convolution of f and g.

Convolution is more often taken over an infinite range,

$$f \star g = \int_{-\infty}^{+\infty} f(\tau)g(t-\tau)\,\mathrm{d}\tau = \int_{-\infty}^{+\infty} g(\tau)f(t-\tau)\,\mathrm{d}\tau \tag{5.5}$$

(Bracewell, 1965, p. 25) with the variable (in this case t) implied, and also occasionally written as $f \otimes g$.

An important result concerning Gaussian functions is that the convolution of two Gaussians

$$f = \frac{1}{\sigma_1\sqrt{2\pi}}\,\mathrm{e}^{-(t-\mu_1)^2/(2\sigma_1^2)}$$
$$g = \frac{1}{\sigma_2\sqrt{2\pi}}\,\mathrm{e}^{-(t-\mu_2)^2/(2\sigma_2^2)}$$

is another Gaussian

$$f \star g = \frac{1}{\sqrt{2\pi\left(\sigma_1^2+\sigma_2^2\right)}}\,\mathrm{e}^{-[t-(\mu_1+\mu_2)]^2/[2(\sigma_1^2+\sigma_2^2)]}$$

5.6.2 Some properties of convolution

Let f, g, and h be arbitrary functions and let a be a constant. Convolution satisfies the properties

$$f \star g = g \star f$$
$$f \star (g \star h) = (f \star g) \star h$$
$$f \star (g + h) = (f \star g) + (f \star h)$$

(Bracewell, 1965, p. 27), as well as

$$a(f \star g) = (af) \star g = f \star (ag)$$

(Bracewell 1965, p. 49).

Taking the derivative of a convolution gives

$$(f \star g)' = f' \star g = f \star g'$$

(Bracewell, 1965, p. 119).

In probability theory, the probability distribution of the sum of two or more independent random variables is the convolution of their individual distributions:

$$F(t) \star G(t) = \int F(t-x)\,\mathrm{d}G(x) \tag{5.6}$$

5.7 DISTRIBUTIONAL CONVOLUTION

When defining the convolution between two distributions, we cannot follow the same route as that leading to the convolution between two ordinary functions. This is because if f and g are two generic distributions, the product of these distributions may not be defined. In order to extend the convolution process to distributions, we have to introduce a further operation, that is the *direct product* (or *tensor product*) between two distributions. This will be discussed below.

5.7.1 The direct product of distribution

As was mentioned in the preceding section, the direct product of distributions is an operation that arises in the development of convolution. In fact, the definition of convolution is based on that of the direct product, and some properties of the direct product carry over to convolution.

We will follow the same notation as Zemanián (1987), and in order to specify the particular variables that constitute a Euclidean space, we will attach these variables as subscripts to the symbol **R**. For example, $\mathbf{R}_t$ is the one-dimensional Euclidean space consisting of all real values for t; $\mathbf{R}_{x,y}$ is the two-dimensional Euclidean space composed of all real pairs (x, y). Similarly, when such subscripts appear on the symbols for spaces of functions or distributions, they will denote the independent variables on which the elements of these spaces are defined. Thus, $\mathcal{D}_\tau$ is the space $\mathcal{D}$ of testing functions that are defined over $\mathbf{R}_\tau$, and $\mathcal{S}'_{t,\tau}$ is the space $\mathcal{S}'$ of distributions of slow growth defined over $\mathbf{R}_{t,\tau}$.

Let us consider two distributions, $f(t)$ in $\mathcal{D}'_t$ and $g(\tau)$ in $\mathcal{D}'_\tau$. The *direct product* or *tensor product* is an operation that combines these two distributions to obtain another distribution in $\mathcal{D}'_{t,\tau}$, which is denoted by $f(t) \times g(\tau)$, in the following way: if $\varphi(t, \tau)$ is an element of $\mathcal{D}_{t,\tau}$, then $\langle g(\tau), \varphi(t, \tau)\rangle$ is clearly a function of t. It is possible to demonstrate that it is a testing function in $\mathcal{D}_t$ (Zemanian, 1987, Corollary 2.7-2a). Upon applying $f(t)$ to this testing function, we obtain the definition of the direct product:

$$\langle f(t) \times g(\tau), \varphi(t, \tau)\rangle \equiv \langle f(t), \langle g(\tau), \varphi(t, \tau)\rangle\rangle \tag{5.7}$$

However, this is only a definition; the use of the direct product is established by means of the following:

Theorem 5.7.1 (*Zemanian, 1987, p. 115*) *The direct product $f(t) \times g(t)$ of two distributions $f(t)$ and $g(t)$ is a distribution in $\mathcal{D}'_{t,\tau}$.*

The direct product is an operation with respect to which the property of being a slow growth distribution is preserved. So the direct product of two distributions of slow growth is another distribution of slow growth. Also, it is possible to verify that the direct product of two distributions is a commutative operation.

As for the support of the direct product, the following theorem holds:

Theorem 5.7.2 (*Zemanian, 1987, p. 118*) *The support of the direct product of two distributions is the Cartesian product of their supports.*

5.7.2 The convolution of distributions

As we have already mentioned, we cannot define the convolution between two distributions following the same route that was used for the function convolution. So let us try to achieve our objective by viewing the resulting function $h(t)$ defined by

$$\langle h, \varphi \rangle = \langle f \star g, \varphi \rangle = \int_{-\infty}^{+\infty} \mathrm{d}t \int_{-\infty}^{+\infty} f(\tau) g(t-\tau) \varphi(\tau) \, \mathrm{d}\tau$$

as a regular distribution.

If we still assume that $f(t)$ and $g(\tau)$ are continuous functions with bounded supports and let φ be in $\mathcal{D}$, we can state that the integrand of the above integral is continuous and has a bounded support on the (t, τ) plane. So we can indifferently write the above integral as a double integral. By applying the change of variable $\tau = x$ and $t = x + y$ and noting that the corresponding Jacobian determinant is equal to 1, we obtain

$$\langle f \star g, \varphi \rangle = \int_{-\infty}^{+\infty} \int_{-\infty}^{+\infty} f(x) g(y) \varphi(x+y) \, \mathrm{d}x \, \mathrm{d}y \tag{5.8}$$

The last expression has a form that is similar to that of the direct product of two regular distributions. Thus, the rule that defines the convolution $f \times g$ of two distributions $f(t)$ and $g(t)$ is suggested by this expression to be

$$\begin{aligned} \langle f \star g, \varphi \rangle &\equiv \langle f(t) \times g(\tau), \varphi(t+\tau) \rangle \\ &\equiv \langle f(t), \langle g(\tau), \varphi(t+\tau) \rangle \rangle \end{aligned} \tag{5.9}$$

However, a problem arises in this case. Even though the function $\varphi(t+\tau)$ is infinitely smooth, it is not a testing function, since its support is not bounded in the (t, τ) plane. In fact, consider a function $\varphi(x)$ which is different from zero only in a bounded set, say $x \in [a, b]$, when we consider the same φ as function of $x + y$ where both x and y are in $\mathbb{R}$, then the support of φ is the region of the plane that satisfies the equation $a < x + y < b$. Actually this is an infinite strip of finite width that runs parallel to the line $x + y = 0$ (see Figure 5.3).

However, a meaning can still be assigned to the right-hand side of (5.9) if the supports of f and g are suitably restricted.

In particular, if the support of $f(t) \times g(\tau)$ intersects the support of $\varphi(t+\tau)$ in a bounded set, say Ω, we can replace the right-hand side of (5.9) by

$$\langle f(t) \times g(\tau), \lambda(t, \tau) \varphi(t+\tau) \rangle \tag{5.10}$$

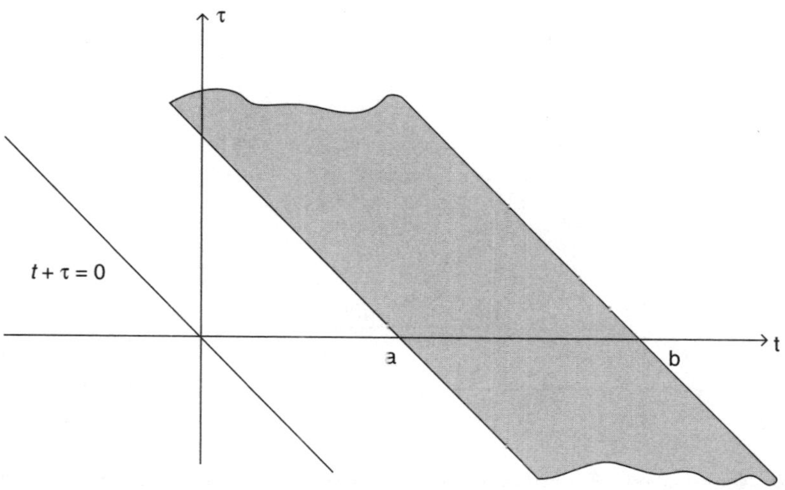

Figure 5.3 The region of the plane which satisfy the equation $a < x + y < b$

where $\lambda(t, \tau)$ is some testing function in $\mathcal{D}_{t,\tau}$ that is equal to 1 over some neighbourhood of Ω. Since $\lambda(t, \tau)\varphi(t + \tau)$ will also be a testing function in $\mathcal{D}_{t,\tau}$, (5.9) and therefore (5.10) enables us to define $f \star g$ in this case as a functional over all $\varphi \in \mathcal{D}$. This replacement is legitimate because the values of a testing function outside some neighbourhood of the support of $f(t) \times g(t)$ can be altered at will without affecting the value assigned by $f(t) \times g(t)$ to that testing function. Yet, we have to determine the conditions under which the intersection of the supports of $f(t) \times g(t)$ and $\varphi(t + \tau)$ is always bounded for all φ in $\mathcal{D}$ and whether $f \star g$ is a distribution. This is resolved by the following:

Theorem 5.7.3 (***Zemanian, 1987, p. 124***) *Let f and g be two distributions over $\mathbb{R}^1$ and let their convolution $f \star g$ be defined by (5.9). Then $f \star g$ will exist as a distribution over $\mathbb{R}^1$ under any one of the following conditions:*

(a) *either f or g has a bounded support;*
(b) *both f and g have supports bounded on the left;*
(c) *both f and g have supports bounded on the right;*

Proof. Let Ω_f and Ω_g be the supports of $f(t)$ and $g(t)$ respectively. Under condition (a), $\Omega_f \times \Omega_g$ is contained in either a horizontal or a vertical strip of finite width in the (t, τ) plane. Under condition (b), $\Omega_f \times \Omega_g$ is contained in a quarter-plane lying above some horizontal line and to the right of some vertical line in the (t, τ) plane (see Figure 5.4). Finally, under condition (c), $\Omega_f \times \Omega_g$ is contained in a quarter-plane lying below some horizontal line and to the left of some vertical line in the (t, τ) plane. Under every one of these conditions, the intersection of $\Omega_f \times \Omega_g$ with the support of $\varphi(t + \tau)$, where $\varphi \in \mathcal{D}$, will be a bounded set. Hence the definition of convolution is applicable, and specifies $f \star g$ as a functional on $\mathcal{D}$. □

The convolution of two distributions is a commutative operation.

We should mention that if the supports of the distributions are not restricted but, instead, sufficiently strong restrictions are placed on the behaviour of the distributions as their arguments approach infinity, then the convolution of distributions can still be defined.

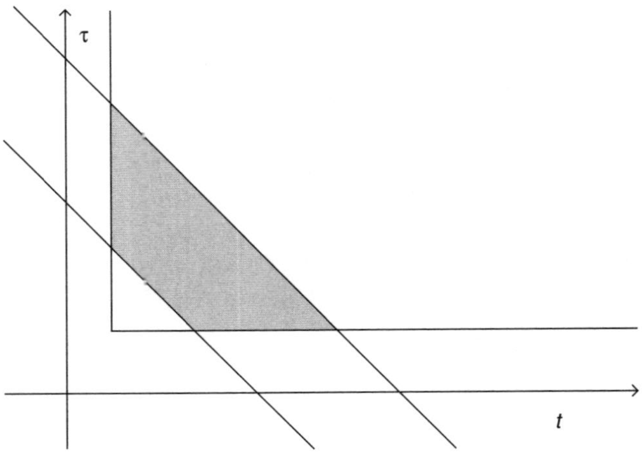

Figure 5.4 The situation described in the above theorem when both distributions are bounded from the left

Example 5.7.1 *The convolution of the delta functional with any distribution yields that distribution again; the convolution of the mth derivative of the delta functional with any distribution yields the mth derivative of that distribution. Let us verify these results. Since convolution is commutative, we can write*

$$\langle \delta \star f, \varphi \rangle = \langle f \star \delta, \varphi \rangle$$

but this is equivalent to

$$\langle f \star \delta, \varphi \rangle = \langle f(t), \langle \delta(\tau), \varphi(t + \tau) \rangle \rangle = \langle f(t), \varphi(t) \rangle$$

so

$$\delta \star f = f$$

from a distributional point of view. In a similar way we can demonstrate the second result:

$$\begin{aligned} \langle \delta^{(m)} \star f, \varphi \rangle &= \langle f \star \delta^{(m)}, \varphi \rangle = \langle f(t), \langle \delta^{(m)}(\tau), \varphi(t + \tau) \rangle \rangle \\ &= \langle f(t), (-1)^m \varphi^{(m)}(t) \rangle = \langle f^{(m)}(t), \varphi(t) \rangle \end{aligned}$$

5.8 THE CONVOLUTION OF DISTRIBUTIONS IN $\mathcal{S}$

Let us now come to the application of the generalized functions we discussed in Chapter 1. Notice that, for our purposes, the standard setting presented above is not sufficient. In fact, we need to define the convolution of two distributions in a suitable way.

Let $f(x)$ and $g(y)$ be two functions in $\mathcal{S}'_x$ and $\mathcal{S}'_y$ respectively. According to the case of $\mathcal{D}$, it is natural to define the convolution $f \star g$ as a distribution on $\mathcal{S}$ through the direct product of two regular distributions as

$$\begin{aligned}\langle f \star g, \varphi\rangle &= \int_{-\infty}^{+\infty}\int_{-\infty}^{+\infty} f(\tau)g(t-\tau)\varphi(t)\,dt\,d\tau \\ &= \int_{-\infty}^{+\infty}\int_{-\infty}^{+\infty} f(x)g(y)\varphi(x+y)\,dx\,dy \\ &= \langle f(x)\times g(y), \varphi(x+y)\rangle \end{aligned} \tag{5.11}$$

Nevertheless, even if φ is a function in $\mathcal{S}$, the function $\varphi(x, y)$ is not a testing function in $\mathcal{S}_{x,y}$, that is, the set of rapid descent testing functions defined in $\mathbb{R}^2$. In fact, $\varphi(x+y)$ satisfies

$$|x+y|^m \left|\frac{\partial^{k_1}\partial^{k_2}\varphi(x+y)}{\partial x^{k_1}\partial y^{k_2}}\right| = |x+y|^m \left|\varphi^{(k)}(x+y)\right| \le C_{mk}$$

for $k_1 + k_2 = k$ instead of

$$\sqrt{(x^2+y^2)^m}\left|\frac{\partial^{k_1}\partial^{k_2}\varphi(x+y)}{\partial x^{k_1}\partial y^{k_2}}\right| \le C_{mk_1k_2}$$

which is required in order to have $\varphi(x+y) \in \mathcal{S}_{x,y}$.

So, we consider a new set of testing functions of $\mathbb{R}^2$: let $\hat{\mathcal{S}}_{x,y}$ be the set of all complex-valued functions $\psi(x, y)$ that satisfy the infinite set of inequalities

$$|x+y|^m \left|\frac{\partial^{k_1}\partial^{k_2}\psi(x,y)}{\partial x^{k_1}\partial y^{k_2}}\right| \le C_{mk_1k_2} \tag{5.12}$$

over all $(x, y) \in \mathbb{R}^2$, where the quantity of the right-hand side is a constant with respect to (x, y) but depends upon the choice of the m, k_1, k_2. The definition is analogous to the one that characterizes $\mathcal{S}_{x,y}$ with $|x+y|$ instead of the norm $\sqrt{x^2+y^2}$. Since $|x+y| \le \sqrt{2}\sqrt{x^2+y^2}$, we have $\mathcal{S}_{x,y} \subset \hat{\mathcal{S}}_{x,y}$.

Notice that $\hat{\mathcal{S}}_{x,y}$ is a vector space and that, if $\varphi \in \mathcal{S}$ is a complex-valued function of a one-dimensional variable, then $\varphi(x+y) \in \hat{\mathcal{S}}_{x,y}$.

In analogy to what was done in $\mathcal{S}_{x,y}$, $\{\psi_\nu(x, y)\}_{\nu=1}^{+\infty}$ is said to converge in $\hat{\mathcal{S}}_{x,y}$ if every function $\psi_\nu \in \hat{\mathcal{S}}_{x,y}$ and if, for all non-negative integers m, k_1, k_2, the sequence

$$\left\{|x+y|^m \frac{\partial^{k_1}\partial^{k_2}\psi_\nu(x,y)}{\partial x^{k_1}\partial y^{k_2}}\right\}_{\nu=1}^{+\infty}$$

converges uniformly over all of $\mathbb{R}^2$. Clearly the convergence in $\mathcal{S}_{x,y}$ implies the convergence in $\hat{\mathcal{S}}_{x,y}$.

We denote by $\hat{\mathcal{S}}'_{x,y}$ the set of all distributions on $\hat{\mathcal{S}}_{x,y}$, that is the set of all functions f assigning a number $\langle f, \psi\rangle$ to all $\psi \in \hat{\mathcal{S}}_{x,y}$ that are linear and continuous with respect to the topology defined in $\hat{\mathcal{S}}_{x,y}$. Obviously, $\hat{\mathcal{S}}'_{x,y} \subset \mathcal{S}'_{x,y}$.

Let assume now that $f : \mathbb{R} \to \mathbb{R}$ such that $\int_{-\infty}^{+\infty} |f(x)|\,dx < +\infty$ and $\theta(y) = \mathbb{I}_{[0,+\infty)}(y)$, the *Heaviside step function*. Obviously $f \in \mathcal{S}'_x$ and $\theta \in \mathcal{S}'_y$.

Lemma 5.8.1 *Let $\psi \in \hat{\mathcal{S}}_{x,y}$, then*

$$\int_{-\infty}^{+\infty}\int_{-\infty}^{+\infty} |f(x)\theta(y)\psi(x,y)|\,dy\,dx < +\infty$$

Proof. By (5.12), since $\psi \in \hat{\mathcal{S}}_{x,y}$, there exists $C_{000} > 0$ such that $|\psi(x, y)| \leq C_{000}$, $(x, y) \in \mathbb{R}^2$. Now,

$$\int_{-\infty}^{+\infty}\int_{-\infty}^{+\infty} |f(x)\theta(y)\psi(x, y)|\,\mathrm{d}y\,\mathrm{d}x = \int_{-\infty}^{+\infty}\int_{-\infty}^{+\infty} \theta(y)|f(x)||\psi(x, y)|\,\mathrm{d}y\,\mathrm{d}x = \int_{-\infty}^{+\infty}\int_{0}^{+\infty} |f(x)||\psi(x, y)|\,\mathrm{d}y\,\mathrm{d}x$$

Let, for a fixed $K > 0$,

$$\mathcal{D}_K = \left\{(x, y) \in \mathbb{R}^2 : y \geq 0, |x + y| \leq K\right\}$$

and

$$\overline{\mathcal{D}}_K = \left\{(x, y) \in \mathbb{R}^2 : y \geq 0, |x + y| > K\right\}$$

We have

$$\int_{-\infty}^{+\infty}\int_{0}^{+\infty} |f(x)||\psi(x, y)|\,\mathrm{d}y\,\mathrm{d}x = \iint_{\mathcal{D}_K} |f(x)||\psi(x, y)|\,\mathrm{d}y\,\mathrm{d}x + \iint_{\overline{\mathcal{D}}_K} |f(x)||\psi(x, y)|\,\mathrm{d}y\,\mathrm{d}x \tag{5.13}$$

But

$$\begin{aligned}\iint_{\mathcal{D}_K} |f(x)||\psi(x, y)|\,\mathrm{d}y\,\mathrm{d}x &= \int_{-\infty}^{+\infty} |f(x)| \left(\int_{\max\{-K-x,0\}}^{\max\{K-x,0\}} |\psi(x, y)|\,\mathrm{d}y\right)\mathrm{d}x \\ &\leq \int_{-\infty}^{+\infty} |f(x)| \left(\int_{\max\{-K-x,0\}}^{\max\{K-x,0\}} C_{000}\,\mathrm{d}y\right)\mathrm{d}x \\ &= C_{000}\int_{-\infty}^{+\infty} [\max\{K - x, 0\} - \max\{-K - x, 0\}]|f(x)|\,\mathrm{d}x \\ &\leq C_{000}\int_{-\infty}^{+\infty} [K - x - (-K - x)]|f(x)|\,\mathrm{d}x \\ &= 2KC_{000}\int_{-\infty}^{+\infty} |f(x)|\,\mathrm{d}x < +\infty \end{aligned} \tag{5.14}$$

On the other hand, assuming that $m > 1$,

$$\begin{aligned}&\iint_{\overline{\mathcal{D}}_K} |f(x)||\psi(x, y)|\,\mathrm{d}y\,\mathrm{d}x \leq \iint_{\overline{\mathcal{D}}_K} |f(x)|\frac{C_{m00}}{|x + y|^m}\,\mathrm{d}y\,\mathrm{d}x \\ &= C_{m00}\int_{-\infty}^{+\infty} |f(x)| \left(\int_{0}^{\max\{-x-K,0\}} \frac{1}{|x + y|^m}\,\mathrm{d}y + \int_{\max\{-x+K,0\}}^{+\infty} \frac{1}{|x + y|^m}\,\mathrm{d}y\right)\mathrm{d}x\end{aligned} \tag{5.15}$$

But, if $x \leq -K$,

$$\int_0^{\max\{-x-K,0\}} \frac{1}{|x+y|^m}\,\mathrm{d}y = \int_0^{-x-K} \frac{1}{(-x-y)^m}\,\mathrm{d}y$$
$$= \frac{1}{m-1}\frac{1}{K^{m-1}} - \frac{1}{m-1}\frac{1}{(-x)^{m-1}} \tag{5.16}$$

while, if $x > -K$,

$$\int_0^{\max\{-x-K,0\}} \frac{1}{|x+y|^m}\,\mathrm{d}y = 0 \tag{5.17}$$

Moreover, if $x > K$, assuming that $m > 1$,

$$\int_{\max\{-x+K,0\}}^{+\infty} \frac{1}{|x+y|^m}\,\mathrm{d}y = \int_0^{+\infty} \frac{1}{(x+y)^m}\,\mathrm{d}y$$
$$= \frac{1}{(m-1)}\frac{1}{x^{m-1}} \tag{5.18}$$

while, if $x \leq K$,

$$\int_{\max\{-x+K,0\}}^{+\infty} \frac{1}{|x+y|^m}\,\mathrm{d}y = \int_{K-x}^{+\infty} \frac{1}{(x+y)^m}\,\mathrm{d}y$$
$$= \frac{1}{(m-1)}\frac{1}{K^{m-1}} \tag{5.19}$$

Substituting (5.16), (5.17), (5.18) and (5.19) in (5.15) we get

$$\iint_{\overline{\mathcal{D}}_K} |f(x)||\psi(x,y)|\,\mathrm{d}y\,\mathrm{d}x$$
$$\leq \frac{C_{m00}}{m-1}\int_{-\infty}^{+\infty} |f(x)|\left[\left(\frac{1}{K^{m-1}} - \frac{1}{(-x)^{m-1}}\right)\mathbb{I}_{\{x\leq -K\}} + \frac{1}{x^{m-1}}\mathbb{I}_{\{x>K\}} + \frac{1}{K^{m-1}}\mathbb{I}_{\{-K<x\leq K\}}\right]\mathrm{d}x$$
$$\leq \frac{C_{m00}}{m-1}\int_{-\infty}^{+\infty} |f(x)|\left[\left(\frac{1}{K^{m-1}} + \frac{1}{(-x)^{m-1}}\right)\mathbb{I}_{\{x\leq -K\}} + \frac{1}{x^{m-1}}\mathbb{I}_{\{x>K\}} + \frac{1}{K^{m-1}}\mathbb{I}_{\{-K<x\leq K\}}\right]\mathrm{d}x$$
$$\leq \frac{C_{m00}}{m-1}\int_{-\infty}^{+\infty} |f(x)|\left[\frac{2}{K^{m-1}}\mathbb{I}_{\{x\leq -K\}} + \frac{1}{K^{m-1}}\mathbb{I}_{\{x>K\}} + \frac{1}{K^{m-1}}\mathbb{I}_{\{-K<x\leq K\}}\right]\mathrm{d}x$$
$$\leq \frac{2C_{m00}}{(m-1)K^{m-1}}\int_{-\infty}^{+\infty} |f(x)|\,\mathrm{d}x < +\infty \tag{5.20}$$

By (5.14), (5.20) and (5.13), the thesis follows. □

If $\{\psi_\nu(x,y)\}_{\nu=1}^{+\infty}$ is a sequence in $\hat{\mathcal{S}}_{x,y}$ converging in $\hat{\mathcal{S}}_{x,y}$ to zero, since $\psi_\nu(x,y)$ converges uniformly on $\mathbb{R}^2$,

$$\int_{-\infty}^{+\infty}\int_{-\infty}^{+\infty} |f(x)\theta(y)\psi_\nu(x,y)|\,\mathrm{d}y\,\mathrm{d}x \underset{\nu\to+\infty}{\to} 0$$

This way $f(x)\theta(y) \in \hat{\mathcal{S}}'_{x,y}$ and the convolution $f \star \theta$ can be defined as a distribution on $\mathcal{S}$ through (5.11).

Thinking of f as a probability density, the above arguments allow us to define the convolution of a probability $\mathbb{P}$ having f as its density and the function θ by

$$\mathbb{P} \star \theta \equiv f \star \theta$$

Nevertheless, the convolution between a probability $\mathbb{P}$ and the function θ can be defined as well, using analogous arguments, by

$$\begin{aligned}
\langle \mathbb{P} \star \theta, \varphi \rangle &= \int_{-\infty}^{+\infty} \mathrm{d}t \int_{-\infty}^{+\infty} \mathbb{P}(\mathrm{d}\tau)\theta(t-\tau)\varphi(t)\,\mathrm{d}t \\
&= \int_{-\infty}^{+\infty} \int_{-\infty}^{+\infty} \mathbb{P}(\mathrm{d}x)\theta(y)\varphi(x+y)\,\mathrm{d}y
\end{aligned}$$

6

The Fourier Transform

6.1 INTRODUCTION

In this chapter we develop the main results concerning the Fourier transform, needed for the results that were presented in Chapter 1. First of all, we will recall the classical properties of ordinary Fourier transformation of functions. After that, we will introduce the Fourier transform from the distributional point of view. The chapter closes with a number of useful examples written in the form of exercises with solutions, some of which were used in the computations of Chapter 1.

6.2 THE FOURIER TRANSFORMATION OF FUNCTIONS

6.2.1 Fourier series

A Fourier series is an expansion of a periodic function $f(x)$ in terms of an infinite sum of sines and cosines. As such, a Fourier series exploits the orthogonality relationships of sine and cosine functions. The field of computation and study of a Fourier series is known as *harmonic analysis* and is extremely useful as a way to break up an arbitrary periodic function into a set of simple terms that can be plugged in, solved individually, and recombined to obtain the solution of the original problem – or an approximation of it at any degree of accuracy that may be considered desirable for practical purposes. Examples of successive approximations of a common function using a Fourier series are illustrated in Figure 6.1.

In particular, since the superposition principle holds for solutions of linear homogeneous ordinary differential equations, if one such equation can be solved in the case of a single sinusoid, the solution for an arbitrary function is immediately available by expressing the original function as a Fourier series and then plugging in the solution for each sinusoidal component. In some special cases where the Fourier series can be summed in closed form, this technique can even yield analytic solutions.

The computation of the (usual) Fourier series is based on the integral identities

$$\int_{-\pi}^{\pi} \sin(mx)\sin(nx)\,\mathrm{d}x = \pi\,\delta_{mn} \tag{6.1}$$

$$\int_{-\pi}^{\pi} \cos(mx)\cos(nx)\,\mathrm{d}x = \pi\,\delta_{mn} \tag{6.2}$$

$$\int_{-\pi}^{\pi} \sin(mx)\cos(nx)\,\mathrm{d}x = 0 \tag{6.3}$$

$$\int_{-\pi}^{\pi} \sin(mx)\,\mathrm{d}x = 0 \tag{6.4}$$

$$\int_{-\pi}^{\pi} \cos(mx)\,\mathrm{d}x = 0 \tag{6.5}$$

for $m, n \neq 0$, where δ_{mn} is the Kronecker δ.

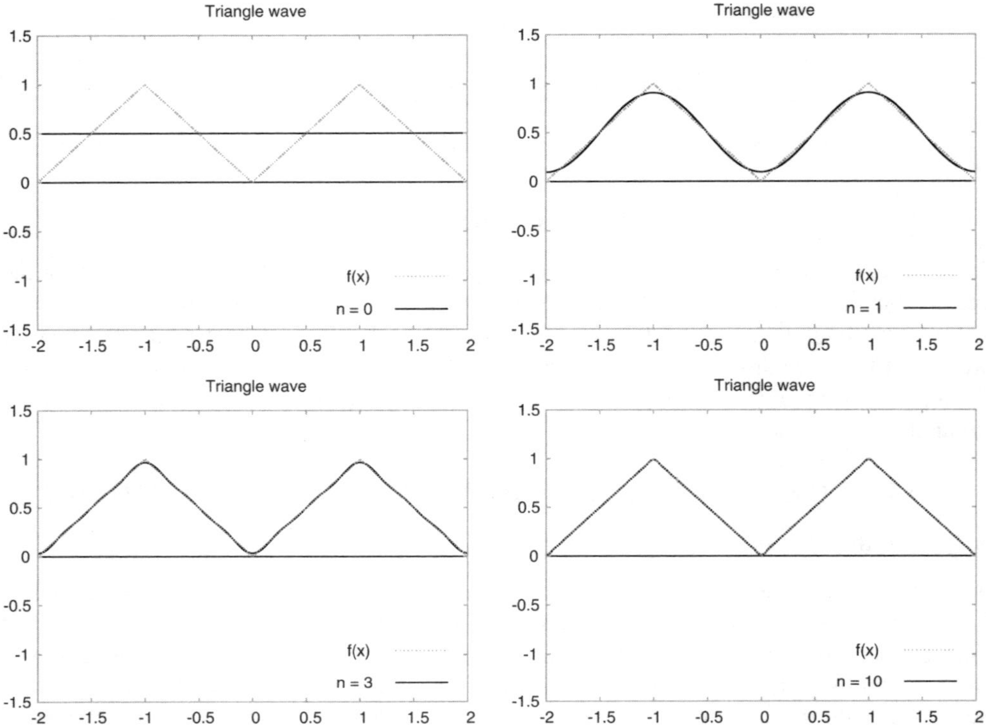

Figure 6.1 Some examples of successive approximations of a common function using Fourier series.

Using the method for a generalized Fourier series, the usual Fourier series involving sines and cosines is obtained by taking $f_1(x) = \cos x$ and $f_2(x) = \sin x$. Since these functions form a complete orthogonal system over $[-\pi, \pi]$, the Fourier series of a function $f(x)$ is given by

$$f(x) = \frac{1}{2}a_0 + \sum_{n=1}^{+\infty} a_n \cos(nx) + \sum_{n=1}^{+\infty} b_n \sin(nx) \tag{6.6}$$

where

$$a_0 = \frac{1}{\pi}\int_{-\pi}^{\pi} f(x)\,\mathrm{d}x \tag{6.7}$$

$$a_n = \frac{1}{\pi}\int_{-\pi}^{\pi} f(x)\cos(nx)\,\mathrm{d}x \tag{6.8}$$

$$b_n = \frac{1}{\pi}\int_{-\pi}^{\pi} f(x)\sin(nx)\,\mathrm{d}x \tag{6.9}$$

and $n = 1, 2, 3, \ldots$. Notice that the coefficient of the constant term a_0 was written in a special form compared to the general form for a generalized Fourier series in order to preserve symmetry with the definitions of a_n and b_n.

A Fourier series converges to the function $\overline{f}$ (equal to the original function at points of continuity or to the average of the two limits at points of discontinuity)

$$\overline{f} = \frac{1}{2}\left[\lim_{x\to x_0^-} f(x) + \lim_{x\to x_0^+} f(x)\right] \quad \text{for } -\pi < x_0 < \pi \tag{6.10}$$

$$\overline{f} = \frac{1}{2}\left[\lim_{x\to \pi^+} f(x) + \lim_{x\to \pi^-} f(x)\right] \quad \text{for } x_0 = -\pi, \pi \tag{6.11}$$

if the function satisfies so-called Dirichlet conditions. Dini's test gives a condition for the convergence of a Fourier series.

For a function $f(x)$ that is periodic in an interval $[-L, L]$ instead of $[-\pi, \pi]$, a simple change of variables can be used to transform the interval of integration from $[-\pi, \pi]$ to $[-L, L]$. Let

$$x = \frac{\pi x'}{L} \qquad \mathrm{d}x = \frac{\pi\,\mathrm{d}x'}{L} \tag{6.12}$$

Solving for x' gives $x' = Lx/\pi$, and substituting gives

$$f(x') = \frac{1}{2}a_0 + \sum_{n=1}^{+\infty} a_n \cos\left(\frac{n\pi x'}{L}\right) + \sum_{n=1}^{+\infty} b_n \sin\left(\frac{n\pi x'}{L}\right) \tag{6.13}$$

Therefore,

$$a_0 = \frac{1}{L}\int_{-L}^{L} f(x')\,\mathrm{d}x' \tag{6.14}$$

$$a_n = \frac{1}{L}\int_{-L}^{L} f(x')\cos\left(\frac{n\pi x'}{L}\right)\mathrm{d}x' \tag{6.15}$$

$$b_n = \frac{1}{L}\int_{-L}^{L} f(x')\sin\left(\frac{n\pi x'}{L}\right)\mathrm{d}x' \tag{6.16}$$

Similarly, if the function is instead defined on the interval $[0, 2L]$, the above equations simply become

$$a_0 = \frac{1}{L}\int_{0}^{2L} f(x')\,\mathrm{d}x' \tag{6.17}$$

$$a_n = \frac{1}{L}\int_{0}^{2L} f(x')\cos\left(\frac{n\pi x'}{L}\right)\mathrm{d}x' \tag{6.18}$$

$$b_n = \frac{1}{L}\int_{0}^{2L} f(x')\sin\left(\frac{n\pi x'}{L}\right)\mathrm{d}x' \tag{6.19}$$

In fact, for $f(x)$ periodic with period $2L$, any interval $(x_0, x_0 + 2L)$ can be used, with the choice being driven just by convenience or personal preference.

If a function is even, so that $f(x) = f(-x)$, then $f(x)\sin(nx)$ is odd. (This follows since $\sin(nx)$ is odd and an even function times an odd function gives an odd function.) Therefore, $b_n = 0$ for all n. Similarly, if a function is odd so that $f(x) = -f(-x)$, then $f(x)\cos(nx)$ is odd. (This follows since $\cos(nx)$ is even and an even function times an odd function gives an odd function.) Therefore, $a_n = 0$ for all n.

The notion of a Fourier series can also be extended to complex coefficients. Consider a real-valued function $f(x)$. Write

$$f(x) = \sum_{n=-\infty}^{+\infty} A_n \mathrm{e}^{inx} \tag{6.20}$$

Now examine

$$\begin{aligned}
\int_{-\pi}^{\pi} f(x)\,\mathrm{e}^{-imx}\,\mathrm{d}x &= \int_{-\pi}^{\pi} \left(\sum_{n=-\infty}^{+\infty} A_n \mathrm{e}^{inx} \mathrm{e}^{-imx}\,\mathrm{d}x \right) \\
&= \sum_{n=-\infty}^{+\infty} A_n \int_{-\pi}^{\pi} \mathrm{e}^{i(n-m)x}\,\mathrm{d}x \\
&= \sum_{n=-\infty}^{+\infty} A_n \int_{-\pi}^{\pi} \cos[(n-m)x] + i \sin[(n-m)x]\,\mathrm{d}x \\
&= \sum_{n=-\infty}^{+\infty} A_n 2\pi\,\delta_{mn} \\
&= 2\pi\,A_m
\end{aligned} \tag{6.21}$$

so

$$A_n = \frac{1}{2\pi} \int_{-\pi}^{\pi} f(x)\,\mathrm{e}^{-inx}\,\mathrm{d}x \tag{6.22}$$

The coefficients can be expressed in terms of those in the Fourier series

$$\begin{aligned}
A_n &= \frac{1}{2\pi} \int_{-\pi}^{\pi} f(x)[\cos(nx) - i \sin(nx)]\,\mathrm{d}x \\
&= \begin{cases}
\dfrac{1}{2\pi} \displaystyle\int_{-\pi}^{\pi} f(x)[\cos(nx) + i \sin(nx)]\,\mathrm{d}x & n < 0 \\
\dfrac{1}{2\pi} \displaystyle\int_{-\pi}^{\pi} f(x)\,\mathrm{d}x & n = 0 \\
\dfrac{1}{2\pi} \displaystyle\int_{-\pi}^{\pi} f(x)[\cos(nx) - i \sin(nx)]\,\mathrm{d}x & n > 0
\end{cases}
\end{aligned} \tag{6.23}$$

and, computing the integrals

$$A_n = \begin{cases}
\dfrac{1}{2}(a_n + i b_n) & \text{for } n < 0 \\
\dfrac{1}{2} a_0 & \text{for } n = 0 \\
\dfrac{1}{2}(a_n - i b_n) & \text{for } n > 0
\end{cases} \tag{6.24}$$

For a function periodic in $[-L/2, L/2]$, these become

$$f(x) = \sum_{n=-\infty}^{+\infty} A_n e^{i(2\pi nx/L)} \tag{6.25}$$

$$A_n = \frac{1}{L}\int_{-L/2}^{L/2} f(x)\, e^{-i(2\pi nx/L)}\, dx \tag{6.26}$$

6.2.2 Fourier transform

The Fourier transform is a generalization of the complex Fourier series in the limit as $L \to \infty$. There are several common conventions in the definition of the Fourier transform, and in this book we will use the following:

$$f(x) = \int_{-\infty}^{+\infty} F(k)\, e^{-2\pi ikx}\, dk \tag{6.27}$$

$$F(k) = \int_{-\infty}^{+\infty} f(x)\, e^{2\pi ikx}\, dx \tag{6.28}$$

where

$$F(k) = F_x[f(x)](k) = \int_{-\infty}^{+\infty} f(x)\, e^{2\pi ikx}\, dx \tag{6.29}$$

is called the forward $(+i)$ Fourier transform, and

$$f(x) = F_k^{-1}[F(k)](x) = \int_{-\infty}^{+\infty} F(k)\, e^{-2\pi ikx}\, dk \tag{6.30}$$

is called the inverse $(-i)$ Fourier transform. The reader should be aware that in many cases it is possible to find the opposite definition (which is actually dominant in computer science). Furthermore, notice that some authors (especially physicists) prefer to write the transform in terms of angular frequency $\omega = 2\pi\nu$ instead of the oscillation frequency ν. However, this destroys the symmetry, resulting in the transform pair

$$H(\omega) = F[h(t)] = \int_{-\infty}^{+\infty} h(t)\, e^{i\omega t}\, dt \tag{6.31}$$

$$h(t) = F^{-1}[H(\omega)] = \frac{1}{2\pi}\int_{-\infty}^{+\infty} H(\omega)\, e^{-i\omega t}\, d\omega \tag{6.32}$$

To restore the symmetry of the transforms, the convention

$$g(y) = F[f(t)] = \frac{1}{\sqrt{2\pi}}\int_{-\infty}^{+\infty} f(t)\, e^{iyt}\, dt \tag{6.33}$$

$$f(t) = F^{-1}[g(y)] = \frac{1}{\sqrt{2\pi}}\int_{-\infty}^{+\infty} g(y)\, e^{-iyt}\, dy \tag{6.34}$$

is sometimes used.

In general, the Fourier transform pair may be defined using two arbitrary constants a and b as

$$F(\omega) = \sqrt{\frac{|b|}{(2\pi)^{1-a}}} \int_{-\infty}^{+\infty} f(t)\, \mathrm{e}^{ib\omega t}\, \mathrm{d}t \tag{6.35}$$

$$f(t) = \sqrt{\frac{|b|}{(2\pi)^{1+a}}} \int_{-\infty}^{+\infty} F(\omega)\, \mathrm{e}^{-ib\omega t}\, \mathrm{d}\omega \tag{6.36}$$

Since any function can be split up into even and odd portions $E(x)$ and $O(x)$,

$$f(x) = \frac{1}{2}[f(x) + f(-x)] + \frac{1}{2}[f(x) - f(-x)] = E(x) + O(x) \tag{6.37}$$

a Fourier transform can always be expressed in terms of the Fourier cosine transform and Fourier sine transform as

$$F_x[f(x)](k) = \int_{-\infty}^{+\infty} E(x)\cos(2\pi kx)\,\mathrm{d}x + i\int_{-\infty}^{+\infty} O(x)\sin(2\pi kx)\,\mathrm{d}x \tag{6.38}$$

A function $f(x)$ has forward and inverse Fourier transforms such that

$$f(x) = \int_{-\infty}^{+\infty} \mathrm{e}^{-2\pi ikx}\left[\int_{-\infty}^{+\infty} f(x)\,\mathrm{e}^{2\pi ikx}\,\mathrm{d}x\right]\mathrm{d}k \tag{6.39}$$

for $f(x)$ continuous at x and

$$f(x) = \frac{1}{2}[f(x_+) + f(x_-)] \tag{6.40}$$

for $f(x)$ discontinuous at x, provided that the following conditions are satisfied:

1. $\int_{-\infty}^{+\infty} |f(x)|\,\mathrm{d}x$ exists.
2. There is a finite number of discontinuities.
3. The function has bounded variation. A sufficient weaker condition is fulfilment of the Lipschitz condition.

The Fourier transform is linear, since if $f(x)$ and $g(x)$ have Fourier transforms $F(k)$ and $G(k)$, then

$$\begin{aligned}\int [af(x) + bg(x)]\mathrm{e}^{2\pi ikx}\,\mathrm{d}x &= a\int_{-\infty}^{+\infty} f(x)\,\mathrm{e}^{2\pi ikx}\,\mathrm{d}x + b\int_{-\infty}^{+\infty} g(x)\,\mathrm{e}^{2\pi ikx}\,\mathrm{d}x \\ &= aF(k) + bG(k)\end{aligned} \tag{6.41}$$

Therefore,

$$F[af(x) + bg(x)] = aF[f(x)] + bF[g(x)] = aF(k) + bG(k)$$

The Fourier transform is also symmetric since $F(k) = F_x[f(x)](k)$ implies $F(-k) = F_x[f(-x)](k)$. Let $f \star g$ denote the convolution, then the transforms of convolutions of functions have particularly nice forms,

$$F[f \star g] = F[f]F[g] \tag{6.42}$$

$$F[fg] = F[f] \star F[g] \tag{6.43}$$

$$F^{-1}[F(f)F(g)] = f \star g \tag{6.44}$$

$$F^{-1}[F(f) \star F(g)] = fg \tag{6.45}$$

The first of these is derived as follows:

$$\begin{aligned} F[f \star g] &= \int_{-\infty}^{+\infty} \int_{-\infty}^{+\infty} \mathrm{e}^{2\pi i k x} f(x') g(x - x') \, \mathrm{d}x' \, \mathrm{d}x \\ &= \int_{-\infty}^{+\infty} \int_{-\infty}^{+\infty} [\mathrm{e}^{2\pi i k x'} f(x') \, \mathrm{d}x'][\mathrm{e}^{2\pi i k (x - x')} g(x - x') \, \mathrm{d}x] \\ &= \left[\int_{-\infty}^{+\infty} \mathrm{e}^{2\pi i k x'} f(x') \, \mathrm{d}x' \right] \left[\int_{-\infty}^{+\infty} \mathrm{e}^{2\pi i k x''} g(x'') \, \mathrm{d}x'' \right] \\ &= F[f]F[g] \end{aligned} \tag{6.46}$$

where $x'' = x - x'$.

There is also a somewhat surprising and extremely important relationship between the autocorrelation and the Fourier transform known as the Wiener–Khintchine theorem. Let $F_x[f(x)](k) = F(k)$, and $f^\dagger$ denote, as usual, the complex conjugate of f, then the Fourier transform of the absolute square of $F(k)$ is given by

$$F_k[|F(k)|^2](x) = \int_{-\infty}^{+\infty} f^\dagger(\tau) f(\tau + x) \, \mathrm{d}\tau \tag{6.47}$$

The Fourier transform of a derivative $f'(x)$ of a function $f(x)$ is simply related to the transform of the function $f(x)$ itself. Consider

$$F_x[f'(x)](k) = \int_{-\infty}^{+\infty} f'(x) \, \mathrm{e}^{2\pi i k x} \, \mathrm{d}x \tag{6.48}$$

Now use integration by parts to obtain

$$F_x[f'(x)](k) = [f(x) \, \mathrm{e}^{2\pi i k x}]_{-\infty}^{+\infty} - \int_{-\infty}^{+\infty} f(x)(2\pi i k \mathrm{e}^{2\pi i k x} \, \mathrm{d}x) \tag{6.49}$$

The first term consists of an oscillating function times $f(x)$. But if the function is bounded so that

$$\lim_{x \to \pm\infty} f(x) = 0$$

then the term vanishes, leaving

$$F_x[f'(x)](k) = 2\pi i k \int_{-\infty}^{+\infty} f(x) \, \mathrm{e}^{2\pi i k x} \, \mathrm{d}x = 2\pi i k F_x[f(x)](k) \tag{6.50}$$

This process can be iterated for the nth derivative to yield

$$F_x[f^{(n)}(x)](k) = (2\pi ik)^n F_x[f(x)](k) \tag{6.51}$$

If $f(x)$ has the Fourier transform $F_x[f(x)](k) = F(k)$, then the Fourier transform has the shift property

$$\begin{aligned} \int_{-\infty}^{+\infty} f(x-x_0)\,\mathrm{e}^{2\pi ikx}\,\mathrm{d}x &= \int_{-\infty}^{+\infty} f(x-x_0)\,\mathrm{e}^{2\pi i(x-x_0)k}\mathrm{e}^{2\pi i(kx_0)}\,\mathrm{d}(x-x_0) \\ &= \mathrm{e}^{2\pi ikx_0}F(k) \end{aligned} \tag{6.52}$$

so $f(x-x_0)$ has the Fourier transform

$$F_x[f(x-x_0)](k) = \mathrm{e}^{2\pi ikx_0}F(k) \tag{6.53}$$

If $f(x)$ has a Fourier transform $F_x[f(x)](k) = F(k)$, then the Fourier transform obeys a similarity theorem.

$$\int_{-\infty}^{+\infty} f(ax)\mathrm{e}^{2\pi ikx}\,\mathrm{d}x = \frac{1}{|a|}\int_{-\infty}^{+\infty} f(ax)\,\mathrm{e}^{2\pi i(ax)(k/a)}\,\mathrm{d}(ax) = \frac{1}{|a|}F(k/a) \tag{6.54}$$

so $f(ax)$ has the Fourier transform

$$F_x[f(ax)](k) = |a|^{-1}F(k/a) \tag{6.55}$$

In the following, for the sake of simplicity, we shall use also the following symbol for the Fourier transform of a function $f(x)$

$$\tilde{f}(k) = F[f(x)]$$

6.2.3 Parseval theorem

Another result very useful in the following discussion is known as

Theorem 6.2.1 (*Parseval's equation*) *If the locally integrable functions $f(t)$ and $g(t)$ are absolutely integrable over $-\infty < t < \infty$, then*

$$\int_{-\infty}^{\infty} f(x)\tilde{g}(x)\,\mathrm{d}x = \int_{-\infty}^{\infty} \tilde{f}(x)g(x)\,\mathrm{d}x$$

6.3 FOURIER TRANSFORM AND OPTION PRICING

The applications to option pricing that can be found in the literature refer to Fourier transforms of functions, and exploit the theory that has been developed up to this point. Here we give a very quick account of the structure of the two main strategies proposed in the literature, the first by Carr and Madan (1999) and the second by Lewis (2001).

6.3.1 The Carr–Madan approach

The Carr–Madan paper is probably the most influential work concerning the application of Fourier transform methods to pricing issues. Rather than adopting the generalized function approach, they work with functions and immediately they have to tackle the issue that a call payoff does not a have a regular Fourier transform, given that is not a summable function.

Therefore they decide to follow a different route. Let us consider the price of a call option:

$$\mathcal{C}(w) = \int \mathcal{Q}(\mathrm{d}x)[\mathrm{e}^x - \mathrm{e}^w]^+$$

and ask what is needed to make it into an L^1 function. Certainly there must be a value $\alpha > 0$ such that the modified price $\mathcal{C}(w, \alpha)$ defined as:

$$\mathcal{C}(w, \alpha) := \mathrm{e}^{\alpha w} \int \mathcal{Q}(\mathrm{d}x)[\mathrm{e}^x - \mathrm{e}^w]^+$$

is a summable function.

Once we have selected a suitable α that does the job we can proceed to take the Fourier transform

$$\begin{aligned} G(k, \alpha) &:= \mathcal{F}[\mathcal{C}(w, \alpha)] \\ &= \int \mathrm{d}w\, \mathrm{e}^{i2\pi wk} \mathrm{e}^{\alpha w} \int \mathcal{Q}(\mathrm{d}x)[\mathrm{e}^x - \mathrm{e}^w]^+ \end{aligned}$$

The dumping factor $\mathrm{e}^{\alpha w}$ justifies the change of integration order, so we get

$$\begin{aligned} G(k, \alpha) &= \int \mathcal{Q}(\mathrm{d}x) \int_{-\infty}^{x} \mathrm{d}w\, \mathrm{e}^{i2\pi(k - i\alpha/2\pi)w} [\mathrm{e}^x - \mathrm{e}^w] \\ &= \int \mathcal{Q}(\mathrm{d}x) \int_{-\infty}^{x} \mathrm{d}w\, \left[\mathrm{e}^x \mathrm{e}^{i2\pi(k - i\alpha/2\pi)w} - \mathrm{e}^{i2\pi(k - i(1+\alpha)/2\pi)w}\right] \end{aligned}$$

Performing the innermost integral we end up with

$$G(k, \alpha) = \frac{-i}{2\pi}\left[\frac{1}{k - i\alpha/2\pi} - \frac{1}{k - i(\alpha+1)/2\pi}\right] \int \mathcal{Q}(\mathrm{d}x) \mathrm{e}^{i2\pi[k - i(\alpha+1)/2\pi]x}$$

Let us notice that the condition on α to make $\mathcal{C}(w, \alpha)$ into an L^1 function is just $\alpha > 0$. The remaining integral

$$\int \mathcal{Q}(\mathrm{d}x)\, \mathrm{e}^{i2\pi[k - i(\alpha+1)/2\pi]x}$$

is certainly related to the characteristic function of the risk-neutral martingale distribution. In fact, as long as the expectation

$$\int \mathcal{Q}(\mathrm{d}x)\, \mathrm{e}^{(\alpha+1)x} < \infty$$

we have

$$\int \mathcal{Q}(\mathrm{d}x)\, \mathrm{e}^{i2\pi[k - i(\alpha+1)/2\pi]x} = \phi_X\left(k - i\frac{\alpha+1}{2\pi}\right)$$

and

$$G(k, \alpha) = \left(\frac{-i}{2\pi}\right) \phi_X\left(k - i\frac{(\alpha+1)}{2\pi}\right)\left[\frac{1}{k - i\alpha/2\pi} - \frac{1}{k - i(\alpha+1)/2\pi}\right]$$

Clearly:

$$|G(k,\alpha)| \le \left(\frac{1}{2\pi}\right)\left|\phi_X\left(k - i\frac{(\alpha+1)}{2\pi}\right)\right|\left|\left[\frac{1}{k-i\alpha/2\pi} - \frac{1}{k-i(\alpha+1)/2\pi}\right]\right|$$
$$= \left(\frac{1}{4\pi^2}\right)\left|\phi_X\left(k - i\frac{(\alpha+1)}{2\pi}\right)\right|\left(\frac{1}{(k^2-\alpha(\alpha+1)/4\pi^2)^2 + k^2(2\alpha+1)^2/4\pi^2}\right)^{1/2}$$
$$\le \left(\frac{1}{4\pi^2}\right)\mathbb{E}[e^{(\alpha+1)x}]\left(\frac{1}{(k^2-\alpha(\alpha+1)/4\pi^2)^2 + k^2(2\alpha+1)^2/4\pi^2}\right)^{1/2}$$

therefore, $G(k,\alpha) \in L^1$ and we can invert it:

$$\mathcal{C}(w,\alpha) = \int dk\, e^{-i2\pi kw} G(k,\alpha)$$

and recover the price of the call option as

$$\mathcal{C}(w) = \frac{-i\, e^{-\alpha w}}{2\pi}\int dk\, e^{-i2\pi kw}\phi_X\left(k - i\frac{(\alpha+1)}{2\pi}\right)\left[\frac{1}{k-i\alpha/2\pi} - \frac{1}{k-i(\alpha+1)/2\pi}\right] \tag{6.56}$$

6.3.2 The Lewis approach

Another popular approach is due to Lewis (2001). Rather that starting from the price of an option, Lewis goes back to the payoff, noticing that if there is an interval $\mathcal{S}_X := (a, b)$ such that for $\alpha \in \mathcal{S}_X$ we have

$$g(x,\alpha) := e^{-\alpha x}[e^x - e^w]^+ \in L^1$$

we can define the Fourier transform:

$$\hat{g}(k,\alpha) = \int dx\, e^{i2\pi kx} g(x,\alpha)$$

For the case of the call option we have

$$\mathcal{S}_X = \{\alpha > 1\}$$

and

$$\hat{g}(k,\alpha) = \int dx\, e^{i2\pi(k+i\alpha/2\pi)x}(e^x - e^w)\theta(x-w)$$
$$= \int_w^{+\infty} dx\, e^{i2\pi(k+i(\alpha-1)/2\pi)x} - e^w \int_w^{+\infty} dx\, e^{i2\pi(k+i\alpha/2\pi)x}$$

Performing the integrals we obtain

$$\hat{g}(k,\alpha) = e^{i2\pi(k+i(\alpha-1)/2\pi)w}\left(\frac{-i}{2\pi}\right)\left[\frac{1}{(k+i\alpha/2\pi)} - \frac{1}{(k+i(\alpha-1)/2\pi)}\right] \tag{6.57}$$

Clearly $\hat{g}(k,\alpha) \in L^1$ and we have no problem in inverting the transform.

Let's assume now that we have a value β such that

$$\mathbb{E}[e^{\lambda x}] < \infty, \quad \forall \lambda < \beta$$

This implies that the characteristic function $\phi_X(z)$ is analytic in the strip $\mathcal{S}_W : \{0 \le \mathrm{Im}(z) \le \beta\}$

Let's go back now to the pricing equation:

$$\mathcal{C}(w) = \int Q(\mathrm{d}x)(\mathrm{e}^x - \mathrm{e}^w)^+$$

if we have the condition

$$\mathcal{S}_X \cup \mathcal{S}_W \neq \emptyset$$

that is $\beta > \alpha$, we can write the pricing equation as

$$\mathcal{C}(w) = \int [Q(\mathrm{d}x)\,\mathrm{e}^{\lambda x}][\mathrm{e}^{-\lambda x}(\mathrm{e}^x - \mathrm{e}^w)^+]$$

where

$$\int [Q(\mathrm{d}x)\,\mathrm{e}^{\lambda x}] < \infty \quad \text{and} \quad \int \mathrm{d}x\,[\mathrm{e}^{-\lambda x}(\mathrm{e}^x - \mathrm{e}^w)^+] < \infty$$

Under these conditions Parseval's theorem holds and we can write

$$\mathcal{C}(w) = \int \mathrm{d}k\,\phi_X\left(k - i\frac{\lambda}{2\pi}\right)\hat{g}(k,\lambda) \tag{6.58}$$

where, as usual,

$$\phi_X(k) := \mathbb{E}[\mathrm{e}^{i2\pi kx}]$$

Replacing expression (6.57) into (6.58) we get

$$\mathcal{C}(w) = \mathrm{e}^{-(\lambda-1)w}\int \mathrm{d}k\,\mathrm{e}^{i2\pi kw}\phi_X\left(k - i\frac{\lambda}{2\pi}\right)\left(\frac{-i}{2\pi}\right)$$
$$\times\left[\frac{1}{(k + i\lambda/2\pi)} - \frac{1}{(k + i(\lambda-1)/2\pi)}\right]$$

6.4 FOURIER TRANSFORM FOR GENERALIZED FUNCTIONS

We now extend the application of Fourier transform to the realm of generalized functions, which we use to recover option prices.

6.4.1 The Fourier transforms of testing functions of rapid descent

First, we quote two important results which are essential for the definition of the Fourier transform of generalized functions (distributions) of slow growth, which will be discussed in the next section. We refer the reader to the book by Zemanian (1987) for proofs of the following theorems.

Theorem 6.4.1 *If $\phi(t)$ is in $\mathcal{S}$ then its Fourier transform*

$$\tilde{\phi}(\omega) \equiv F[\phi(t)] \equiv \int_{-\infty}^{+\infty}\phi(t)\,\mathrm{e}^{i\omega t}\,\mathrm{d}t$$

is also in $\mathcal{S}$

Theorem 6.4.2 *The Fourier transformation and its inverse are continuous linear mappings of $\mathcal{S}$ onto itself.*

6.4.2 The Fourier transforms of distribution of slow growth

Parseval's equation provides a definition for the Fourier transforms of distributions of slow growth. If the locally integrable function $f(t)$ is absolutely integrable for $-\infty < t < +\infty$, and if φ is a testing function of rapid descent, then their respective Fourier transforms $\tilde{f}$ and $\tilde{\phi}$ certainly exist and one form of Parseval's theorem reads

$$\int_{-\infty}^{+\infty} \tilde{f}(\omega)\varphi(\omega)\,\mathrm{d}\omega = \int_{-\infty}^{+\infty} f(\omega)\tilde{\varphi}(\omega)\,\mathrm{d}\omega \tag{6.59}$$

In our usual notation we can write

$$\langle \tilde{f}, \varphi \rangle = \langle f, \tilde{\varphi} \rangle \tag{6.60}$$

We may generalize equation (6.60) by letting f be any distribution of slow growth. As φ traverses $\mathcal{S}$, (6.60) will define $\tilde{f}$ as a functional on $\mathcal{S}$. In simple words, the Fourier transform $\tilde{f}$ of a distribution f of slow growth is defined as that functional which assigns to each φ in $\mathcal{S}$ the same number as that which f assigns to the Fourier transform $\tilde{\varphi}$ of φ.

The following result holds:

Theorem 6.4.3 *If f is a distribution of slow growth, then its Fourier transform $\tilde{f}$ is also a distribution of slow growth.*

Relation (6.60) also serves as a definition of the inverse Fourier transform of distributions of slow growth. If we set $F[f] = g$ and $F[\varphi] = \psi$, we may rewrite (6.60) as

$$\langle F^{-1}[g], \psi \rangle = \langle g, F^{-1}[\psi] \rangle \tag{6.61}$$

where $g \in \mathcal{S}'$ and $\psi \in \mathcal{S}$. Thus the inverse Fourier transform of an arbitrary distribution g in $\mathcal{S}'$ is that functional which assigns to each ψ in $\mathcal{S}$ the same number as that which g assigns to the inverse Fourier transform of ψ. $F^{-1}[g]$ can again be shown to be a distribution of slow growth.

Since, with $f \in \mathcal{S}'$ and $\psi \in \mathcal{S}$

$$\langle F^{-1}[F[f]], \phi \rangle = \langle F[f], F^{-1}[\varphi] \rangle = \langle f, F[F^{-1}[\varphi]] \rangle = \langle f, \varphi \rangle$$

it follows that $F^{-1}[F[f]] = f$. Similarly, $F[F^{-1}[f]] = f$. Thus, the Fourier transform and its inverse provide one-to-one mappings of the space $\mathcal{S}'$ onto itself. It also follows that $F[f] = 0$ if and only if $f = 0$ (here f is the zero distribution but when f is taken to be a function, its values may be different from zero on a set of measure zero).

It is worth mentioning that our present definition of the Fourier transformation is not applicable when f is an arbitrary distribution in $\mathcal{D}'$. This is because $F[\varphi]$ will not be in $\mathcal{D}$ when $\varphi \in \mathcal{D}$ and $\varphi \neq 0$. Consequently the right-hand side of equation (6.60) may be meaningless. As was mentioned in the introduction to this chapter, by employing another space of testing functions and its dual space of continuous linear functionals, it becomes possible to construct the Fourier transform of any generalized function in $\mathcal{D}'$.

The ordinary Fourier transform is a special case of the distributional Fourier transform. An important advantage of distribution theory is represented by the following result:

Theorem 6.4.4 *The Fourier transform and its inverse are continuous linear mapping of S' onto itself. Consequently, if a series*

$$\sum_{\nu=1}^{\infty} g_\nu$$

converges in S' to g, then the Fourier transform may be applied to this series term-by-term to obtain

$$\tilde{g} = \sum_{\nu=1}^{\infty} \tilde{g}_\nu$$

where the last series again converges in S'.

Such term-by-term transformation is not in general permissible in classical analysis.

6.5 EXERCISES

We now provide some examples in the form of exercises. Some of them are useful to understand and perform the computations that have been given in Chapter 1.

Exercise 6.5.1 *Compute $F(x)$*

Solution. From the relation

$$F[\mathbf{1}] = \delta$$

we can write

$$F[x\mathbf{1}] = \frac{i}{2\pi}\delta'$$

Exercise 6.5.2 *Compute the Fourier transform of the distributions δ^+, δ^- defined by:*

$$\delta^+(x) \equiv \frac{i}{2\pi} g^+(x), \quad \delta^-(x) \equiv \frac{-i}{2\pi} g^-(x)$$

Solution. From the definition:

$$\langle F\delta^+, \gamma\rangle = \langle \delta^+, F\gamma\rangle, \quad \langle F\delta^-, \gamma\rangle = \langle \delta^-, F\gamma\rangle$$

For the first part we have

$$\begin{aligned}
\langle \delta^+, F\gamma\rangle &= \frac{i}{2\pi}\lim_{\epsilon\to 0^+}\int \mathrm{d}x\,\mathrm{d}\lambda \frac{\gamma(\lambda)}{x+i\epsilon}\mathrm{e}^{2\pi i\lambda x} \\
&= \frac{i}{2\pi}\lim_{\epsilon\to 0^+}\int_{-\infty}^{0} \mathrm{d}\lambda\,\gamma(\lambda)\int \mathrm{d}x \frac{1}{x+i\epsilon}\mathrm{e}^{-2\pi i|\lambda|x} \\
&+ \frac{1}{2\pi i}\lim_{\epsilon\to 0^+}\int_{0}^{+\infty} \mathrm{d}\lambda\,\gamma(\lambda)\int \mathrm{d}x \frac{1}{x+i\epsilon}\mathrm{e}^{2\pi i|\lambda|x} \\
&= \lim_{\epsilon\to 0^+}\int_{-\infty}^{0} \mathrm{d}\lambda\,\gamma(\lambda)\mathrm{e}^{-2\pi\epsilon\lambda} = \int_{-\infty}^{0} \mathrm{d}\lambda\,\gamma(\lambda)
\end{aligned}$$

It follows that

$$F\delta^+ = \theta(-\lambda)$$

For the second part we have

$$\begin{aligned}
\langle \delta^-, F\gamma \rangle &= -\frac{i}{2\pi} \lim_{\epsilon\to 0^+} \int \mathrm{d}x\, \mathrm{d}\lambda \frac{\gamma(\lambda)}{x - i\epsilon} \mathrm{e}^{2\pi i \lambda x} \\
&= -\frac{i}{2\pi} \lim_{\epsilon\to 0^+} \int_{-\infty}^{0} \mathrm{d}\lambda\, \gamma(\lambda) \int \mathrm{d}x \frac{1}{x - i\epsilon} \mathrm{e}^{-2\pi i |\lambda| x} \\
&\quad - \frac{i}{2\pi} \lim_{\epsilon\to 0^+} \int_{0}^{+\infty} \mathrm{d}\lambda\, \gamma(\lambda) \int \mathrm{d}x \frac{1}{x - i\epsilon} \mathrm{e}^{2\pi i |\lambda| x} \\
&= \lim_{\epsilon\to 0^+} \int_0^{\infty} \mathrm{d}\lambda\, \gamma(\lambda) \mathrm{e}^{2\pi \epsilon \lambda} = \int_0^{\infty} \mathrm{d}\lambda \gamma(\lambda)
\end{aligned}$$

It follows that

$$F\delta^- = \theta(\lambda)$$

Exercise 6.5.3 *Let $\Phi(x)$ be a probability measure, and define*

$$\phi(t) = \int \mathrm{d}F\, \mathrm{e}^{2\pi i t x}$$

its characteristic function. Express $\Phi(x)$ as an integral function of $\phi(t)$.

Solution. Clearly we have

$$\int \mathrm{d}\Phi\, \theta(x - y) = \Phi(y)$$

on the other hand

$$\begin{aligned}
\int \mathrm{d}\Phi\, \theta(x - y) &= \int \mathrm{d}\Phi\, F\delta^+ \\
&= \frac{i}{2\pi} \lim_{\epsilon\to 0^+} \int \mathrm{d}\Phi \frac{\mathrm{d}\lambda}{\lambda + i\epsilon} \mathrm{e}^{2\pi i \lambda (x-y)} \\
&= \frac{i}{2\pi} \lim_{\epsilon\to 0^+} \int \mathrm{d}\lambda \frac{\phi(\lambda)}{\lambda + i\epsilon} \mathrm{e}^{-2\pi i \lambda y} \\
&= \frac{i}{2\pi} \int \mathrm{d}\lambda \phi(\lambda) \left[p.v. \frac{1}{\lambda} - i\pi\delta(\lambda) \right] \mathrm{e}^{-2\pi i \lambda y} \\
&= \frac{\phi(0)}{2} - \frac{1}{2\pi i} \int \frac{\mathrm{d}\lambda}{\lambda} \left[\phi(\lambda) \mathrm{e}^{-2\pi i \lambda y} - \phi(0) \right]
\end{aligned}$$

From the definition of the characteristic function we have $\phi(0) = 1$, therefore we can conclude that

$$\Phi(y) = \frac{1}{2} + \frac{1}{2\pi} \int \frac{\mathrm{d}\lambda}{i\lambda} \left[1 - \phi(\lambda) \mathrm{e}^{-2\pi i \lambda y} \right]$$

besides, from

$$\Phi(y) = -\frac{1}{2\pi i} \lim_{\epsilon\to 0^+} \int \mathrm{d}\lambda \frac{\phi(\lambda)}{\lambda + i\epsilon} \mathrm{e}^{-2\pi i \lambda y}$$

we conclude that

$$F\Phi = -\frac{1}{2\pi i} \lim_{\epsilon\to 0^+} \frac{\phi(\lambda)}{\lambda + i\epsilon}$$

Exercise 6.5.4 *Compute* $F[\theta(x) - \theta(-x)]$.

Solution. since

$$\theta(x) - \theta(-x) = 2\theta(x) - 1$$

we have

$$F[\theta(x) - \theta(-x)] = 2\delta^{-} - \delta$$

Exercise 6.5.5 *Compute* $F[|x|]$.

Solution. Since

$$|x| = x[\theta(x) - \theta(-x)]$$

we have

$$F|x| = F[x(\theta(x) - \theta(-x))] = \frac{i}{2\pi}\frac{\mathrm{d}}{\mathrm{d}t}[2\delta^{+} - \delta]$$

6.6 FOURIER OPTION PRICING WITH GENERALIZED FUNCTIONS

We now go through the derivation of option prices which was discussed in Chapter 1. Neglecting normalization and scaling factors, the payoff for a call option is given by

$$\mathcal{C}(w) = [\,\mathrm{e}^{x} - \mathrm{e}^{w}\,]^{+}$$

while the price of that same option is

$$\mathcal{C}(w) = \int Q(\mathrm{d}x)[\,\mathrm{e}^{x} - \mathrm{e}^{w}\,]^{+} \tag{6.62}$$

where $Q(\mathrm{d}x)$ is the risk-neutral martingale measure.

As usual we write the payoff as

$$\mathcal{C}(w) = [\,\mathrm{e}^{x} - \mathrm{e}^{w}\,]\theta(x - w)$$

and the price will be written as

$$\mathcal{C}(w) = \int Q(\mathrm{d}x)[\,\mathrm{e}^{x} - \mathrm{e}^{w}\,]\theta(x - w)$$

Following the lines of Chapter 1 we introduce a new measure

$$Q^{*}(\mathrm{d}x) = Q(\mathrm{d}x)\,\mathrm{e}^{x}$$

and split the call price into two parts:

$$\mathcal{C}(w) = \int Q^{*}(\mathrm{d}x)\theta(x - w) - \mathrm{e}^{w}\int Q(\mathrm{d}x)\theta(x - w)$$

In this form we recognize two convolutions except for the sign of the argument in the θ function. Let's introduce the reflection operator s as

$$s : \mathcal{R} \to \mathcal{R}, \quad s : x \to -x$$

so that, given a function f, we can write

$$f \circ s : x \to f(-x)$$

(not to be confused with $s \circ f : x \to -f(x)$).

A simple property of the Fourier transform is the fact that

$$\mathcal{F}(f \circ s) = \overline{\mathcal{F}} f,$$

and using the reflection operator s we can write

$$\int Q^*(\mathrm{d}x)\theta(x - w) = (\theta \circ s) \star Q^*, \qquad \int Q(\mathrm{d}x)\theta(x - w) = (\theta \circ s) \star Q$$

For convolution of distributions we have seen that

$$f \star g = \overline{\mathcal{F}}(\mathcal{F} f \mathcal{F} g)$$

therefore

$$\begin{aligned}\int Q(\mathrm{d}x)\theta(x - w) &= (\theta \circ s) \star Q \\ &= \overline{\mathcal{F}}[\mathcal{F}(\theta \circ s)\mathcal{F}(Q)] \\ &= \overline{\mathcal{F}}[\overline{\mathcal{F}}(\theta)\mathcal{F}(Q)]\end{aligned}$$

In conclusion

$$\int Q(\mathrm{d}x)\theta(x - w) = \int \mathrm{d}k\, \mathrm{e}^{-i2\pi k w} \delta^+(k)\phi_X(k)$$

Following the same line of reasoning we get

$$\int Q^*(\mathrm{d}x)\theta(x - w) = \int \mathrm{d}k\, \mathrm{e}^{-i2\pi k w} \delta^+(k)\phi_X\left(k - \frac{i}{2\pi}\right)$$

and the price of the call option is given by

$$\mathcal{C}(w) = \int \mathrm{d}k\, \mathrm{e}^{-i2\pi k w} \delta^+(k)\phi_X\left(k - \frac{i}{2\pi}\right) - \mathrm{e}^{w} \int \mathrm{d}k\, \mathrm{e}^{-i2\pi k w} \delta^+(k)\phi_X(k) \tag{6.63}$$

7

Fourier Transforms at Work

7.1 INTRODUCTION

In this chapter we apply the pricing formula proposed in Chapter 1 to real-world data. First of all, since this book was shot like a police story, starting with the ending, we take a few words to show how we got to that final scene, and we collect the hints that make the overall picture clear.

As in Duffie *et al.* (2000), the main character of our movie is the pricing kernel of the options written on some underlying asset S_t, that is, the value of the digital option. Differently from the above paper, we are able to split the Fourier transform of our digital options into the payoff and density specific to the model involved. In this sense, our story establishes a link to the approaches proposed by Carr and Madan (1999) and Lewis (2001). Differently from those, the core of our story is the Fourier transform of the payoff of the digital option. This is well defined because we work in the framework of generalized functions, that is functionals, instead of functions. So, this gives us a smooth way to substitute the Fourier transform of the payoff in the price of the digital option. The route is not so smooth when we try to define the convolution – in the generalized function sense – of the payoff and the digital options. Luckily, this convolution is well defined under a very mild condition, which corresponds to the requirement that the probability distribution must have a finite first moment. This is not a very stringent requirement for an application of pricing in an arbitrage-free setting, where it corresponds to assuming that the price of the underlying asset exists.

This way, we are back to the happy ending from which we started. We have a general pricing formula for European options, with strike K and maturity T:

$$\mathcal{O}(S_t; m, T, \omega) = \frac{1}{2}\omega S_t(1-m) + S_t\frac{i}{2\pi}\left(\mathrm{d}(k,0)m - \mathrm{d}\left(k, \frac{i}{2\pi}\right)\right) \tag{7.1}$$

In the formula, $m \equiv B(t,T)K/S_t$ denotes moneyness, $k \equiv \log(m)$, ω is a binary variable taking value 1 for call options and -1 for put, and

$$\mathrm{d}(k,\alpha) \equiv \int \frac{\mathrm{d}u}{u}(\mathrm{e}^{-2\pi iuk}\phi_X(u-\alpha) - 1) \tag{7.2}$$

is what we call the **characteristic integral** summarizing the price of all options. This is actually linked to the Hilbert transform of the characteristic function. Notice that the price is entirely defined by the moneyness parameter m and the characteristic function ϕ_X, representing the probability distribution of the increments of the logarithm of price between time t and maturity T.

It is now time to show that this is reality and not merely a movie, and to apply the pricing formula to actual price data. Of course, the first thing we want to recover is the Black–Scholes formula, once we plug the characteristic function of the Gaussian distribution into the **characteristic integral**. Then, we would like to move on and apply the model to produce smiles under general assumptions concerning the dynamics of log-price, with jumps of finite and infinite activity, and stochastic volatility. Finally, we want to carry the formula to actual market data, to calibrate the volatility surface, back out the underlying asset dynamics, and price exotic options accordingly.

7.2 THE BLACK–SCHOLES MODEL

We start by reproducing the Black–Scholes model. It suffices to reproduce the price of a digital call, that we recall is

$$\mathcal{C}_{\text{CoN}}(k) = N\left(\frac{\log(S_t/(B(t,T)K)) - (\sigma^2/2)T}{\sigma\sqrt{T}}\right)$$

The dynamics of X_t is

$$X_t = -\frac{\sigma^2}{2}t + \sigma W_t$$

where W_t is a Wiener process. As $X_t \stackrel{d}{=} N\left(-\frac{\sigma^2}{2}t, \sigma\sqrt{t}\right)$, we know that the characteristic function is

$$\phi_{X_T}(u) = \exp\left(-2\pi i u\frac{\sigma^2}{2}T - \frac{(2\pi u\sigma)^2 T}{2}\right)$$

and the **characteristic integral** is

$$\mathrm{d}(k,\alpha) = \int \frac{\mathrm{d}u}{u}\left[\exp\left(-2\pi i(u-\alpha)\left[\frac{\sigma^2}{2}T + k\right] - \frac{(2\pi(u-\alpha)\sigma)^2 T}{2}\right) - 1\right]$$

Let

$$\mu = 2\pi\left[-k - \frac{\sigma^2}{2}T\right]$$
$$Z = 2\pi\sigma\sqrt{T}$$

then

$$\mathrm{d}(k,\alpha) = \int \frac{\mathrm{d}u}{u}\left[\exp\left(i(u-\alpha)\mu - \frac{(u-\alpha)^2 Z^2}{2}\right) - 1\right]$$

Now, the pricing equation for a digital call option becomes:

$$\mathcal{C}_{\text{CoN}}(k) = \frac{1}{2} + \frac{1}{2\pi i}\int \frac{\mathrm{d}u}{u}\left[\exp\left(iu\mu - \frac{u^2 Z^2}{2}\right) - 1\right]$$

The exponential $\exp(iu\mu)$ can safely be expanded in a uniformly convergent power series. Besides, we rescale $Zu \to u$ and we get:

$$\mathcal{C}_{\text{CoN}}(k) = \frac{1}{2} + \frac{1}{2\pi i}\sum_{n=1}^{\infty}\frac{i^n}{n!}\left(\frac{\mu}{Z}\right)^n \int \mathrm{d}u\, u^{n-1}\exp\left(-\frac{u^2}{2}\right) + R(Z)$$

where

$$R(Z) = \frac{1}{2\pi i}\int \frac{\mathrm{d}u}{u}\left[\exp\left(-\frac{u^2Z^2}{2}\right) - 1\right]$$

This term is clearly zero since it is a trivial matter to check that

$$\frac{\mathrm{d}R(Z)}{\mathrm{d}Z} = 0, \quad R(0) = 0$$

It is convenient to shift back to zero the summation index n; moreover, we can notice that the integral in u contributes something different from zero only for even values of the power:

$$\mathcal{C}_{\text{CoN}}(k) = \frac{1}{2} + \frac{1}{2\pi i}\sum_{n=0}^{\infty}\frac{i^{2n+1}}{(2n+1)!}\left(\frac{\mu}{Z}\right)^{2n+1}\int \mathrm{d}u\, u^{2n}\exp\left(-\frac{u^2}{2}\right)$$

We recall the result:

$$\int \mathrm{d}u\, u^{2n}\exp\left(-\frac{u^2}{2}\right) = (2n-1)!!\sqrt{2\pi}, \quad (2n+1)! = 2n!!(2n+1)!! = 2^n n!(2n+1)!!$$

where it follows that

$$\begin{aligned}\mathcal{C}_{\text{CoN}}(k) &= \frac{1}{2} + \frac{1}{i\sqrt{2\pi}}\sum_{n=0}^{\infty}\frac{i^{2n+1}}{(2n+1)!}\left(\frac{\mu}{Z}\right)^{2n+1}(2n-1)!! \\ &= \frac{1}{2} + \frac{1}{\sqrt{2\pi}}\sum_{n=0}^{\infty}\frac{(-1)^n}{2^n n!}\left(\frac{\mu}{Z}\right)^{2n+1}\frac{1}{(2n+1)}\end{aligned}$$

Notice also that

$$\frac{x^{2n+1}}{2n+1} = \int_0^x \mathrm{d}y\, y^{2n}$$

It follows that

$$\begin{aligned}\mathcal{C}_{\text{CoN}}(k) &= \frac{1}{2} + \frac{1}{\sqrt{2\pi}}\int_0^{\mu/Z}\mathrm{d}x\sum_{n=0}^{\infty}\frac{(-x^2)^n}{2^n n!} \\ &= \frac{1}{2} + \int_0^{\mu/Z}\frac{\mathrm{d}x}{\sqrt{2\pi}}\exp\left(-\frac{x^2}{2}\right) \\ &= \int_{-\infty}^{\mu/Z}\frac{\mathrm{d}x}{\sqrt{2\pi}}\exp\left(-\frac{x^2}{2}\right)\end{aligned}$$

Since:

$$\frac{\mu}{Z} = \frac{-\kappa - \frac{1}{2}\sigma^2 T}{\sigma\sqrt{T}}$$

by the definition of K we end up with

$$\mathcal{C}_{\text{CoN}}(k) = N\left(\frac{\log(S_t(B(t,T)K)) - (\sigma^2/2)T}{\sigma\sqrt{T}}\right)$$

as it should be.

7.3 FINITE ACTIVITY MODELS

We now include finite activity jumps in the model above. Let $N(t)$ be a Poisson process, with intensity λ counting the number of events occurring before time t. The dynamics of X_t is then described by

$$X_t = \mu t + \sigma W(t) + \sum_{i=1}^{N(t)} J_i$$

where $W(t)$ is a standard Brownian motion, μ is a chosen drift term and the $\{J_i\}$ are i.i.d. random variables. The last term is a compound Poisson process and we know that

$$\phi_{X_t}(u) = \exp\left(i2\pi u\mu t - 2(\pi u\sigma)^2 t - \lambda t(1 - \phi_J(u))\right)$$

where

$$\phi_J(u) = \mathbb{E}\left[e^{i2\pi u J}\right]$$

As stated in Proposition 4.5.1, the process $e^{\alpha X_t}/\zeta(\alpha, t)$, with $\zeta(\alpha, t) = \mathbb{E}[e^{\alpha X_t}]$, is a martingale. But, by Proposition 2.4.7

$$\zeta(\alpha, t) = \exp\left[\alpha\mu t + \frac{(\alpha\sigma)^2}{2}t - \lambda t\left(1 - \phi_J\left(\frac{\alpha}{i2\pi}\right)\right)\right]$$

So we can consider the martingale

$$\zeta(1, t)^{-1}\, e^{X_t}$$

as a representation of the process Z_t

$$Z(t) = -\frac{\sigma^2}{2}t + \sigma W(t) + \sum_{j=0}^{N(t)} J_j + t\lambda\left(1 - \phi_J\left(\frac{1}{2\pi i}\right)\right)$$

which denotes the dynamics of the logarithm of asset price under the risk-neutral measure. The relevant characteristic function would then be

$$\phi_Z(u) = \exp\left[-2\pi iu\frac{\sigma^2}{2}T - \frac{(2\pi u\sigma)^2 T}{2} - \lambda T(1 - \phi_J(u)) + 2\pi iuT\lambda\left(1 - \phi_J\left(\frac{1}{2\pi i}\right)\right)\right]$$

and the specific shape will be fully specified by the characteristic function of the dimension of jumps $\phi_J(u)$. We give below two specific instances of jump distribution.

7.3.1 Discrete jumps

The simplest model we can imagine is one in which jumps may take two states. So, we denote the dimension of these jumps j_1, j_2 and the corresponding probability $p, q = 1 - p$.

For this process we have

$$\begin{aligned}\kappa &= pu + qd \\ \phi_J(u) &= p\, e^{-2\pi iuj_1} + q\, e^{-2\pi iuj_2} \\ \frac{d}{du}\phi_J(u) &= -2\pi i\left(j_1 p\, e^{-2\pi ikj_1} + j_2 q\, e^{-2\pi ikj_2}\right)\end{aligned}$$

In Figures 7.1 and 7.2 we report examples of smiles generated by the discrete jump models. The idea is to take j_1 as the upward jump and j_2 as the downward jump. Figure 7.1. shows

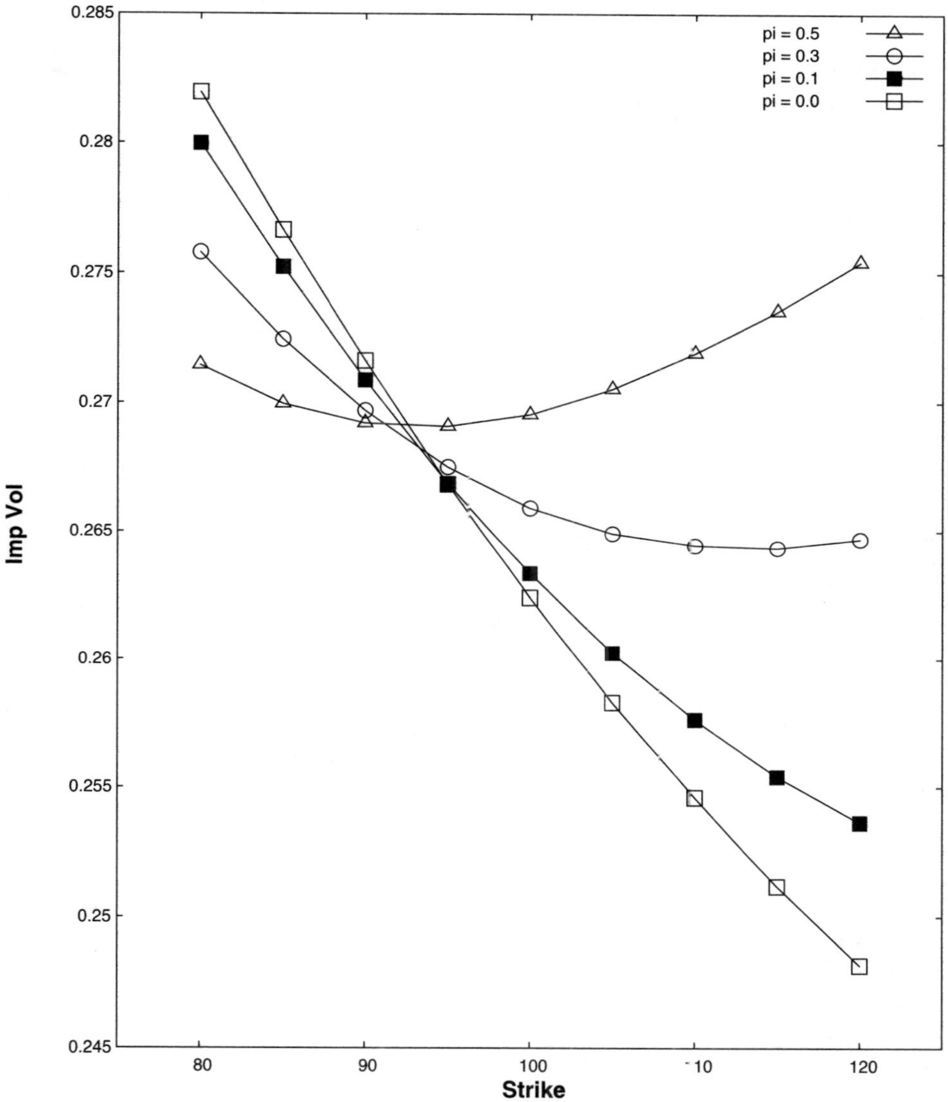

Figure 7.1 The dependency on the probability of the j_1 jump of the smile in the discrete jump model. The parameters kept fixed the size of jumps $j_1 = 0.3, \quad j_2 = -0.3$

that as the probability of the upward jump decreases, the smile gets more and more skewed. Figure 7.2 shows that the skew is shifted up by the size of the jump.

7.3.2 The Merton model

In the Merton (1976) model the logarithm of jumps are normally distributed. Formally, we have

$$J \stackrel{d}{=} N(a, b)$$

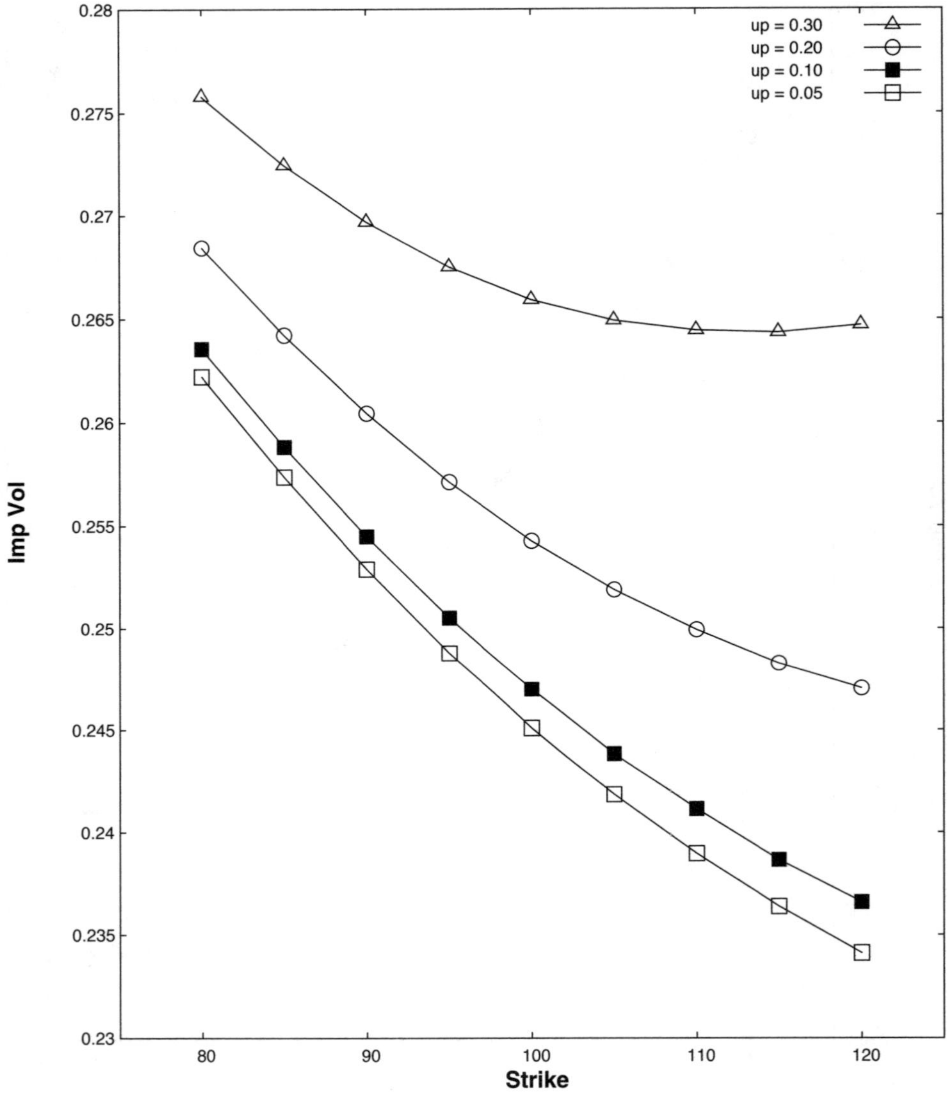

Figure 7.2 The dependency on the size of the j_1 jump of the smile in the discrete jump model. The parameters kept fixed the size of jumps $p = 0.3, \quad j_2 = -0.3$

and the characteristic function for jumps is

$$\psi_J(u) = \exp\left(i2\pi ua - 2(\pi ub)^2\right)$$

Figures 7.3 and 7.4 report smiles for different values of the mean and variance of log-jumps.

7.4 INFINITE ACTIVITY MODELS

We now turn to some models with infinite activity, namely the Variance Gamma model and the CGMY model, with $Y > 0$.

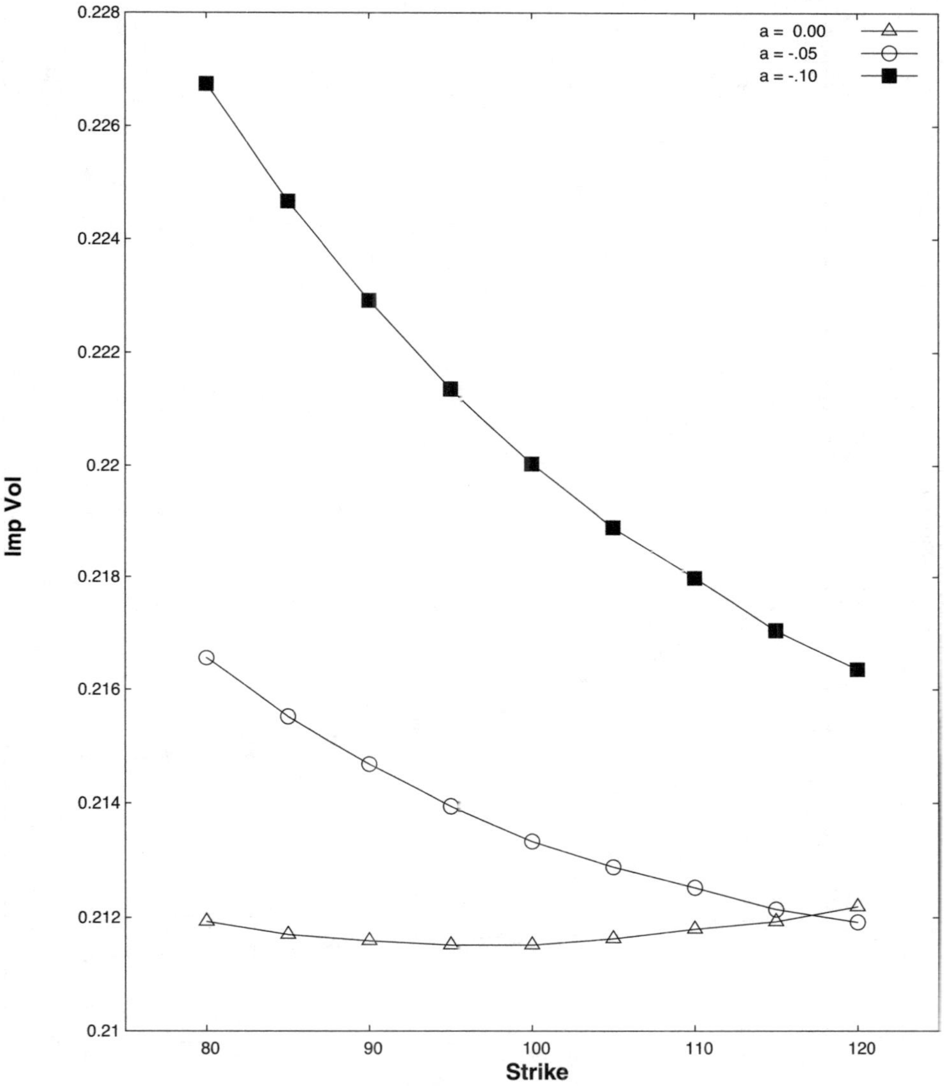

Figure 7.3 The dependency of the smile on the mean size of jumps in the Merton jump-diffusion model. The parameter kept fixed is jump volatility: $b = 0.1$

7.4.1 The Variance Gamma model

Let X_t and Y_t be two Gamma processes; more precisely, $X_t \overset{d}{=} \Gamma(ct, 1/m)$ and $Y_t \overset{d}{=} \Gamma(ct, 1/g)$. Then, as shown in Example 2.3.1,

$$V_t^\gamma \cong X_t - Y_t$$

is a Variance Gamma process and

$$\phi_{V^\gamma}(u) = \left(\frac{1}{1 - i2\pi mu}\right)^{ct} \left(\frac{1}{1 + i2\pi gu}\right)^{ct} = \left(\frac{1}{1 + 4\pi^2 u^2 gm - i2\pi(m-g)u}\right)^{ct}$$

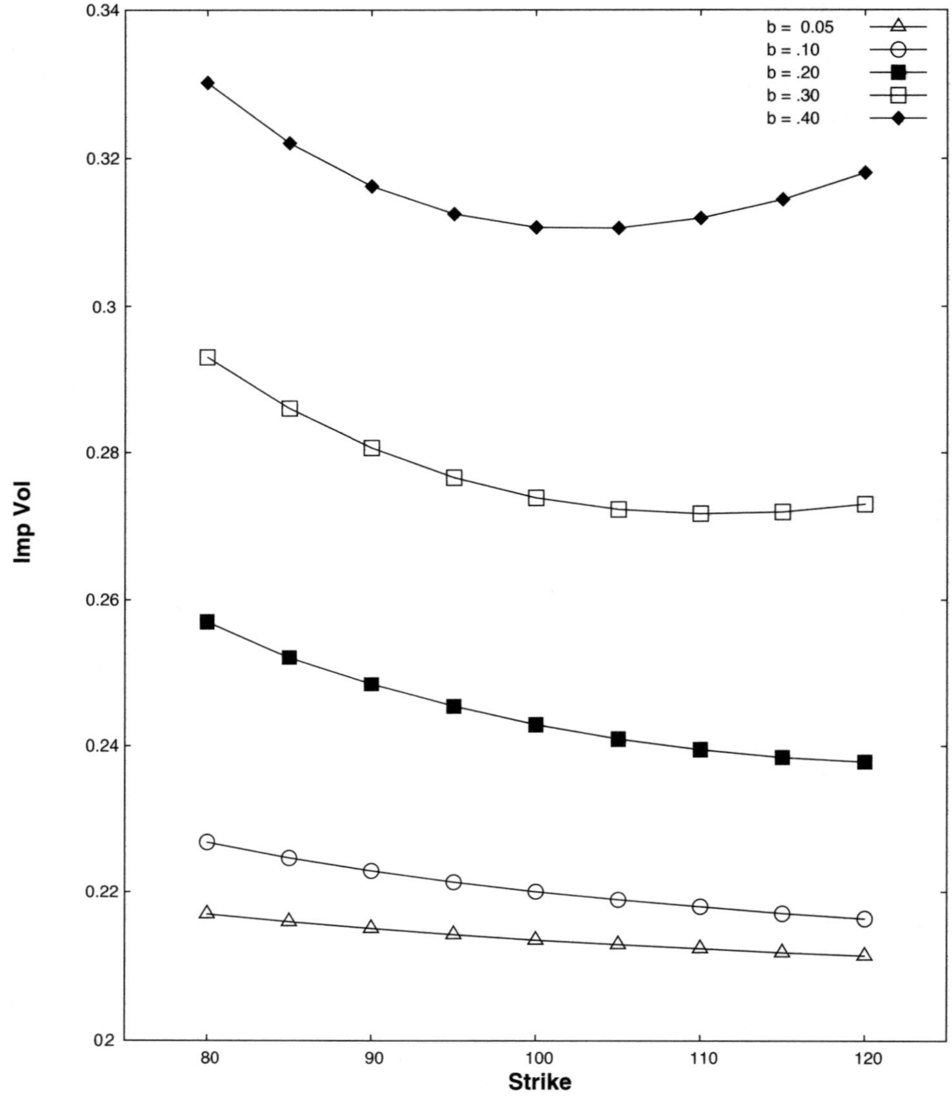

Figure 7.4 The dependency on the b parameter (jumps volatility) of the smile in the Merton jump-diffusion model. The parameter kept fixed is jump mean: $a = -0.1$

As shown in Example 3.2.4, the Variance Gamma process can be represented as a time changed Brownian motion with drift, that is

$$Z_t = \theta\gamma(t) + \sigma W_{\gamma(t)}$$

where $\gamma(t)$ is a Gamma process such that $\gamma(t) \stackrel{d}{=} \Gamma(t/\nu, 1/\nu)$ and

$$\phi_{Y_t}(u) = \left(\frac{1}{1 + 2\nu(\pi u\sigma)^2 - i2\pi u\nu\theta}\right)^{t/\nu}$$

If we cast Example 3.2.4 in the present notation we have the identifications:

$$\nu = \frac{1}{c}, \quad \nu\theta = (m - g), \quad \frac{\nu\sigma^2}{2} = gm$$

$$c = \frac{1}{\nu}$$
$$g = \sqrt{\frac{\nu\sigma^2}{2} + \frac{\nu^2\theta^2}{4}} - \frac{\nu\theta}{2}$$
$$m = \sqrt{\frac{\nu\sigma^2}{2} + \frac{\nu^2\theta^2}{4}} + \frac{\nu\theta}{2}$$

As in the finite activity case, in order to construct a martingale process for asset prices we have to compute

$$\zeta(\alpha, t) = \mathbb{E}[e^{\alpha V_t^Y}]$$

Again, thanks to Proposition 2.4.7, we obtain

$$\zeta(\alpha, t) = \left(\frac{1}{1 - \alpha^2 gm - \alpha(m - g)}\right)^{ct}$$

and we shall consider the martingale process:

$$\begin{aligned}\zeta^{-1}(1, t)\, e^{V_t^Y} &= \exp\left(c\,t \log(1 - gm - (m - g)) + V_t^Y\right)\\ &= \exp\left(\frac{t}{\nu}\log\left(1 - \nu\theta - \frac{\nu\sigma^2}{2}\right) + V_t^Y\right)\end{aligned}$$

The characteristic function associated to it is:

$$\begin{aligned}\phi_Z(u) &\equiv \exp\left(\frac{i2\pi ut}{\nu}\log\left(1 - \nu\theta - \frac{\nu\sigma^2}{2}\right)\right)\mathbb{E}\left[e^{i2\pi u V_t^Y}\right]\\ &= \exp\left(\frac{i2\pi ut}{\nu}\log\left(1 - \nu\theta - \frac{\nu\sigma^2}{2}\right)\right)\left(\frac{1}{1 + 2\nu(\pi k\sigma)^2 - i2\pi k\nu\theta}\right)^{t/\nu}\end{aligned}$$

Figures 7.5 and 7.6 describe the behaviour of smiles with respect to the parameters θ governing skewness and ν governing kurtosis (in a symmetric smile with $\theta = 0$).

7.4.2 The CGMY model

Consider now a CGMY process. As described in Chapter 2, this is an extension of the Variance Gamma model. It includes another parameter, called Y. Applying again Proposition 2.4.7 in order to construct a martingale process, we have

$$\phi_Z(u) = \exp\left(\frac{t}{\nu}(\varphi(i2\pi u) - i2\pi u\varphi(1))\right)$$

with

$$\varphi(x) = \Gamma(-Y)\left[(M - x)^Y - M^Y + (G + x)^Y - G^Y\right]$$

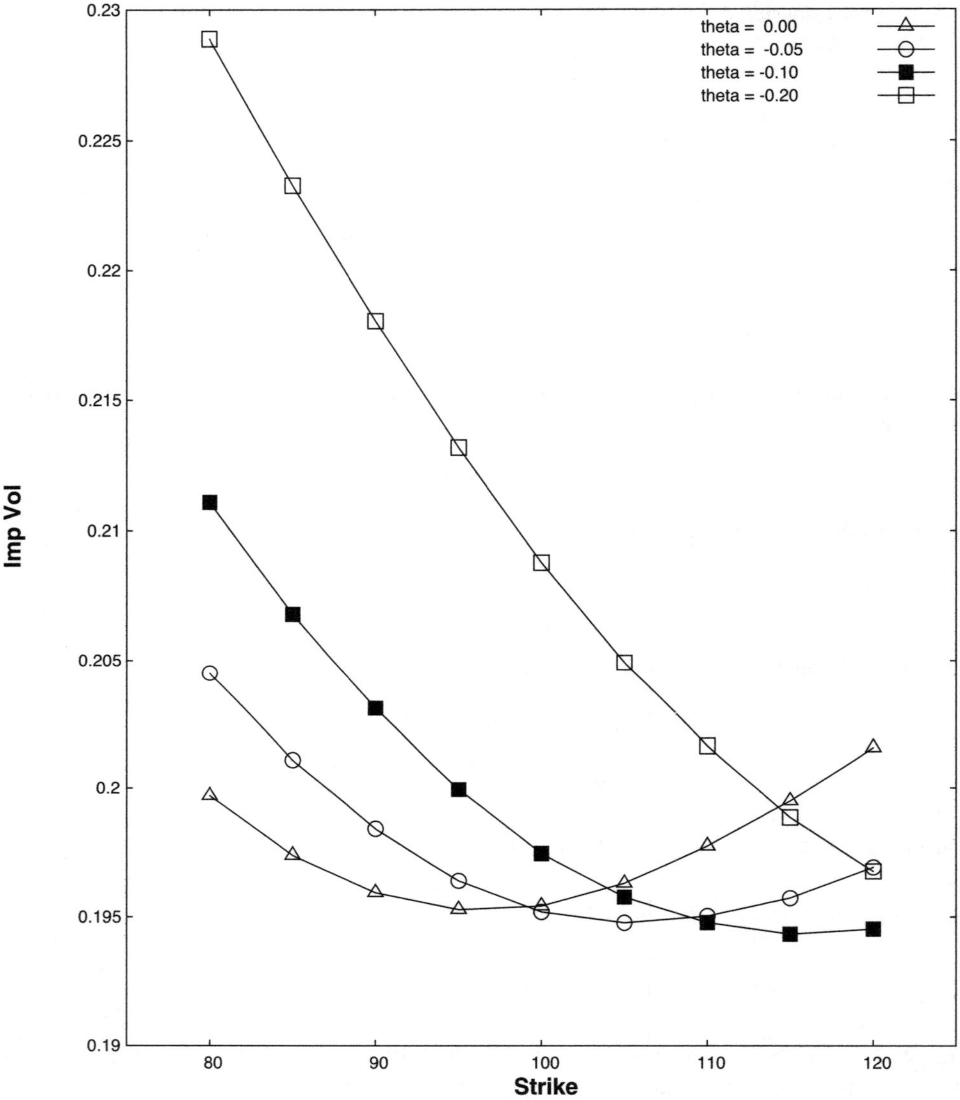

Figure 7.5 The dependency on θ of the smile in the Variance Gamma model. The parameters kept fixed are: $\sigma = 0.2, \quad \nu = 0.2$

Figure 7.7 shows the behaviour of smiles with several values of the parameter Y. We concentrate on values greater than zero, corresponding to infinite activity. We see that an increase of the parameter brings about an upward shift of the smile.

7.5 STOCHASTIC VOLATILITY

Let us consider the following risk-neutral diffusion for the log of a price process:

$$dX_t = -\frac{\alpha^2(t, \nu_t)}{2} dt + \alpha(t, \nu_t)\, dW_t, \quad X_0 = 0 \tag{7.3}$$

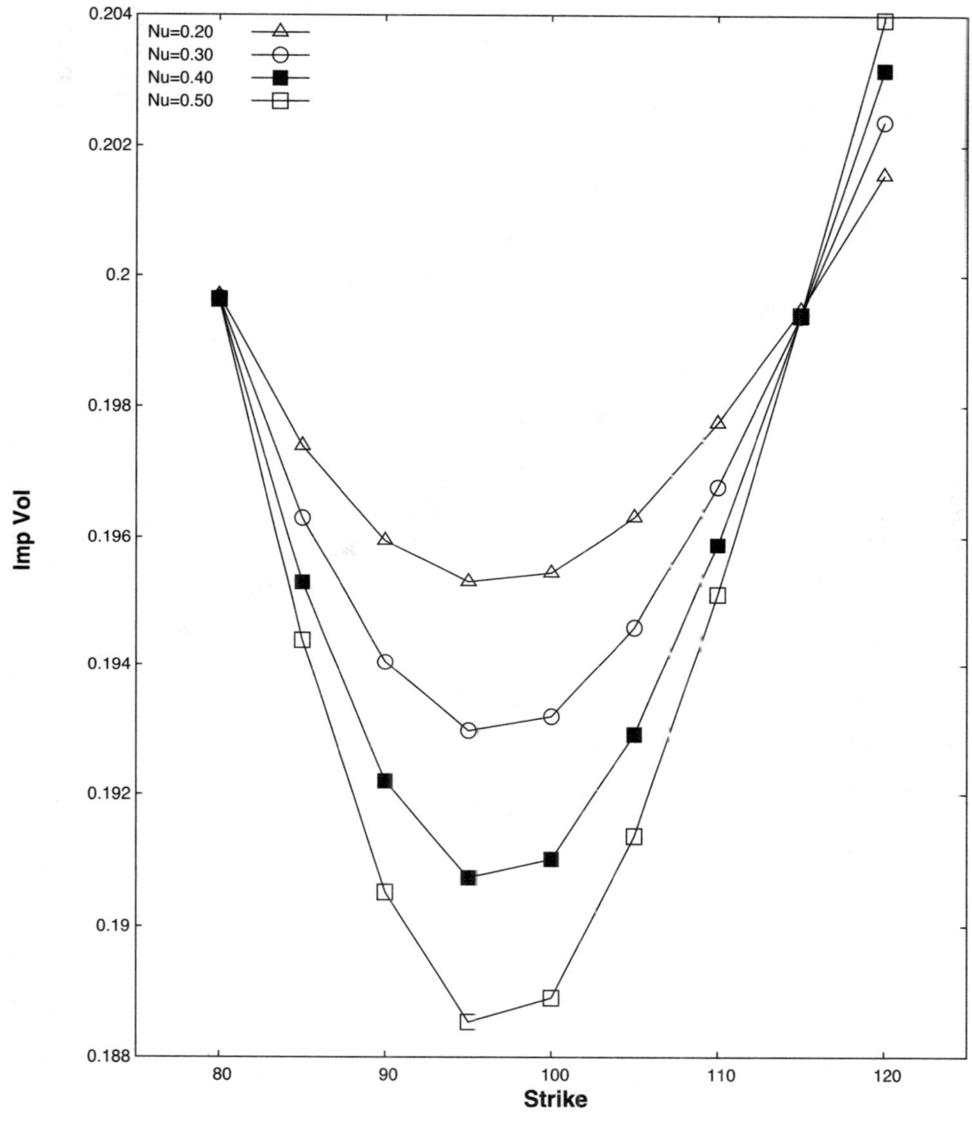

Figure 7.6 The dependency on ν of the smile in the Variance Gamma model. The parameters kept fixed are: $\sigma = 0.2, \quad \theta = -0.0$

where α is a deterministic function of (t, ν), differentiable all the times we need it, and ν is an exogenous process described, in the same martingale measure, by

$$d\nu_t = \mu_\nu(t, \nu)\, dt + \sigma_\nu(t, \nu)\, dY_t$$

where the Brownian motionss W_t and Y_t are correlated:

$$E[dW, dY] = \rho$$

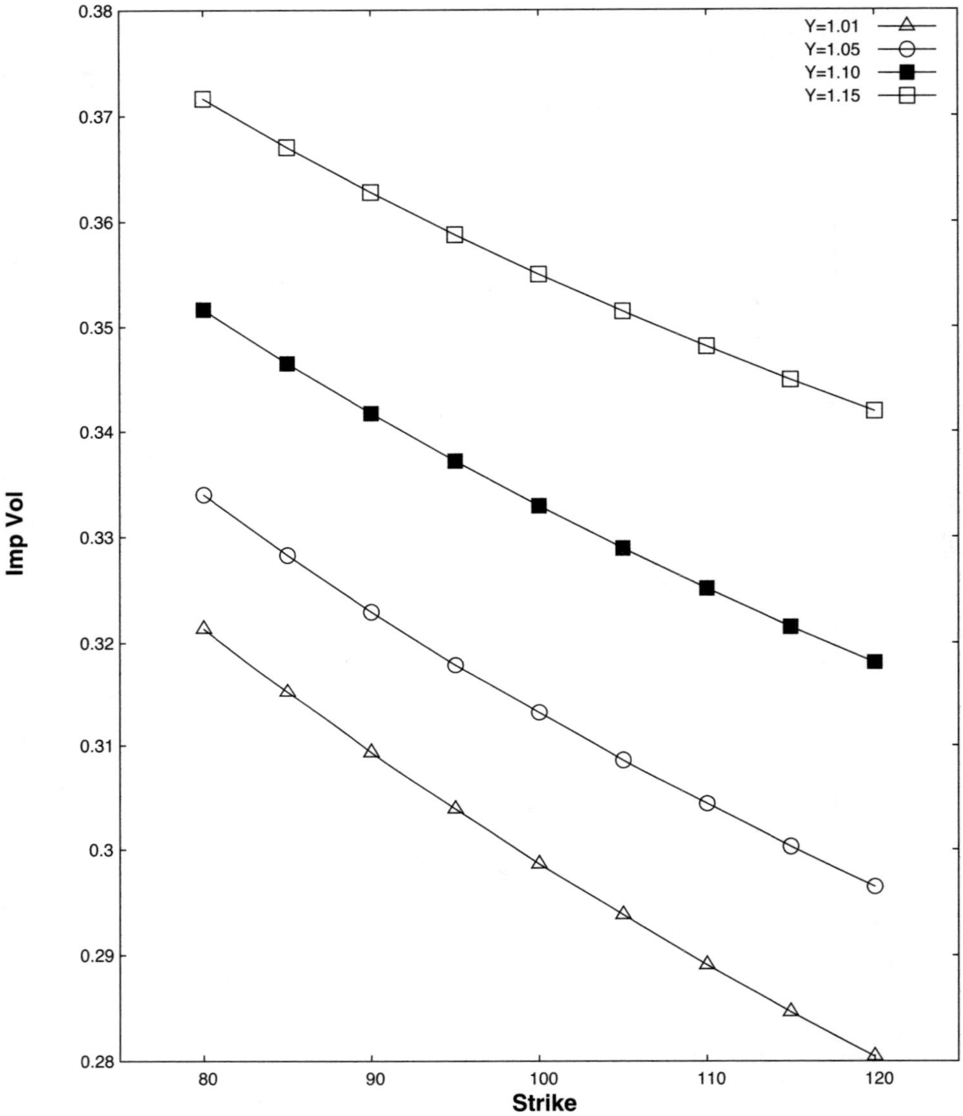

Figure 7.7 The dependency of the smile on the parameter Y in the CGMY model. The parameters kept fixed are: $\sigma = 0.0312, \quad \nu = 2.386, \quad \theta = -0.0938, \quad \eta = 0.0428$

(for instance, $W_t = \rho Y_t + \sqrt{1-\rho^2} L_t$ for a third Brownian motion L_t independent of Y_t). We are interested in computing the moment generating function $\phi_X(v)$ of the random variable X_T defined by:

$$\Phi_X(v) \equiv E_W \left[\mathrm{e}^{v X_T} \right]$$

for some complex value of v. Ultimately, we will be interested in the characteristic function where $v = i2\pi u$.

Let us fix a trajectory

$$\nu_T = \int_0^T [\mu_\nu(s,\nu)\,\mathrm{d}s + \sigma_\nu(s,\nu)\,\mathrm{d}Y_s]$$

for volatility. Then

$$\begin{aligned}\Phi_X(\nu) &\equiv E_W\left[\mathrm{e}^{\nu X_T}\right]\\ &= E_Y\left[E_L\left[\mathrm{e}^{\nu X_T}|\nu_T\right]\right] = E_Y\left[\phi_X(\nu, Y)\right]\end{aligned}$$

where

$$\begin{aligned}\phi_X(\nu, Y) &= E_L\left[\exp\left(-\frac{\nu}{2}\int_0^T \alpha^2(t,\nu_t)\,\mathrm{d}t + \nu\int_0^T \alpha(t,\nu_t)\,\mathrm{d}W_t\right)\right]\\ &= \exp\left(-\frac{\nu}{2}\int_0^T \alpha^2(t,\nu_t)\,\mathrm{d}t + \rho\nu\int_0^T \alpha(t,\nu_t)\,\mathrm{d}Y_t + \frac{(1-\rho^2)\nu^2}{2}\int_0^T \alpha^2(s,\nu_s)\,\mathrm{d}s\right)\\ &= \exp\left(\frac{\nu^2-\nu}{2}\int_0^T \alpha^2(t,\nu_t)\,\mathrm{d}t + \rho\nu\int_0^T \alpha(t,\nu_t)\,\mathrm{d}Y_t - \frac{\rho^2\nu^2}{2}\int_0^T \alpha^2(s,\nu_s)\,\mathrm{d}s\right)\end{aligned}$$

It follows that:

$$\begin{aligned}\Phi_X(\nu) &= E_Y\left[\exp\left(\frac{\nu^2-\nu}{2}\int_0^T \alpha^2(t,\nu_t)\,\mathrm{d}t + \rho\nu\int_0^T \alpha(t,\nu_t)\,\mathrm{d}Y_t - \frac{\rho^2\nu^2}{2}\int_0^T \alpha^2(s,\nu_s)\,\mathrm{d}s\right)\right]\\ &= E_{\overline{Y}}\left[\exp\left(\frac{\nu^2-\nu}{2}\int_0^T \alpha^2(t,\nu_t)\,\mathrm{d}t\right)\right]\end{aligned}$$

where ν follows the process:

$$\mathrm{d}\nu_t = [\mu_\nu(t,\nu) + \rho\nu\alpha(t,\nu_t)\sigma_\nu(t,\nu_t)]\mathrm{d}t + \sigma_\nu(t,\nu)\,\mathrm{d}\overline{Y}_t$$

7.5.1 The Heston model

The risk-neutral martingale measure for the Heston model is defined by:

$$\begin{aligned}\mathrm{d}X_t &= -\frac{\sigma_X^2\nu_t}{2}\mathrm{d}t + \sigma_X\sqrt{\nu_t}\,\mathrm{d}W_t, \quad X_0 = 0\\ \mathrm{d}\nu_t &= \lambda(\overline{\nu} - \nu_t)\,\mathrm{d}t + \eta\sqrt{\nu_t}\,\mathrm{d}Y_t, \quad \nu_0 = \sigma^2\end{aligned}$$

Clearly σ_X is fully redundant in the sense that it can be reabsorbed into a redefinition of η and $\overline{\nu}$. This can be easily seen by defining a $\nu_t' = \sigma_X^2\nu_t$.

We connect to the formalism of the previous section assigning:

$$\begin{aligned}\alpha(t,\nu_t) &= \sqrt{\nu_t}\\ \mu_\nu(t,\nu_t) &= \lambda(\overline{\nu} - \nu_t)\,\mathrm{d}t\\ \sigma_\nu(t,\nu_t) &= \eta\sqrt{\nu_t}\end{aligned}$$

The ν process therefore reads

$$\begin{aligned}
d\nu_t &= \lambda[\overline{\nu} - \nu_t]dt + \rho \upsilon \eta \nu_t\, dt + \eta\sqrt{\nu_t}\, d\overline{Y}_t \\
&= (\lambda - \rho\upsilon\eta)[\frac{\lambda\overline{\nu}}{\lambda - \rho\upsilon\eta} - \nu_t]dt + \eta\sqrt{\nu_t}\, d\overline{Y}_t
\end{aligned} \tag{7.4}$$

We map the model into a complex CIR model by setting:

$$\kappa = \lambda - \rho\upsilon\eta, \quad \theta = \frac{\lambda\overline{\nu}}{\kappa} \tag{7.5}$$

where κ and θ denote the mean reversion and the long run equilibrium parameters of the process respectively. Then, the computation of the characteristic function for the Heston model reduces to the computation of the expectation

$$\mathbb{E}\left[\exp\left(-\Lambda \int_0^T \nu_t\, dt\right)\right] \tag{7.6}$$

where

$$\Lambda = -\frac{\upsilon - \upsilon^2}{2} = -i2\pi u\left(\frac{1 - i2\pi u}{2}\right)$$

and

$$\begin{aligned}
d\nu_t &= \kappa[\theta - \nu_t]dt + \eta\sqrt{\nu_t}\, d\overline{Y}_t \\
\kappa &= \lambda - i2\pi\rho u\eta, \quad \theta = \frac{\lambda\overline{\nu}}{\kappa}
\end{aligned}$$

7.5.2 Vanilla options in the Heston model

The derivation of the characteristic function in the Heston model requires the computation of the expectation of

$$\exp\left(-\Lambda \int_0^T \nu_t\, dt\right)$$

within the *square-root* model. This is quite standard, even though we report a formal derivation in Appendix G. While referring the reader there for details, here we simply report the characteristic function to be used in the pricing formula. We have

$$\begin{aligned}
dX_t &= (r - q)\, dt - \frac{\nu_t}{2} dt + \sqrt{\nu_t}\, dW_t, \quad X_0 = 0 \\
d\nu_t &= \lambda(\overline{\nu} - \nu_t)\, dt + \eta\sqrt{\nu_t}\, dY_t, \quad \nu_0 = \sigma^2 \\
\mathbb{E}[dW_t\, dY_t] &= \rho\, dt \\
\mathbb{E}\left[e^{i2\pi u X_T}\right] &= e^{i2\pi u(r-q)T + A_T(0,\Lambda) - B_T(0,\Lambda)\nu_0} \\
B_T(t, \Lambda) &= z_p \frac{1 - e^{-\gamma(T-t)}}{1 - g\, e^{-\gamma(T-t)}} \\
A_T(t, \Lambda) &= -\lambda\overline{\nu}\left[z_p(T - t) - \frac{2}{\eta^2}\log\left(\frac{1 - g}{1 - g\, e^{-\gamma(T-t)}}\right)\right] \\
z_p &= \frac{\gamma - \kappa}{\eta^2}
\end{aligned}$$

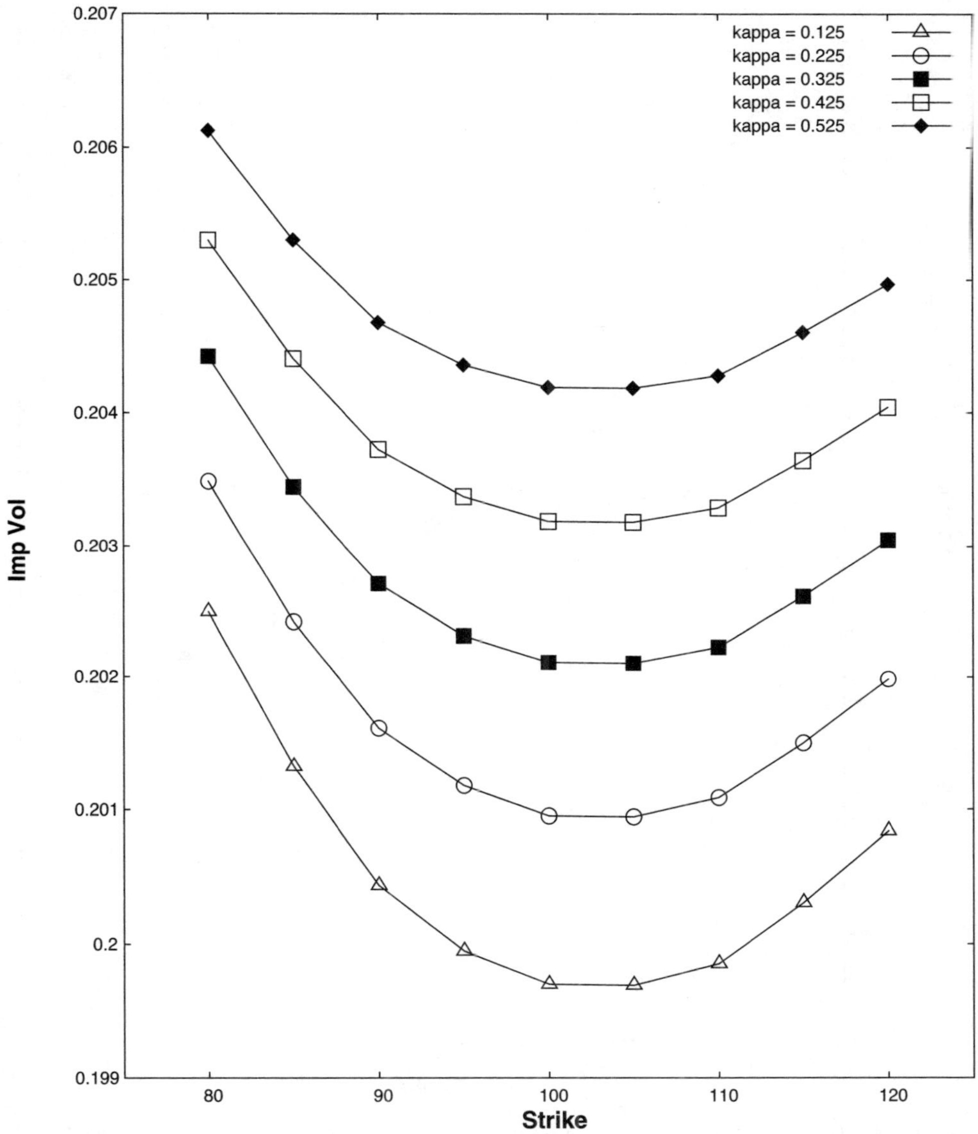

Figure 7.8 The dependency of the smile on the κ parameter in the Heston model. The parameters kept fixed are: $\theta = 0.05, \quad \eta = 0.10, \quad \rho = 0.00$

$$
\begin{aligned}
g &= -\frac{\gamma - \kappa}{\gamma + \kappa} \\
\gamma &= \sqrt{\kappa^2 + 2\Lambda\eta^2} \\
\kappa &= \lambda - i2\pi\rho u\eta \\
\gamma &= \sqrt{\kappa^2 + 2\Lambda\eta^2} \\
\Lambda &= -i2\pi u\left(\frac{1 - i2\pi u}{2}\right)
\end{aligned}
$$

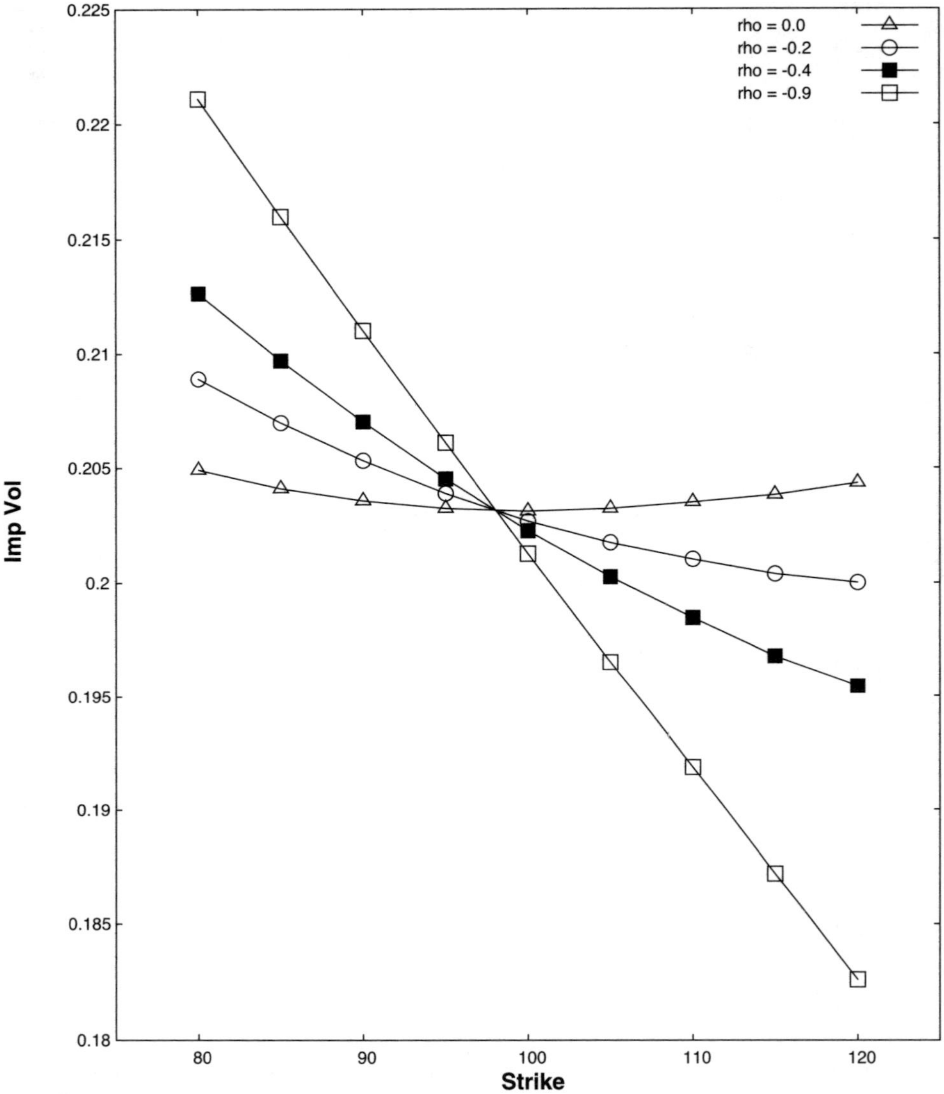

Figure 7.9 The dependency of the smile on the correlation ρ parameter in the Heston model. The fixed parameters are: $\kappa = 0.425, \quad \theta = 0.05, \quad \eta = 0.1$

The Black–Scholes limit

A smooth Black–Scholes limit is achieved by setting $\eta = 0$, while κ, ν_0, η are freely chosen. In this case the system is described by:

$$\begin{aligned}
S_t &= S_0\, \mathrm{e}^{X_t} \\
\mathrm{d}X_t &= -\frac{\nu(t)}{2}\mathrm{d}t + \sqrt{\nu_t}\,\mathrm{d}W_t, \qquad X_0 = 0 \\
\mathrm{d}\nu_t &= \kappa(\overline{\nu} - \nu_t)\,\mathrm{d}t \quad \nu_0 = \sigma^2
\end{aligned}$$

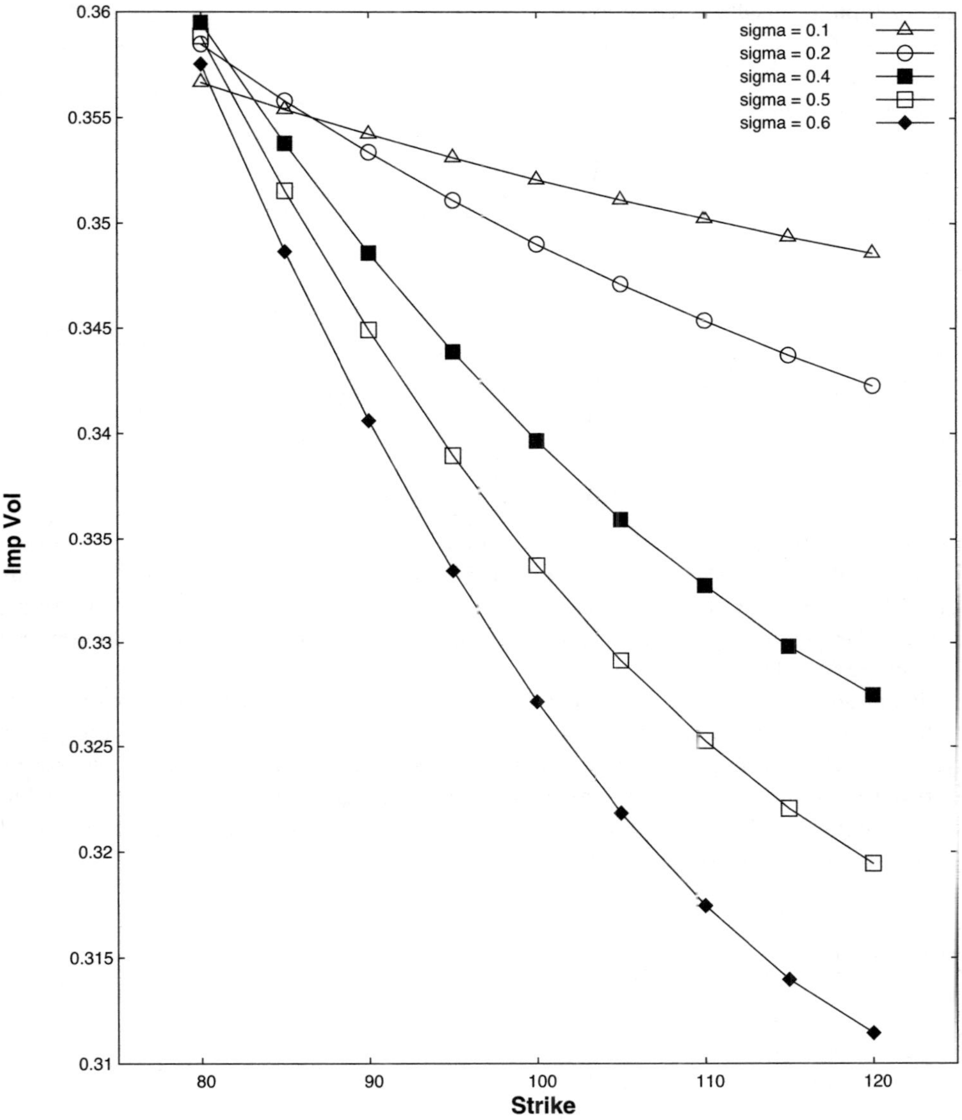

Figure 7.10 The dependency of the smile on vol of vol η parameter in the Heston model. The parameters kept fixed are: $\rho = -0.40$, $\kappa = 0.425$, $\theta = 0.5$

Under these conditions the variable X_T is a normal variate with mean

$$M(T) = \left(r - \frac{1}{2T} \int_0^T \mathrm{d}s\, v_s \right) T$$

and variance

$$\sigma^2(T) = \frac{1}{T} \int_0^t v_s\, \mathrm{d}s$$

The equation for ν_t admits the solution:

$$\int_{\nu_0}^{\nu_t} \frac{\mathrm{d}\nu_s}{\kappa(\overline{\nu} - \nu_s)} = t$$

$$-\frac{1}{\kappa}\log\left(\frac{\overline{\nu} - \nu_t}{\overline{\nu} - \nu_0}\right) = t$$

$$\nu_t = \overline{\nu} + (\overline{\nu} - \nu_0)\,\mathrm{e}^{-\kappa t}$$

and

$$\int_0^T \nu_t \,\mathrm{d}t = \overline{\nu} T + \frac{\overline{\nu} - \nu_0}{\kappa}\left(1 - \mathrm{e}^{-\kappa T}\right)\left(1 - \mathrm{e}^{-\kappa T}\right)$$

7.6 FFT AT WORK

We now apply the Fourier transform pricing formula to a set of market data. We start from the volatility surface of a rather liquid market index, the German equity index DAX. Data have not been treated in any way as one can easily grasp from the severe roughness of the surface, probably due to liquidity factors on some maturities and strike prices. A plot of the surface is given in Figure 7.11.

Say our goal is to compute a path-dependent contingent claim where it is important to use all the information contained in the whole surface and not just at some time horizon. We will try to back out the parameters of the model from market data (market *calibration*) and then use the parameters to price exotic options by simulation. In particular, since we will be looking at Asian options expiring at $T = 1.0$ year, we are going to calibrate our models on the whole surface for all the time horizons up to one year.

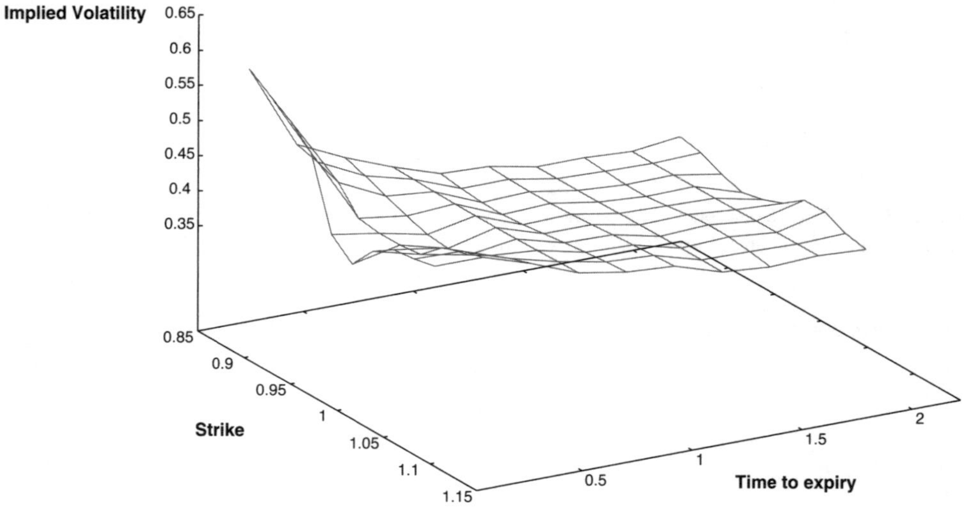

Figure 7.11 The DAX volatility surface we have been working with

7.6.1 Market calibration

We first describe the calibration process. Let $k_1^i, \ldots, k_{n_i}^i$ be the strikes corresponding to the time horizon t_i, and $\mathcal{O}(t_i, k_n^i)$ the prices implied by the surface at hand. Using FFT we compute the set of prices $\mathcal{O}(t_i, \overline{\mathbf{k}}, \mathbf{\Theta})$, where $\mathbf{\Theta}$ is a set of parameters characterizing the model, and $\overline{\mathbf{k}}$ the strikes handed back by the FFT integration procedure. In order to match the observed values we use linear interpolation and call $\mathbf{O}(t_i, k_{i_n}, \mathbf{\Theta})$ the interpolated price. The calibration is performed by minimizing, with respect to the set $\mathbf{\Theta}$, the function:

$$\sum_i \sum_{n=1}^{n_i} |\mathcal{O}(t_i, k_n^i) - \mathbf{O}(t_i, k_n^i, \mathbf{\Theta})|^2$$

We look at different models that we expect able to represent the implicit information hidden in a volatility surface. In particular, the models analyzed are Variance Gamma (VG), a two-state discrete jump model (DJ), the jump diffusion Merton model (MJ) and the Heston model (Heston). Each model is characterized by a set of parameters, and the values giving the best fit are given by:

- **VG:** $\theta = -0.5114, \quad \nu = 0.0329, \quad \sigma = 0.4450$
- **DJ:** $\sigma = 0.4380, \quad \lambda = 0.3397 \begin{cases} \pi_1 = 0.323 & j_1 = -0.031, \\ \pi_2 = 0.677 & j_2 = -0.133 \end{cases}$
- **MJ:** $\sigma = 0.4406, \quad \lambda = 0.4931, \quad a = -0.041, \ b = 0.041$
- **Heston:** $\rho = -0.8950, \quad \kappa = 1.488, \quad \theta = 0.161, \quad \sigma = 0.276, \quad \overline{v} = 0.210$

In Table 7.1 we report the first four moments associated to the p.d.f. implied by each model. Gauging the model that provides the best fit is beyond the scope of this book. Nevertheless, in order to give a rough idea of the goodness of fit, we report scatter plots that superpose the observed smiles and those fitted by the models. These are displayed in Figures 7.12 to 7.15.

7.6.2 Pricing exotics

We now apply the parameters calibrated above to price exotic claims by simulation (some techniques for simulation were described in Chapter 3). For the sake of illustration, we evaluate a standard Asian option whose payoff is given by

$$\mathcal{A} = \left[\frac{1}{N} \sum_{i=1}^{N} S(t_i) - K \right]^+$$

Table 7.1 Moments of the distribution calibrated on market data. DAX equity index

Model	Mean	Volatility	Skewness	Kurtosis
VG	−0.0509	0.3214	−0.1547	3.2132
DJ	−0.0490	0.3131	−0.0090	3.0038
MJ	−0.0489	0.3128	−0.0022	3.0007
Heston	−0.0489	0.3207	−0.4787	3.3427

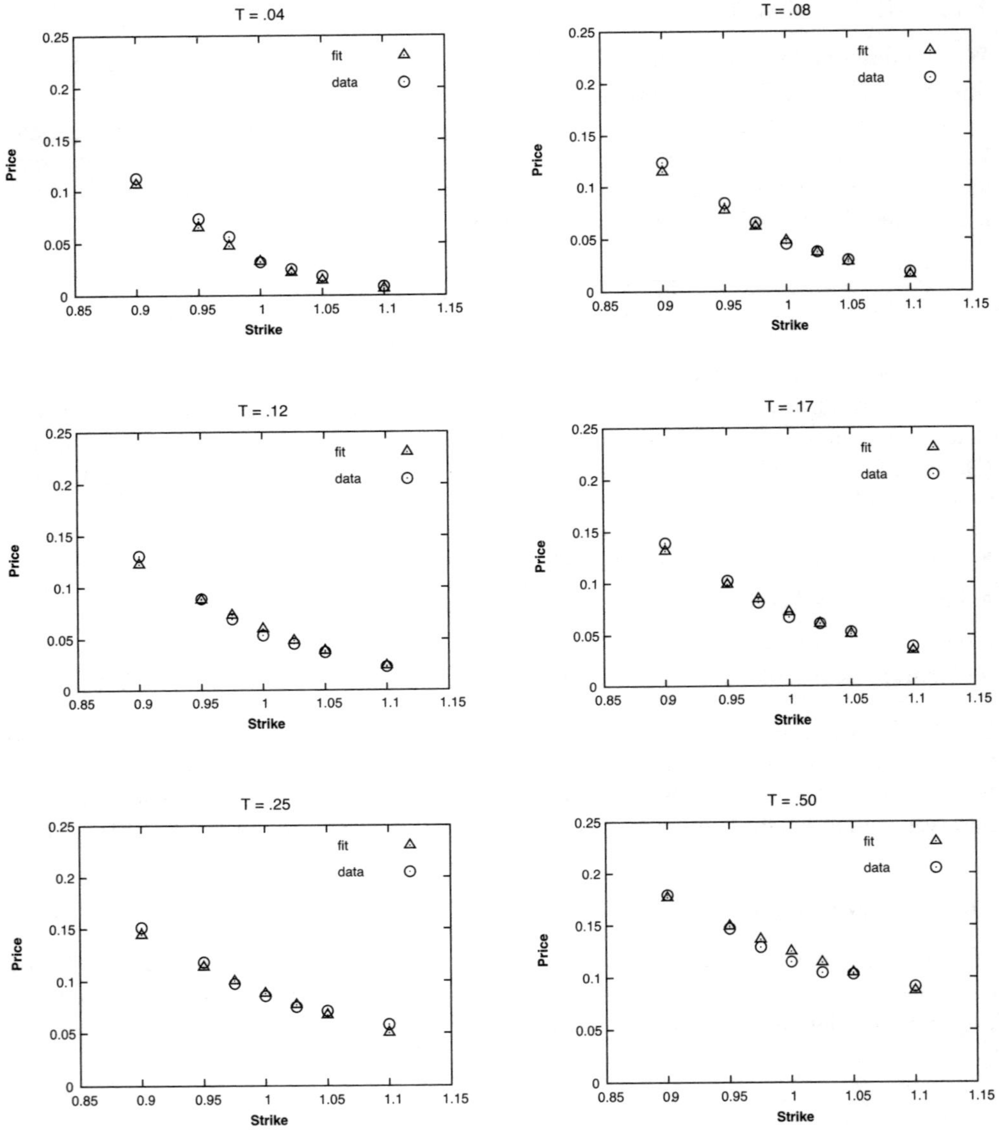

Figure 7.12 Prices vs fitted prices at different time horizons for the VG model. The parameters determined by the calibration procedure are: $\theta = -0.5114,\ \nu = 0.0329,\ \sigma = 0.4450$

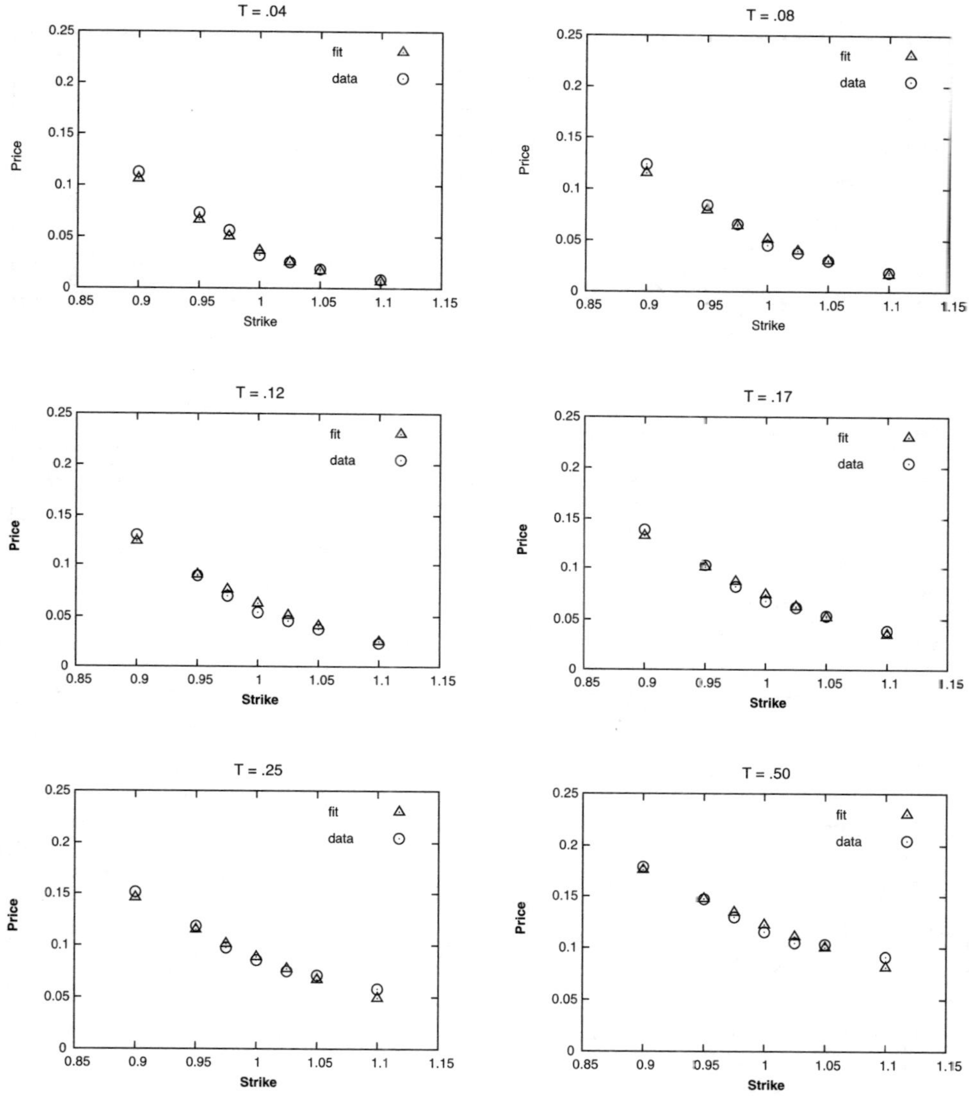

Figure 7.13 Prices vs fitted prices at different time horizons for the Heston model. The parameters determined by the calibration procedure are: $\rho = -0.8950$, $\kappa = 1.488$, $\theta = 0.161$, $\sigma = 0.276$, $\overline{v} = 0.210$

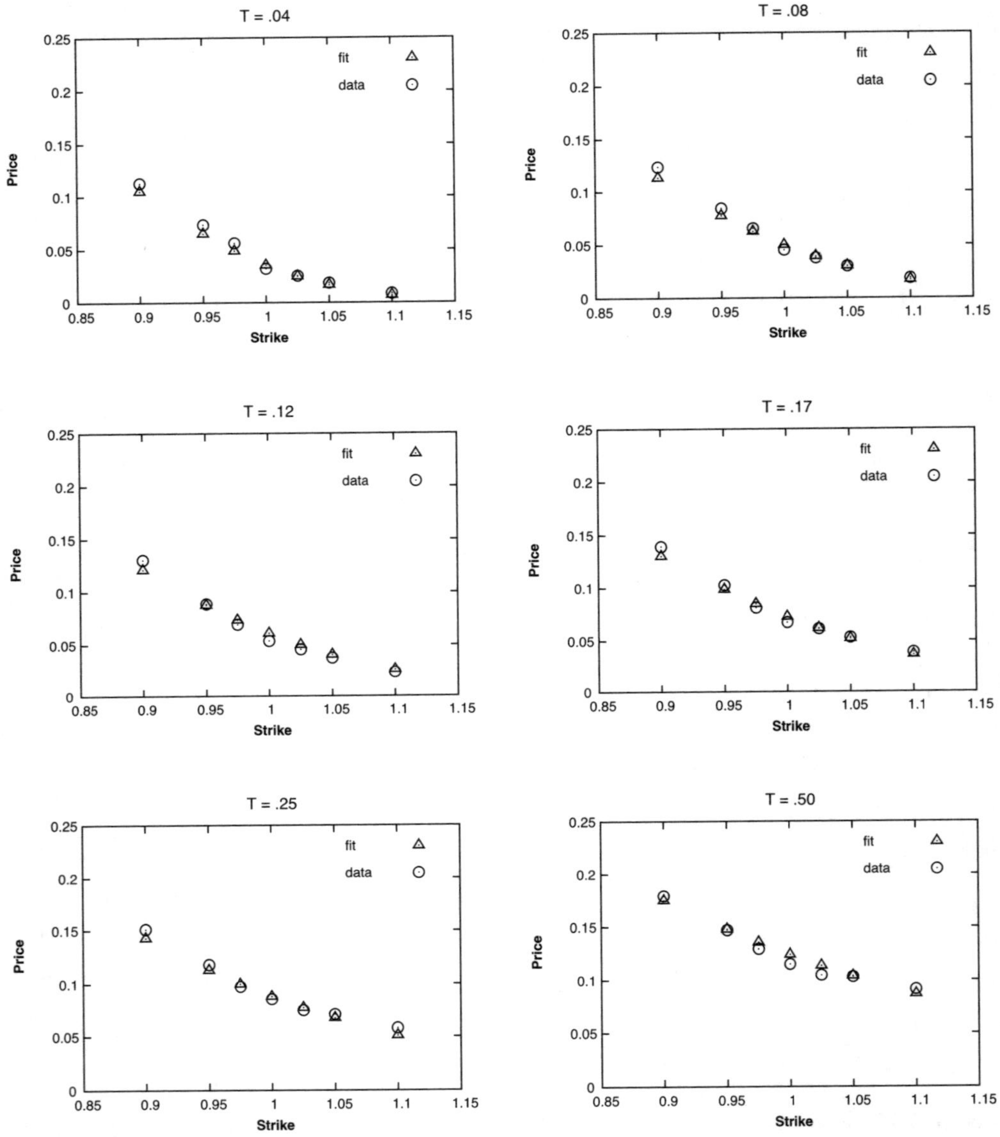

Figure 7.14 Prices vs fitted prices at different time horizons for the jump diffusion Merton model. The parameters determined by the calibration procedure are: $\sigma = 0.4406$, $\lambda = 0.4931$, $a = -0.041$, $b = 0.041$

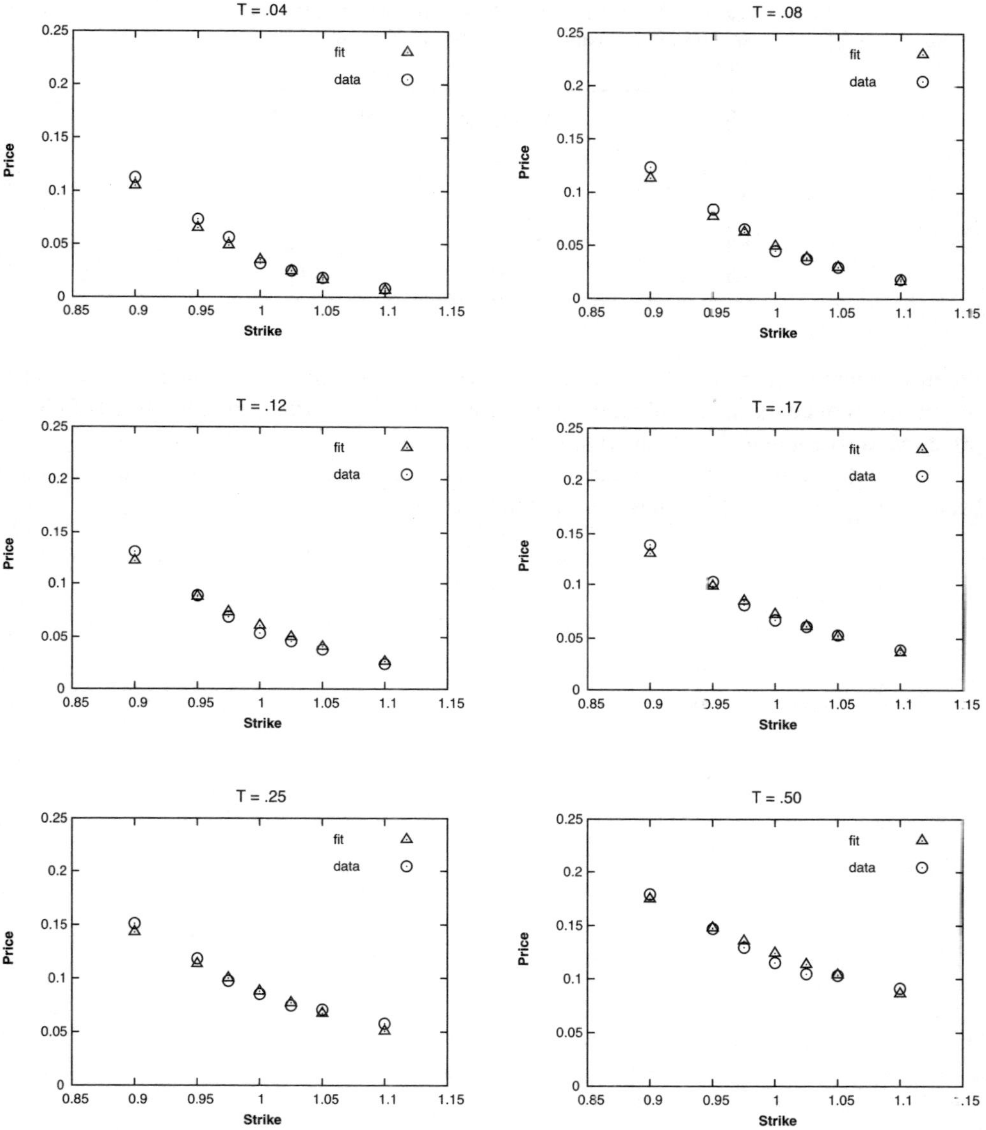

Figure 7.15 Prices vs fitted prices at different time horizons for the two-state jump diffusion model. The parameters determined by the calibration procedure are: $\sigma = 0.4380$, $\lambda = 0.3397$, $\pi_1 = 0.323$, $j_1 = -0.031$, $\pi_2 = 0.677$, $j_2 = -0.133$

Table 7.2 Asian and average strike Asian options for several models calibrated on the DAX volatility surface

Model	Asian	Asian Av. Strike
Heston	8.6740	5.5156
VG	8.6098	5.7789
DJ	8.5814	5.7554
NJ	8.5853	5.7548
ATM Bs	8.0138	5.2950

We also compute the price of the same claim within the BS model parameterized in such a way as to produce the same volatility of the original model. Furthermore, the Black–Scholes prices are also computed with an ATM forward calibration.

The results are summarized in Table 7.2. No statistical error is quoted, given that we have been running several million iterations and the measured error is $<1e^{-04}$.

Appendices

A

Elements of Probability

A.1 ELEMENTS OF MEASURE THEORY

Definition A.1.1 *Given a set Ω, a family $\mathcal{F}$ of subsets of Ω is a σ-algebra if*

(1) $\emptyset, \ \Omega \in \mathcal{F}$;
(2) *if $A \in \mathcal{F}$ then $A^c \in \mathcal{F}$ (where $A^c = \Omega \setminus A$);*
(3) *if $\{A_n\}_{n\geq 1}$ is such that $A_n \in \mathcal{F}$ for all $n \geq 1$, then $\bigcup_{n\geq 1} A_n \in \mathcal{F}$.*

The elements of $\mathcal{F}$ are called *measurable sets* and the pair $(\Omega, \mathcal{F})$ a *measurable space*. By the De Morgan formula and (2) and (3) above

$$\bigcap_{n\geq 1} A_n = \left(\bigcup_{n\geq 1} A_n^c\right)^c \in \mathcal{F}$$

Obviously the family $\mathcal{P}(\Omega)$ of all subsets of Ω is always a σ-algebra.

If $\mathcal{A}$ is a family of subsets of Ω, the smallest σ-algebra containing $\mathcal{A}$ is called the *σ-algebra generated by $\mathcal{A}$*. If $\Omega = \mathbb{R}$, the smallest σ-algebra $\mathcal{B}$ containing all subintervals of $\mathbb{R}$ is called the *Borel σ-algebra* and its elements *Borel sets*.

A measure is a function that associates to each measurable set a real number. More precisely:

Definition A.1.2 *Let $(\Omega, \mathcal{F})$ be a measurable space. A measure is a function $\mu : \mathcal{F} \to [0, +\infty]$ such that*

(1) $\mu(\emptyset) = 0$;
(2) *if $\{A_n\}_{n\geq 1}$ is a disjoint sequence of sets in $\mathcal{F}$ then*

$$\mu\left(\bigcup_{n\geq 1} A_n\right) = \sum_{n\geq 1} \mu(A_n)$$

The measure μ is *finite* or *infinite* (say also *integrable* or *not integrable*) as $\mu(\Omega) < +\infty$ or $\mu(\Omega) = +\infty$. Any measurable set $A \in \mathcal{F}$ such that $\mu(A) = 0$ is called a *negligible* or a *null* set.

Let $\Omega = \mathbb{R}$. A measure μ on the Borel sets $\mathcal{B}$ is said *locally finite* if for every compact $B \in \mathcal{B}$, $\mu(B) < +\infty$. Locally finite measures are also called *Radon measures*. The Lebesgue measure is clearly a Radon measure.

Definition A.1.3 *Let $(\Omega, \mathcal{F})$ be a measurable space. A real-valued function $f : \Omega \to \mathbb{R}$ is called measurable if for every Borel set $A \in \mathcal{B}$*

$$f^{-1}(A) = \{\omega \in \Omega : f(\omega) \in A\} \in \mathcal{F}$$

Let $(\Omega, \mathcal{F})$ be a measurable space provided by the measure μ; two measurable functions f and g are said *equal μ-almost everywhere* if and only if $\mu(\{x \in \Omega : f(x) \neq g(x)\}) = 0$ (and we shall write $f = g$ μ-a.e).

If $\mu(\Omega) = 1$, μ is a *probability* measure and is denoted by $\mathbb{P}$. In particular, if $\mathcal{F}$ is a σ-algebra in Ω and $\mathbb{P}$ a probability on $\mathcal{F}$, the triple $(\Omega, \mathcal{F}, \mathbb{P})$ is called a *probability space*. A *support* of $\mathbb{P}$ is any event $A \in \mathcal{F}$ such that $\mathbb{P}(A) = 1$.

The following are trivial consequences of the above definitions. Let $(\Omega, \mathcal{F}, \mathbb{P})$ be a probability space, then

(1) if $A \in \mathcal{F}$, then $\mathbb{P}(A^c) = 1 - \mathbb{P}(A)$;
(2) if $A, B \in \mathcal{F}$, $A \subset B$, then $\mathbb{P}(A) \leq \mathbb{P}(B)$;
(3) if $A, B \in \mathcal{F}$, then $\mathbb{P}(A \cup B) = \mathbb{P}(A) + \mathbb{P}(B) - \mathbb{P}(A \cap B)$.

Definition A.1.4 *Let $(\Omega, \mathcal{F}, \mathbb{P})$ be a probability space. If $A, B \in \mathcal{F}$ with $\mathbb{P}(A) > 0$, the quantity*

$$\mathbb{P}(B|A) = \frac{\mathbb{P}(A \cap B)}{\mathbb{P}(A)}$$

is called the "conditional probability of B with respect to A".

Intuitively the conditional probability $\mathbb{P}(B|A)$ is the probability that B occurs knowing that A has occurred.

Definition A.1.5 *Let $(\Omega, \mathcal{F}, \mathbb{P})$ be a probability space. $A, B \in \mathcal{F}$ are "independent" if and only if*

$$\mathbb{P}(A \cap B) = \mathbb{P}(A)\mathbb{P}(B)$$

Definition A.1.6 *Let $(\Omega, \mathcal{F}, \mathbb{P})$ be a probability space. $A_1, A_2, \ldots \in \mathcal{F}$ are "independent" if and only if for every k and every $i_1, i_2, \ldots, i_k$*

$$\mathbb{P}(A_{i_1} \cap \ldots \cap A_{i_k}) = \mathbb{P}(A_{i_1}) \cdots \mathbb{P}(A_{i_k})$$

Let $(\Omega, \mathcal{F}, \mathbb{P})$ be a probability space. A real-valued *random variable* is a measurable function $X : \Omega \to \mathbb{R}$. By definition, the function

$$A \to \mathbb{P}(\{\omega \in \Omega : X(\omega) \in A\})$$

is well defined for every real Borel set A. This function is called the *law* or the *distribution* of X. In the sequel we shall use the more concise notation $\mathbb{P}(X \in A)$ to denote $\mathbb{P}(\{\omega \in \Omega : X(\omega) \in A\})$.

The function $F_X : \mathbb{R} \to [0, 1]$, defined as $F_X(t) = \mathbb{P}(X \leq t) = \mathbb{P}(X \in (-\infty, t])$, is called the *cumulative distribution function of X*.

Let X and Y be two random variables defined on the same probability space $(\Omega, \mathcal{F}, \mathbb{P})$; we shall say that they are *equal $\mathbb{P}$-almost surely* (and we shall write $X = Y$ a.s.) if and only if $\mathbb{P}(X = Y) = 1$.

The random variables $X_1, \ldots, X_n$ defined on the same probability space $(\Omega, \mathcal{F}, \mathbb{P})$ are *independent* if and only if

$$\mathbb{P}(X_1 \in A_1, \ldots, X_n \in A_n) = \mathbb{P}(X_1) \cdots \mathbb{P}(A_n)$$

for all $A_1, \ldots, A_n$ real Borel sets.

A.1.1 Integration

Let $(\Omega, \mathcal{F})$ be a measurable space provided by the measure μ. If f is a non-negative simple function, that is $f = \sum_{i=1}^{n} x_i \mathbf{1}_{A_i}$ with $x_i \geq 0$ for all $i = 1, \ldots, n$ and $\{A_i\}$ is a finite decomposition of Ω into elements of $\mathcal{F}$, then the integral is defined by

$$\int_{\Omega} f(x)\mu(\mathrm{d}x) = \sum_{i=1}^{n} x_i \mu(A_i)$$

Now, if f is a non-negative measurable function, the integral is defined by

$$\int_{\Omega} f(x)\mu(\mathrm{d}x) = \sup\left\{\int_{\Omega} g(x)\mu(\mathrm{d}x) : g \text{ simple function } g \leq f\right\}$$

For a general measurable function f consider its positive part

$$f^{+}(x) = \begin{cases} f(x) & \text{if } f(x) \geq 0 \\ 0 & \text{otherwise} \end{cases}$$

and its negative part

$$f^{-}(x) = \begin{cases} -f(x) & \text{if } f(x) \leq 0 \\ 0 & \text{otherwise} \end{cases}$$

These functions are non-negative and measurable and $f = f^{+} - f^{-}$. The general integral is defined by

$$\int_{\Omega} f(x)\mu(\mathrm{d}x) = \int_{\Omega} f^{+}(x)\mu(\mathrm{d}x) - \int_{\Omega} f^{-}(x)\mu(\mathrm{d}x) \tag{A.1}$$

unless $\int_{\Omega} f^{+}(x)\mu(\mathrm{d}x) = \int_{\Omega} f^{-}(x)\mu(\mathrm{d}x) = -\infty$, in which case f has no integral.

If $\int_{\Omega} f^{+}(x)\mu(\mathrm{d}x)$ and $\int_{\Omega} f^{-}(x)\mu(\mathrm{d}x)$ are both finite (that is, $\int_{\Omega} |f(x)|\mu(\mathrm{d}x) < +\infty$), then f is integrable and has (A.1) as its *definite integral*. If $\int_{\Omega} f^{+}(x)\mu(\mathrm{d}x) = +\infty$ and $\int_{\Omega} f^{-}(x)\mu(\mathrm{d}x) < +\infty$ (or $\int_{\Omega} f^{-}(x)\mu(\mathrm{d}x) = +\infty$ and $\int_{\Omega} f^{+}(x)\mu(\mathrm{d}x) < +\infty$), then f is not integrable but is, in accordance with (A.1), assigned $+\infty$ (or $-\infty$) as its definite integral.

Main properties of the integral:

1. If f and g are two integrable functions, then, if $f = g$ μ-a.e., $\int_{\Omega} f(x)\mu(\mathrm{d}x) = \int_{\Omega} g(x)\mu(\mathrm{d}x)$.
2. *Monotonicity.* If f and g are two integrable functions, with $f \geq g$ μ-a.e., then $\int_{\Omega} f(x)\mu(\mathrm{d}x) \geq \int_{\Omega} g(x)\mu(\mathrm{d}x)$.
3. *Linearity.* If f and g are two integrable functions and $a, b \in \mathbb{R}$, then $af + bg$ is integrable and $\int_{\Omega}(af(x) + bg(x))\mu(\mathrm{d}x) = a\int_{\Omega} f(x)\mu(\mathrm{d}x) + b\int_{\Omega} g(x)\mu(\mathrm{d}x)$.
4. $\left|\int_{\Omega} f(x)\mu(\mathrm{d}x) - \int_{\Omega} g(x)\mu(\mathrm{d}x)\right| \leq \int_{\Omega} |f(x) - g(x)|\mu(\mathrm{d}x)$.

Expected values and moments

If $(\Omega, \mathcal{F}, \mathbb{P})$ is a probability space and X a random variable, the integral $\int_{\Omega} X(\omega)\mathbb{P}(d\omega)$, when defined, is called the *expected value of* X and is denoted by $\mathbb{E}^{\mathbb{P}}[X]$ (when there is no ambiguity the probability can be removed and the expected value indicated by $\mathbb{E}[X]$). Clearly the expected value only depends on the distribution of a random variable: so random variables with the same distribution share the same expected value.

If, for a positive integer k, $\mathbb{E}[|X|^k] < +\infty$, the expected value $\mathbb{E}[X^k]$ is called the k-moment of X. Since $|x|^j \leq 1 + |x|^k$ for $j \leq k$, if X has a finite k-moment, it has a j finite moment as well. Clearly, moments, being expectations, are uniquely determined by distributions. The moments of a random variable may or may not exist depending on how fast the distribution of X decays at infinity.

Definition A.1.7 *The set of all random variables for which $\mathbb{E}[|X|^p] < +\infty$ is denoted by L^p and is equipped with the distance $d(X, Y) = (\mathbb{E}[|X - Y|^p])^{1/p}$.*

The k-centred moment of a random variable X is defined as the k-moment of $X - \mathbb{E}[X]$, that is

$$\mathbb{E}[(X - \mathbb{E}[X])^k]$$

The second moment of a random variable X is called the variance, $\text{Var}(X) = \mathbb{E}[(X - \mathbb{E}[X])^2] = \mathbb{E}[X^2] - \mathbb{E}[X]^2$.

Scale-free versions of centred moments can be obtained by suitable normalizing. For example,

$$s(X) = \frac{\mathbb{E}[(X - \mathbb{E}[X])^3]}{\text{Var}^{3/2}(X)}$$

is called the *skewness* of X: if $s(X) > 0 (< 0)$, X is said to be positively (negatively) skewed:

$$k(X) = \frac{\mathbb{E}[(X - \mathbb{E}[X])^4] - 3\text{Var}(X)}{\text{Var}^2(X)}$$

is called the excess kurtosis of X. X is leptokurtic (that is fat tailed) if $k(X) > 0$. By definition the skewness and the kurtosis are invariant with respect to a change of scale:

$$\forall c > 0, \qquad s(cX) = s(X), \quad k(cX) = k(X)$$

Since, for X normally distributed, we have $s(X) = k(X) = 0$ these quantities can be seen as measures of deviation from normality.

A.1.2 Lebesgue integral

The development of the integral concept in most introductory analysis courses is centered almost exclusively on the Riemann integral. This relatively intuitive approach begins by taking a partition $P = \{x_0, \ldots, x_n\}$ of the domain of a real-valued function f. Given P we form the sum

$$\sum_{i=1}^{n} (x_i - x_{i-1}) f(\xi_i)$$

where $x_{i-1} < \xi_i < x_i$.

The integral of f, if it exists, is the limit of the sum as $n \to \infty$.

Although the Riemann integral suffices in most daily situations, it suffers from several difficulties: for example, let us consider an extreme case by looking at the bizarre function $f(x)$ defined to be 1 for every rational number in [0, 1] and 0 for every non-rational number. Now, since there are "very few" rational numbers, only a countable number in fact, we strongly suspect that the integral of this function would be zero. However, if we form the upper and lower Riemann integrals by partitioning [0, 1] into small segments Δx_i and write

$$\int_{SUP} f(x)\,\mathrm{d}x = \sum_i \Delta x_i \max[f(x)], \quad x_i \leq x \leq x_i + \Delta x_i$$

$$\int_{INF} f(x)\,\mathrm{d}x = \sum_i \Delta x_i \min[f(x)], \quad x_i \leq x \leq x_i + \Delta x_i$$

in the usual way, we see that no matter how small the subinterval is, Δx_i, the maximum of $f(x)$ on this interval is always 1 and the minimum is always 0. Thus

$$\int_{SUP} f(x)\,\mathrm{d}x = 1$$

and

$$\int_{INF} f(x)\,\mathrm{d}x = 0$$

so the Riemann integral does not exist.

This is a particular example but, from a general point of view, the class of Riemann integrable functions is relatively small. Another problem, related to the previous one, is that the Riemann integral does not have satisfactory limit properties. That is, given a sequence of Riemann integrable functions $\{f_n\}$with a limit function $f = \lim_{n\to\infty} f_n$, it does not necessarily follow that the limit function f is Riemann integrable

An equally intuitive method of integration was presented by Lebesgue in 1902. Rather then partitioning the domain of the function, as in the Riemann integral, Lebesgue chose to partition the range. Thus for each interval in the partition, rather than searching for the value of the function between the end points of the interval in the domain, he considered how much of the domain is mapped by the function to some value between two end points in the range.

Partitioning the range of a function and counting the resultant rectangles becomes tricky since we must employ some way of determining (or measuring) how much of the domain is sent to a particular portion of a partition of the range. Measure theory addresses just this problem and, as usual, we refer the interested reader to the relevant bibliography.

As it turns out, the Lebesgue integral solves many of the problems left by the Riemann integral. For example, in the theory of Lebesgue integration, the integral of the above-described function does exist, and equals zero. We say that $f(x) = 0$ except on a set of points of *measure zero*, or $f(x) = 0$ *almost everywhere*. The intuitive content of this sentence is the following: if we have a countable number of points on the real line and are given a small strip of paper of length ϵ, then we can paste a small piece of the strip over each element of the set by dividing it into a countable number of pieces of width $\epsilon/2^n$. Since $\sum_{n=1}^{\infty} \epsilon/2^n = \epsilon$, we use up only our given strip in the process. But since the original strip can be arbitrarily small, the set of points on which $f(x)$ is non-zero is negligible with respect to the set on which it is zero, despite the fact that every real number is arbitrarily close to some rational number. Thus rational numbers are a set of measure zero on the real line.

A.1.3 The characteristic function

Definition A.1.8 *The characteristic function of a random variable X is defined for real t by*

$$\phi_X(t) = \mathbb{E}\left[e^{itX}\right]$$

The characteristic function in non-probabilistic contexts is called the Fourier transform (see Chapter 6). The characteristic function has three fundamental properties:

1. If X and Y are independent random variables. $\phi_{X+Y}(t) = \phi_X(t)\phi_Y(t)$.
2. The characteristic function uniquely determines the distribution.
3. From the pointwise convergence of a sequence of characteristic functions, it follows the convergence of the corresponding distributions; more precisely, $X_n \xrightarrow{d} X$ if and only if $\phi_{X_n}(t) \to \phi_X(t)$ for all t.

The moments of a random variables are related to the derivatives at 0 of its characteristic function: if $\mathbb{E}[|X|^n] < +\infty$, then ϕ_X has n continuous derivatives at 0 and

$$\mathbb{E}[X^k] = \frac{1}{i^k}\frac{\partial^k \phi_X(0)}{\partial t^k}, \qquad \forall k = 1, \ldots n \tag{A.2}$$

On the other hand, if ϕ_X has n continuous derivatives at 0, then $\mathbb{E}[|X|^n] < +\infty$ and (A.2) holds.

A.1.4 Relevant probability distributions

Binomial distribution

The binomial distribution with parameters $n \in \mathbb{N} \setminus \{0\}$ and $p \in [0, 1]$ is defined through its density

$$p(x) = \begin{cases} \binom{n}{x} p^x(1-p)^{n-x} & x = 0, 1, \ldots, n \\ 0 & \text{otherwise} \end{cases}$$

This is denoted as $B(n, p)$ and is the distribution of a random variable X with values on $\{0, 1, \ldots, n\}$. If $\{X_k\}_{k=1,\ldots,n}$ is a family of independent and identically distributed random variables with $\mathbb{P}(X_k = 1) = p$ and $\mathbb{P}(X_k = 0) = 1 - p$, then $X = \sum_{k=1}^n X_k$ is binomially distributed.

$\mathbb{E}[X] = np$ and $\text{Var}(X) = np(1-p)$.

Poisson distribution

The Poisson distribution with parameter $\lambda > 0$ is defined through its density

$$p(x) = \begin{cases} e^{-\lambda}\lambda^x/x! & x = 0, 1, \ldots \\ 0 & \text{otherwise} \end{cases}$$

It is denoted as $\text{Poi}(\lambda)$ and it is the distribution of a random variable with values on $\mathbb{N}$. As shown in Example 2.2.3, it is obtained as the limit of binomial distributions $B(n, \lambda/n)$ as $n \to +\infty$.

In the book, we shall use the convention that $X \stackrel{d}{=} \mathrm{Poi}(0)$ means $\mathbb{P}(X = 0) = 1$ and $X \stackrel{d}{=} \mathrm{Poi}(\infty)$ means $\mathbb{P}(X = +\infty) = 1$.

For $\lambda > 0$, $\mathbb{E}[X] = \lambda$ and $\mathrm{Var}(X) = \lambda$.

The characteristic function of the Poisson distribution with parameter λ is

$$\phi(t) = \exp\left(\lambda(e^{it} - 1)\right)$$

If $X_1 \stackrel{d}{=} \mathrm{Poi}(\lambda_1)$, $X_2 \stackrel{d}{=} \mathrm{Poi}(\lambda_2)$ are independent, then

$$\begin{aligned}\phi_{X_1+X_2}(t) &= \phi_{X_1}(t)\phi_{X_2}(t) = \exp\left(\lambda_1(e^{it} - 1)\right)\exp\left(\lambda_2(e^{it} - 1)\right) \\ &= \exp\left((\lambda_1 + \lambda_2)(e^{it} - 1)\right)\end{aligned}$$

and $X_1 + X_2 \stackrel{d}{=} \mathrm{Poi}(\lambda_1 + \lambda_2)$. As a consequence, if $X \stackrel{d}{=} \mathrm{Poi}(\lambda)$, for every integer $n \geq 1$, $X = X_1 + X_2 + \cdots + X_n$ where $X_1, \ldots, X_n$ are independent and identically distributed random variables with law $\mathrm{Poi}(\lambda/n)$.

Normal distribution

The Normal distribution with parameters $\mu \in \mathbb{R}$ and $\sigma^2 > 0$ is defined through its density

$$f(x) = \frac{1}{\sqrt{2\pi}\sigma} e^{-\frac{(x-\mu)^2}{2\sigma^2}}, \qquad x \in \mathbb{R}$$

It is denoted as $N(\mu, \sigma^2)$ and it is the distribution of a random variable with values on $\mathbb{R}$. If $X \stackrel{d}{=} N(0, 1)$, then $\sigma X + \mu \stackrel{d}{=} N(\mu, \sigma^2)$.

Since $e^{-\frac{x^2}{2}}$ is a symmetric function around $x = 0$, if $X \stackrel{d}{=} N(0, 1)$ then also $-X \stackrel{d}{=} N(0, 1)$ and this can be expressed by saying that the distribution $N(0, 1)$ is symmetric. As a consequence, as $x\, e^{-\frac{x^2}{2}}$ is an odd function, if $X \stackrel{d}{=} N(0, 1)$

$$\mathbb{E}[X] = \int_{-\infty}^{+\infty} \frac{1}{\sqrt{2\pi}} x\, e^{-\frac{x^2}{2}}\, dx = 0$$

On the other hand, integrating by parts

$$\mathbb{E}[X^2] = \int_{-\infty}^{+\infty} \frac{1}{\sqrt{2\pi}} x^2\, e^{-\frac{x^2}{2}}\, dx = 1$$

and so $\mathrm{Var}(X) = 1$.

By the linearity of expectations, if $X \stackrel{d}{=} N(\mu, \sigma^2)$ then

$$\mathbb{E}[X] = \mu, \qquad \mathrm{Var}(X) = \sigma^2$$

More generally, if $X \stackrel{d}{=} N(\mu, \sigma^2)$, all odd-order moments are zero, while the even ones are

$$\mathbb{E}[(X - \mu)^{2k}] = \sigma^k \frac{(2k)!}{2^k k!}$$

The characteristic function of the normal distribution with parameters μ and σ^2 is

$$\phi(t) = \exp\left(-\frac{\sigma^2 t^2}{2} + i\mu t\right)$$

If $X_1 \stackrel{d}{=} N(\mu_1, \sigma_1^2)$, $X_2 \stackrel{d}{=} N(\mu_2, \sigma_2^2)$ are independent then

$$\begin{aligned}\phi_{X_1+X_2}(t) = \phi_{X_1}(t)\phi_{X_2}(t) &= \exp\left(-\frac{\sigma_1^2 t^2}{2} + i\mu_1 t\right)\exp\left(-\frac{\sigma_2^2 t^2}{2} + i\mu_2 t\right) \\ &= \exp\left(-\frac{(\sigma_1^2+\sigma_2^2)t^2}{2} + i(\mu_1+\mu_2)t\right)\end{aligned}$$

and $X_1 + X_2 \stackrel{d}{=} N(\mu_1+\mu_2, \sigma_1^2+\sigma_2^2)$. As a consequence, if $X \stackrel{d}{=} N(\mu, \sigma^2)$, for every integer $n \geq 1$, $X = X_1 + X_2 + \cdots + X_n$ where $X_1, \ldots, X_n$ are independent and identically distributed random variables with law $N\left(\frac{\mu}{n}, \left(\frac{\sigma}{\sqrt{n}}\right)^2\right)$.

Exponential distribution

The exponential distribution with parameter $\lambda > 0$ is defined through its density

$$f(x) = \begin{cases} \lambda\, e^{-\lambda x} & x > 0 \\ 0 & \text{otherwise} \end{cases}$$

It is denoted as $\mathcal{E}(\lambda)$ and is the distribution of a random variable with values on $[0, +\infty)$. This distribution is characterized by the *lack of memory property*, meaning that if $X \stackrel{d}{=} \mathcal{E}(\lambda)$,

$$\begin{aligned}\mathbb{P}(X > t+s | X > t) &= \frac{\mathbb{P}(X > t+s, X > t)}{\mathbb{P}(X > t)} = \frac{\mathbb{P}(X > t+s)}{\mathbb{P}(X > t)} \\ &= \frac{\int_{t+s}^{+\infty} \lambda\, e^{-\lambda x}\, dx}{\int_t^{+\infty} \lambda\, e^{-\lambda x}\, dx} = \frac{e^{-\lambda(t+s)}}{e^{-\lambda t}} \\ &= e^{-\lambda s} = \mathbb{P}(X > s)\end{aligned}$$

Moreover, if $X \stackrel{d}{=} \mathcal{E}(\lambda)$

$$\mathbb{E}[X] = \int_0^{+\infty} x\lambda\, e^{-\lambda x}\, dx = \frac{1}{\lambda}$$

$$\mathrm{Var}(X) = \mathbb{E}\left[\left(X - \frac{1}{\lambda}\right)^2\right] = \int_0^{+\infty}\left(x - \frac{1}{\lambda}\right)\lambda\, e^{-\lambda x}\, dx = \frac{1}{\lambda^2}$$

Gamma distribution

The *Gamma function* is a function $\Gamma : \mathbb{R}^+ \to \mathbb{R}^+$ defined as

$$\Gamma(\alpha) = \int_0^{+\infty} x^{\alpha-1}\, e^{-x}\, dx$$

Except for some special cases, the above integral is not explicitly computable. Nevertheless, integrating by parts

$$\Gamma(\alpha+1) = \int_0^{+\infty} x^{\alpha}\, e^{-x}\, dx = -x^{\alpha}\, e^{-x}\Big|_0^{+\infty} + \alpha\int_0^{+\infty} x^{\alpha-1}\, e^{-x}\, dx = \alpha\Gamma(\alpha)$$

and for integers α, inductively we get

$$\Gamma(n) = (n-1)!$$

The Gamma distribution with parameters $\alpha,\ \lambda > 0$ is defined through its density

$$f(x) = \begin{cases} \frac{\lambda^\alpha}{\Gamma(\alpha)} x^{\alpha-1} \mathrm{e}^{-\lambda x} & x > 0 \\ 0 & \text{otherwise} \end{cases}$$

This is denoted by $\Gamma(\alpha, \lambda)$ and is the distribution of a random variable with values on $[0, +\infty)$. If $\alpha = 1$, we recover the exponential distribution with parameter λ. If $X \stackrel{d}{=} N(0, \sigma^2)$, then $X^2 \stackrel{d}{=} \Gamma(\frac{1}{2}, \frac{1}{2\sigma^2})$. The Gamma distributions with an integer α are also called *Erlang laws*. On the other hand, distributions of type $\Gamma\left(\frac{n}{2}, \frac{1}{2}\right)$ are called *chi-square laws with n degrees of freedom* and denoted as $\chi^2(n)$.

If $\beta > 0$ and $X \stackrel{d}{=} \Gamma(\alpha, \lambda)$,

$$\begin{aligned} \mathbb{E}\left[X^\beta\right] &= \frac{\lambda^\alpha}{\Gamma(\alpha)} \int_0^{+\infty} x^{\beta+\alpha-1} \mathrm{e}^{-\lambda x}\, \mathrm{d}x \\ &= \frac{\lambda^\alpha}{\Gamma(\alpha)} \frac{\Gamma(\alpha+\beta)}{\lambda^{\alpha+\beta}} \left\{ \frac{\lambda^{\alpha+\beta}}{\Gamma(\alpha+\beta)} \int_0^{+\infty} x^{\beta+\alpha-1}\, \mathrm{e}^{-\lambda x}\, \mathrm{d}x \right\} = \frac{\Gamma(\alpha+\beta)}{\lambda^\beta \Gamma(\alpha)} \end{aligned}$$

since the quantity inside {} is 1, being the integral of the density of the distribution $\Gamma(\alpha+\beta, \lambda)$. Setting first $\beta = 1$ and then $\beta = 2$ we get

$$\mathbb{E}[X] = \frac{\Gamma(\alpha+1)}{\lambda\Gamma(\alpha)} = \frac{\alpha}{\lambda}$$

$$\mathbb{E}[X^2] = \frac{\Gamma(\alpha+2)}{\lambda^2\Gamma(\alpha)} = \frac{\alpha(\alpha+1)}{\lambda^2}$$

and

$$\mathrm{Var}(X) = \mathbb{E}[X^2] - (\mathbb{E}[X])^2 = \frac{\alpha}{\lambda^2}$$

In particular, if $X \stackrel{d}{=} \chi^2(n)$

$$\mathbb{E}[X] = n \quad \text{and} \quad \mathrm{Var}(X) = 2n$$

The characteristic function of a random variable $X \stackrel{d}{=} \Gamma(\alpha, \lambda)$ is

$$\phi(t) = \frac{\lambda^\alpha}{\Gamma(\alpha)} \int_0^{+\infty} v^{\alpha-1}\, \mathrm{e}^{-(-it+\lambda)v}\, \mathrm{d}v$$

The last complex-valued integral has to be computed by applying contour complex integration techniques (see Appendix C). The explicit evaluation of the characteristic function is the content of the next example where we show that

$$\phi(t) = \frac{1}{(1 - i\lambda^{-1}t)^\alpha}$$

Example A.1.1 *In order to compute the complex-valued integral*

$$\int_0^{+\infty} v^{\alpha-1}\, \mathrm{e}^{-(-it+\lambda)v}\, \mathrm{d}v$$

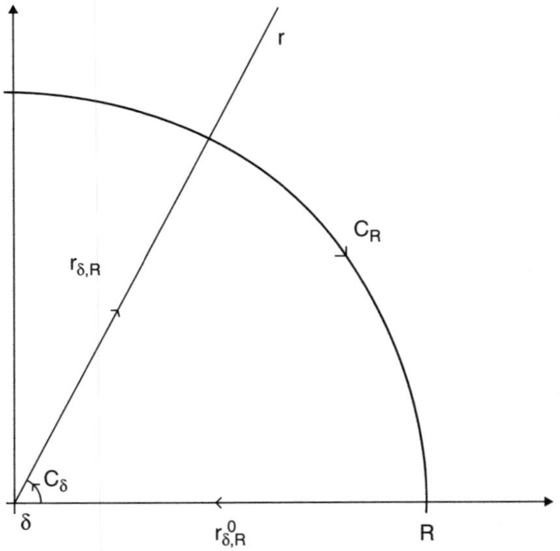

Figure A.1 Path of integration for the Gamma characteristic function

let $r = \{z = x + iy \in \mathbb{C} : y = -\frac{t}{\lambda}x, x \geq 0\}$ *. By the change of variable* $z = v(\lambda - it)$ *we get*

$$\begin{aligned}\phi(t) &= \frac{\lambda^\alpha}{\Gamma(\alpha)} \int_0^{+\infty} v^{\alpha-1} \mathrm{e}^{-(-it+\lambda)v}\,\mathrm{d}v \\ &= \frac{\lambda^\alpha}{\Gamma(\alpha)} \int_r \frac{z^{\alpha-1}}{(\lambda - it)^{\alpha-1}} \mathrm{e}^{-z} (\lambda - it)^{-1}\,\mathrm{d}z = \\ &= \frac{\lambda^\alpha}{\Gamma(\alpha)(\lambda - it)^\alpha} \int_r z^{\alpha-1}\,\mathrm{e}^{-z}\,\mathrm{d}z \qquad \text{(A.3)}\end{aligned}$$

For $\alpha \in (0, 1)$ *the integrand has a singularity in the origin. So, let us now consider, for* $R > \delta > 0$*, the path in Figure A.1.*

Since

$$0 = \int_{r_{\delta,R} \cup C_R \cup r^0_{\delta,R} \cup C_\delta} z^{\alpha-1} \mathrm{e}^{-z}\,\mathrm{d}z$$

$$\int_{r_{\delta,R}} z^{\alpha-1}\,\mathrm{e}^{-z}\,\mathrm{d}z + \int_{C_R} z^{\alpha-1}\,\mathrm{e}^{-z}\,\mathrm{d}z = \int_{r^0_{\delta,R}} z^{\alpha-1}\,\mathrm{e}^{-z}\,\mathrm{d}z + \int_{C_\delta} z^{\alpha-1}\,\mathrm{e}^{-z}\,\mathrm{d}z \qquad \text{(A.4)}$$

Now, substituting $z = R\,\mathrm{e}^{i\theta}$ *in the second integral of the left-hand side we get*

$$\begin{aligned}\int_{C_R} z^{\alpha-1}\,\mathrm{e}^{-z}\,\mathrm{d}z &= \int_0^{\arctan(-\frac{t}{\lambda})} i R^{\alpha-1}\,\mathrm{e}^{i(\alpha-1)\theta}\,\mathrm{e}^{-R\,\mathrm{e}^{i\theta}}\,R\,\mathrm{e}^{i\theta}\mathrm{d}\theta \\ &= \int_0^{\arctan(-\frac{t}{\lambda})} i R^\alpha\,\mathrm{e}^{i\alpha\theta}\,\mathrm{e}^{-R\cos\theta}\,\mathrm{e}^{-Ri\sin\theta}\,\mathrm{d}\theta\end{aligned}$$

But

$$\left|i R^\alpha e^{i\alpha\theta} e^{-R\cos\theta} e^{-iR\sin\theta}\right| = R^\alpha \left|e^{i\alpha\theta}\right| e^{-R\cos\theta} \left|e^{-iR\sin\theta}\right|$$
$$\leq R^\alpha e^{-R\cos\theta} \leq R^\alpha e^{-R\cos\arctan(-\frac{t}{\lambda})}$$
$$= R^\alpha \exp\left(-\frac{R}{\sqrt{1+\frac{t^2}{\lambda^2}}}\right) \underset{R\to+\infty}{\to} 0$$

and

$$\int_{C_R} z^{\alpha-1} e^{-z}\, dz \underset{R\to+\infty}{\to} 0$$

Now, substituting $z = \delta e^{i\theta}$ in the second integral of the right-hand side of (A.4), by the same arguments we get

$$\int_{C_\delta} z^{\alpha-1} e^{-z} dz = \int_0^{\arctan(-\frac{t}{\lambda})} i\delta^\alpha e^{i\alpha\theta} e^{-\delta\cos\theta} e^{-\delta i\sin\theta} d\theta$$

and

$$\left|\delta^\alpha e^{i\alpha\theta} e^{-\delta\cos\theta} e^{-i\delta\sin\theta}\right| \leq \delta^\alpha \exp\left(-\frac{\delta}{\sqrt{1+\frac{t^2}{\lambda^2}}}\right) \underset{\delta\to0}{\to} 0$$

so that

$$\int_{C_\delta} z^{\alpha-1} e^{-z}\, dz \underset{\delta\to0}{\to} 0$$

Let us now focus on the first integral of the right-hand side of (A.4):

$$\int_{r^0_{\delta,R}} z^{\alpha-1} e^{-z}\, dz = \int_\delta^R x^{\alpha-1} e^{-x}\, dx$$

and so

$$\lim_{\substack{R\to+\infty\\ \delta\to0}} \int_{r^0_{\delta,R}} z^{\alpha-1} e^{-z}\, dz = \int_0^{+\infty} x^{\alpha-1} e^{-x}\, dx = \Gamma(\alpha)$$

This way, from (A.4) it follows that

$$\lim_{\substack{R\to+\infty\\ \delta\to0}} \int_{r_{\delta,R}} z^{\alpha-1} e^{-z}\, dz = \Gamma(\alpha)$$

Now, by (A.3), we get

$$\phi(t) = \frac{\lambda^\alpha}{\Gamma(\alpha)(\lambda - it)^\alpha} \int_r z^{\alpha-1} e^{-z}\, dx =$$
$$= \frac{\lambda^\alpha}{\Gamma(\alpha)(\lambda - it)^\alpha} \lim_{\substack{R\to+\infty\\ \delta\to0}} \int_{r_{\delta,R}} z^{\alpha-1} e^{-z}\, dz =$$
$$= \frac{\lambda^\alpha}{\Gamma(\alpha)(\lambda - it)^\alpha} \Gamma(\alpha) = \frac{\lambda^\alpha}{(\lambda - it)^\alpha} = \frac{1}{(1 - i\lambda^{-1}t)^\alpha}$$

Table A.1 Relevant probability distributions and their characteristic functions.

	Probability distributions	
Distribution	Density	Characteristic Function
Normal	$f(x) = \frac{1}{\sqrt{2\pi}\sigma} e^{-\frac{(x-\mu)^2}{2\sigma^2}}$	$\phi(t) = \exp\left(-\frac{\sigma^2 t^2}{2} + i\mu t\right)$
Poisson	$p(x) = e^{-\lambda}\frac{\lambda^x}{x!}\mathbf{1}_{\mathbb{N}}$	$\phi(t) = \exp\left(\lambda(e^{it} - 1)\right)$
Exponential	$f(x) = \lambda\, e^{-\lambda x}\mathbf{1}_{x>0}$	$\phi(t) = \frac{\lambda}{(\lambda - it)}$
Gamma	$f(x) = \frac{\lambda^\alpha}{\Gamma(\alpha)} x^{\alpha-1} e^{-\lambda x}\mathbf{1}_{x>0}$	$\phi(t) = \frac{1}{(1 - i\lambda^{-1}t)^\alpha}$

If $X_1 \stackrel{d}{=} \Gamma(\alpha_1, \lambda)$, $X_2 \stackrel{d}{=} \Gamma(\alpha_1, \lambda)$ are independent then

$$\phi_{X_1+X_2}(t) = \phi_{X_1}(t)\phi_{X_2}(t) = \frac{1}{(1 - i\lambda^{-1}t)^{\alpha_1}} \frac{1}{(1 - i\lambda^{-1}t)^{\alpha_2}}$$
$$= \frac{1}{(1 - i\lambda^{-1}t)^{\alpha_1+\alpha_2}}$$

and $X_1 + X_2 \stackrel{d}{=} \Gamma(\alpha_1 + \alpha_2, \lambda)$. As a consequence, if $X \stackrel{d}{=} \Gamma(\alpha, \lambda)$, for every integer $n \geq 1$, $X = X_1 + X_2 + \cdots + X_n$ where $X_1, \ldots, X_n$ are independent and identically distributed random variables with law $\Gamma(\frac{\alpha}{n}, \lambda)$.

Moreover, by the above results the distribution $\Gamma(\frac{n}{2}, \frac{1}{2})$ is the law of a random variable Y of type $Y = X_1^2 + \cdots + X_n^2$, where $X_1, \ldots, X_n$ are independent and identically $N(0, 1)$ distributed random variables.

A.1.5 Convergence of sequences of random variables

Let $\{X_n\}_n$ be a sequence of random variables defined on the same probability space of the random variable X. X_n is said to *converge almost surely* to X ($X_n \to X$ a.s.) if

$$\mathbb{P}\left(\lim_{n\to+\infty} X_n = X\right) = 1$$

X_n is said to *converge in* L^p to X ($X_n \stackrel{L^p}{\to} X$) if

$$\lim_{n\to+\infty} \mathbb{E}\left[|X_n - X|^p\right] = 0$$

The above type of convergence is the convergence in the metric space L^p. This space is *complete* (more precisely it is a Banach space) meaning that every Cauchy sequence of random variables X_n (that is such that for all $\epsilon > 0$ there exists an $\overline{n}$ such that for all $n, m > \overline{n}$, $d(X^n, X^m) < \epsilon$) converges in L^p to a random variable $X \in L^p$.

X_n is said to *converge in probability* to X ($X_n \stackrel{\mathbb{P}}{\to} X$) if for each $\epsilon > 0$

$$\lim_{n\to+\infty} \mathbb{P}(|X_n - X| > \epsilon) = 0$$

X_n is said to *converge in distribution or in law* to X ($X_n \xrightarrow{d} X$) if

$$\lim_{n\to+\infty} F_{X_n}(t) = F_X(t)$$

for every t in which F_X is continuous. Unlike the other notion of convergence, it does not require the random variables to be defined on a common probability space.

A.1.6 The Radon–Nikodym derivative

Let $(\Omega, \mathcal{F})$ be a measurable space with two measures μ and ν. If for every $A \in \mathcal{F}$, $\mu(A) = 0 \Rightarrow \nu(A) = 0$, then ν is said to be *absolutely continuous with respect to* μ. This means that all negligible sets for μ are negligible sets for ν as well. The following result, known as *Radon–Nikodym theorem*, characterizes absolute continuity.

Theorem A.1.1 *If ν is absolutely continuous with respect to μ there exists a measurable function $Z : \Omega \to [0, +\infty)$ such that for any $A \in \mathcal{F}$*

$$\nu(A) = \int_A Z(\omega)\mu(\mathrm{d}\omega)$$

Z is called the density or Radon–Nikodym derivative of ν with respect to μ and is usually denoted as $\mathrm{d}\nu/\mathrm{d}\mu$. For every function f integrable with respect to the measure ν,

$$\int_\Omega f(\omega)\nu(\mathrm{d}\omega) = \int_\Omega f(\omega)Z(\omega)\mu(\mathrm{d}\omega) = \int_\Omega f(\omega)\frac{\mathrm{d}\nu}{\mathrm{d}\mu}(\omega)\mu(\mathrm{d}\omega)$$

If μ is also absolutely continuous with respect to ν, then μ and ν are said to be *equivalent*, meaning that they share the same negligible sets. This is obviously equivalent to $\mathrm{d}\nu/\mathrm{d}\mu > 0$.

A.1.7 Conditional expectation

Definition A.1.9 *Let $(\Omega, \mathcal{F}, \mathbb{P})$ be a probability space and $\mathcal{A} \subset \mathcal{F}$ a σ-algebra. There exists a random variable denoted by $\mathbb{E}[X|\mathcal{A}]$ called the "conditional expected value of X given $\mathcal{A}$", which has these two properties:*

(1) $\mathbb{E}[X|\mathcal{A}]$ *is measurable $\mathcal{A}$ and integrable;*
(2) $\mathbb{E}[X|\mathcal{A}]$ *satisfies the functional equation*

$$\int_A \mathbb{E}[X|\mathcal{A}]\mathrm{d}\mathbb{P} = \int_A X\,\mathrm{d}\mathbb{P}, \qquad A \in \mathcal{A}$$

To prove the existence of such a random variable, consider first the case of non-negative X. Define a measure ν on $\mathcal{A}$ by $\nu(A) = \int_A X\mathbb{P}(\mathrm{d}\omega)$. This measure is finite because X is integrable and it is absolutely continuous with respect to $\mathbb{P}$. By the Radon–Nykodym theorem there exists a function Y, measurable $\mathcal{A}$, such that $\nu(A) = \int_A Y(\omega)\mathbb{P}(\mathrm{d}\omega)$. This Y has properties (1) and (2) above. If X is not necessarily non-negative, $\mathbb{E}[X^+|\mathcal{A}] - \mathbb{E}[X^-|\mathcal{A}]$ clearly has the required properties.

In general there will be many such random variables $\mathbb{E}[X|\mathcal{A}]$, any one of which is called a *version* of the conditional expected value. Any two versions are equal with probability 1.

Obviously, $\mathbb{E}[X|\{\emptyset, \Omega\}] = \mathbb{E}[X]$ and that $\mathbb{E}[X|\mathcal{F}] = X$ with probability 1. As $\mathcal{A}$ increases, condition (1) becomes weaker and condition (2) becomes stronger.

The value $\mathbb{E}[X|\mathcal{A}](\omega)$ is to be interpreted as the expected value of X for someone who knows, for each $A \in \mathcal{A}$, whether or not it contains the point ω, which in general remains unknown itself. Condition (1) ensures that $\mathbb{E}[X|\mathcal{A}]$ can in principle be calculated from this partial information alone. Condition (2) can be restated as $\int_A (\mathbb{E}[X|\mathcal{A}] - X)\,\mathrm{d}\mathbb{P} = 0$, if the observer, in possession of the partial information contained in $\mathcal{A}$, is offered the opportunity to bet, paying an entry fee of $\mathbb{E}[X|\mathcal{A}]$ and being returned the amount X. If he adopts the strategy of betting if A occurs, this equation says that the game is fair.

Properties of the conditional expectation

Suppose that X, Y and X_n are integrable.

1. If $X = a$ with probability 1, then $\mathbb{E}[X|\mathcal{A}] = a$.
2. For constant a and b, $\mathbb{E}[aX + bY|\mathcal{A}] = a\mathbb{E}[X|\mathcal{A}] + b\mathbb{E}[Y|\mathcal{A}]$.
3. If $X \leq Y$ with probability 1, then $\mathbb{E}[X|\mathcal{A}] \leq \mathbb{E}[Y|\mathcal{A}]$.
4. $|\mathbb{E}[X|\mathcal{A}]| \leq \mathbb{E}[|X||\mathcal{A}]$.
5. If the limit for n that tends to infinity $X_n = X$ with probability 1, $|X_n| \leq Y$ and Y is integrable, then $\lim_{n\to+\infty} \mathbb{E}[X_n|\mathcal{A}] = \mathbb{E}[X|\mathcal{A}]$ with probability 1.
6. If X is measurable $\mathcal{A}$ and if XY is integrable, then $\mathbb{E}[XY|\mathcal{A}] = X\mathbb{E}[Y|\mathcal{A}]$ with probability 1.
7. If $\mathcal{A}_1 \subset \mathcal{A}_2$ are σ-algebras, then $\mathbb{E}\left[\mathbb{E}[X|\mathcal{A}_2]|\mathcal{A}_1\right] = \mathbb{E}[X|\mathcal{A}_1]$ with probability 1.
8. If X is independent on the partial information provided by $\mathcal{A}$, then $\mathbb{E}[X|\mathcal{A}] = \mathbb{E}[X]$.
9. *Jensen's inequality*. If ϕ is a convex function on the real line and $\phi(X)$ is integrable, then $\phi\left(\mathbb{E}[X|\mathcal{A}]\right) \leq \mathbb{E}\left[\phi(X)|\mathcal{A}\right]$.

A.2 ELEMENTS OF THE THEORY OF STOCHASTIC PROCESSES

A.2.1 Stochastic processes

Let $(\Omega, \mathcal{F}, \mathbb{P})$ be a probability space. A stochastic process is a collection of random variables $(X_t)_{t\in\mathcal{T}}$ with values in a common state space, which we will choose specifically as $\mathbb{R}$ (or $\mathbb{R}^d$). In this book we assume that $\mathcal{T} = [0, +\infty)$ or $\mathcal{T} = [0, T]$ and we interpret the index t as time.

The functions of time $t \to X_t(\omega)$ are the paths or trajectories of the process: thus a stochastic process can be viewed as a random function that is a random variable taking values in a function space. The trajectories can be continuous or with jumps for some $t \geq 0$:

$$\Delta X_t = X_{t^+} - X_{t^-} = \lim_{h\downarrow 0} X_{t+h} - \lim_{h\downarrow 0} X_{t-h} \tag{A.5}$$

If all trajectories are continuous functions of time apart from a negligible set, the process is said to be continuous and its state space is $C(\mathcal{T})$, the space of all real-valued continuous functions.

Nevertheless, most of the processes encountered in this book will not have continuous paths. In this case, we suppose that limits in (A.5) always exist for all $t \geq 0$ and that $X_{t^+} = X_t$, i.e. that the paths are almost surely right-continuous with left limits. These paths are called *cadlag*, which is a French acronym for *continue à droite, limitée à gauche* which means "right-continuous with left limit". The jump at t is denoted by $\Delta X_t = X_t - X_{t^-}$. However, cadlag trajectories cannot jump too wildly. In fact, as a consequence of the existence of limits, in any interval $[0, T]$, for every $b > 0$ the number of jumps greater than b must be finite and the number of jumps is at most countable. This way, in $[0, T]$ every cadlag trajectory has a finite number of large jumps and a possibly infinite but countable set of small jumps. The space of all real valued cadlag functions is denoted by $D(\mathcal{T})$.

The choice, among all the others, of assuming cadlag trajectories for financial modelling is justified by the following arguments. If a cadlag trajectory has a jump at time t then the value of $X_t(\omega)$ is unknown before t following the trajectory up to time t: the discontinuity is a sudden event at time t. By contrast, if the left-limit coincides with $X_t(\omega)$ then, an observer following the trajectory up to time t will approach the value of $X_t(\omega)$. It is natural, in a concrete financial context, to assume jumps to be sudden and unforeseeable events.

Filtrations

While time t is elapsing, the observer increases his or her endowment of information. In fact, some events that are random at time 0 may no longer be random at a certain time $t > 0$. In fact, the set of information available at time t can be sufficient to reveal if the event has occurred or not. In order to model the flow of information, we introduce the notion of filtration.

Definition A.2.1 *Given a probability space $(\Omega, \mathcal{F}, \mathbb{P})$, a filtration is an increasing family of σ-algebras $(\mathcal{F}_t)_{t\in\mathcal{T}}$, such that for all $t \geq s \geq 0$, $\mathcal{F}_s \subseteq \mathcal{F}_t \subseteq \mathcal{F}$.*

A probability space equipped with a filtration is called a *filtered probability space*.

$\mathcal{F}_t$ can be interpreted as the set of all events that occurred within time t and so represents the information known at time t. By definition, an $\mathcal{F}_t$-measurable random variable is a random variable whose value is known at time t.

Given a stochastic process $(X_t)_{t\in\mathcal{T}}$ if X_t is $\mathcal{F}_t$-measurable for every $t \in \mathcal{T}$ we say that the stochastic process is $(\mathcal{F}_t)_{t\in\mathcal{T}}$-adapted. This means that the values of the process at time t are revealed by the known information $\mathcal{F}_t$.

Clearly, the values of the process at time t, X_t, are revealed by the σ-algebra generated by X_t. We shall call the *natural filtration* generated by the process X_t the filtration $\left(\mathcal{F}_t^X\right)_{t\geq 0}$ such that $\mathcal{F}_t^X$ is the smallest σ-algebra with respect to which X_t is adapted completed by the null sets.

The assumption that all negligible sets are contained in each $\mathcal{F}_t$ implies, in particular, that all null sets are in $\mathcal{F}_0$, meaning that the fact that a certain evolution for the process is impossible is already known at time 0.

Stopping times

In a stochastic setting it is natural to deal with events happening at random times. For example, given a stochastic process $(X_t)_{t\geq 0}$, we can be interested in the first time at which the value of the process exceeds a given bound b; more precisely, if

$$\tau_b = \inf\{t \geq 0 : X_t > b\} \tag{A.6}$$

If $X_0 < b$, τ_b is a random variable.

A *random time* τ is a random variable with values in the set of times $\mathcal{T}$. It represents the time at which some event is going to occur. Given a filtration $(\mathcal{F}_t)_{t\in\mathcal{T}}$ one can ask if the information

$\mathcal{F}_t$ available at time t is sufficient to state if the event has already happened ($\tau \leq t$) or not ($\tau > t$).

Definition A.2.2 *Given a filtered probability space, a random variable τ with values in $\mathcal{T}$ is a "stopping time" if for all $t \geq 0$, $\{\tau \leq t\} \in \mathcal{F}_t$.*

The term "stopping time" is due to the notion of *stopped process*: given an adapted stochastic process $(X_t)_{t\in\mathcal{T}}$ and a stopping time τ, the process stopping at τ is defined by

$$X_{t\wedge\tau} = \begin{cases} X_t & \text{if } t < \tau \\ X_\tau & \text{if } t \geq \tau \end{cases}$$

The random time τ_b defined in (A.6) is indeed a stopping time.

Given a filtration $(\mathcal{F}_t)_{t\in\mathcal{T}}$ and a stopping time τ, the known information at time τ is the σ-algebra generated by all adapted process observed up to time τ. More precisely

$$\mathcal{F}_\tau = \{A \in \mathcal{F} : \forall t \in \mathcal{T}, A \cap \{\tau \leq t\} \in \mathcal{F}_t\}$$

A.2.2 Martingales

Let $(\Omega, \mathcal{F}, \mathbb{P})$ be equipped with a filtration $(\mathcal{F}_t)_{t\geq 0}$.

Definition A.2.3 (*Martingale*) *A cadlag process $(X_t)_{t\geq 0}$ is a martingale if is is $(\mathcal{F}_t)_{t\geq 0}$-adapted, $\mathbb{E}[|X_t|]$ is finite for any $t \in [0, T]$ and*

$$\forall s < t, \qquad \mathbb{E}[X_t | \mathcal{F}_s] = X_s \tag{A.7}$$

In other words, the best prediction of a martingale future value is its current value.

An obvious consequence of (A.7) is that a martingale has constant expectation: $\forall t \geq 0$, $\mathbb{E}[X_t] = \mathbb{E}[X_0]$.

There are several important results about martingales. They all come in several different forms. We present the L^2-versions as they are most easily formulated.

Theorem A.2.1 *If $(M_t)_{t\geq 0}$ is a martingale then there exists a unique stochastic process $\left(\hat{M}_t\right)_{t\geq 0}$ having cadlag trajectories whose paths coincide with those of $(M_t)_{t\geq 0}$ with probability* 1.

Definition A.2.4 *A martingale $(M_t)_{t\geq 0}$ with $\mathbb{E}[M_t^2] < +\infty$ is called a "square integrable martingale". The space of all square integrable martingales is denoted by $\mathcal{M}^2$.*

A typical way to construct a martingale is the following: given a random variable H with $\mathbb{E}[|H|] < +\infty$, the process M_t defined by $M_t = \mathbb{E}[H | \mathcal{F}_t]$ is a martingale. Moreover

Theorem A.2.2 (*Closure*) *Let $(M_t)_{t\geq 0}$ be a martingale such that $\sup_{t\geq 0}\mathbb{E}[M_t^2] < +\infty$, then there exists a random variable H with $\mathbb{E}[|Y|] < +\infty$ such that $M_t \underset{t\to+\infty}{\rightarrow} H$ almost surely.*

The following is one of the most useful martingale theorems:

Theorem A.2.3 (*Doob's optional stopping*) *Let $(M_t)_{t\geq 0}$ be a martingale and τ a stopping time. If $\sup_{t\geq 0} \mathbb{E}[M_t^2] < +\infty$, then the stopped process $M_{t\wedge\tau}$ is also a martingale and $\mathbb{E}[M_\tau] = \mathbb{E}[M_0]$.*

Theorem A.2.4 (***Doob's maximal inequality***) *Let $(M_t)_{t\geq 0}$ be a square integrable martingale. Then*

$$\mathbb{E}\left[\sup\{M_s^2 : 0 \leq s \leq t\}\right] \leq 4\mathbb{E}[M_t^2]$$

The space $\mathcal{M}^2$ equipped with the distance $d(M, N) = \sqrt{\sup_{t\geq 0}(M_t - N_t)^2}$ is a complete space, that is, for every Cauchy sequence of square integrable martingales there exists a square integrable martingale to which the sequence converges.

B

Elements of Complex Analysis

B.1 COMPLEX NUMBERS

The purpose of this chapter is to give a review of various properties of the complex numbers that may be a useful background for the mathematical chapter.

B.1.1 Why complex numbers?

We shall start from a very simple question: Why do we need new numbers?

The hardest thing about working with complex numbers is understanding why you might want to. Before introducing complex numbers, let us go back and look at simpler examples of how the need to deal with new numbers may arise.

If you start asking what a number may mean to most people, you discover immediately that the numbers, 1, 2, 3, ..., that is the *Natural* numbers, make sense. They provide a way to answer questions of the form "How many ... ?" One may learn about the operations of addition and subtraction, and find that while subtraction is a perfectly good operation, for subtraction some problems, like 3–5, do not have answers if we only work with *Natural* numbers. Then you find that if you are willing to work with *Integers*, $\ldots, -2, -1, 0, 1, 2, \ldots$, then all subtraction problems do have answers! Furthermore, by considering examples such as temperature scales, or your checking account, you see that negative numbers often make sense.

Now that we have clarified subtraction we will deal with division. Some, in fact most, division problems do not have answers that are *Integers*. For example, $3/2$ is not an *Integer*. We need new numbers! Now we have *Rational* numbers (fractions).

However, this is not the end of the story. There are problems with square roots and other operations, but we will not get into that here. The point is that you have had to expand your idea of number on several occasions, and now we are going to do that again.

The "problem" that leads to complex numbers concerns solutions of equations.

$$x^2 - 1 = 0 \tag{B.1}$$

$$x^2 + 1 = 0 \tag{B.2}$$

Equation (B.1) has two solutions, $x = -1$ and $x = 1$. We know that solving an equation in x is equivalent to finding the x-intercepts of a graph; and, the graph of $y = x^2 - 1$ crosses the x-axis at $(-1, 0)$ and $(1, 0)$.

Equation (B.2) has no solutions, and we can see this by looking at the graph of $y = x^2 + 1$. Since the graph has no x-intercepts, the equation has no solutions. Equation (B.2) has no solutions because -1 does not have a square root. In other words, there is no real number such that if we multiply it by itself we get -1. If equation (B.2) is to be given solutions, then we must create a square root of -1. This is what we are going to do in the next paragraph.

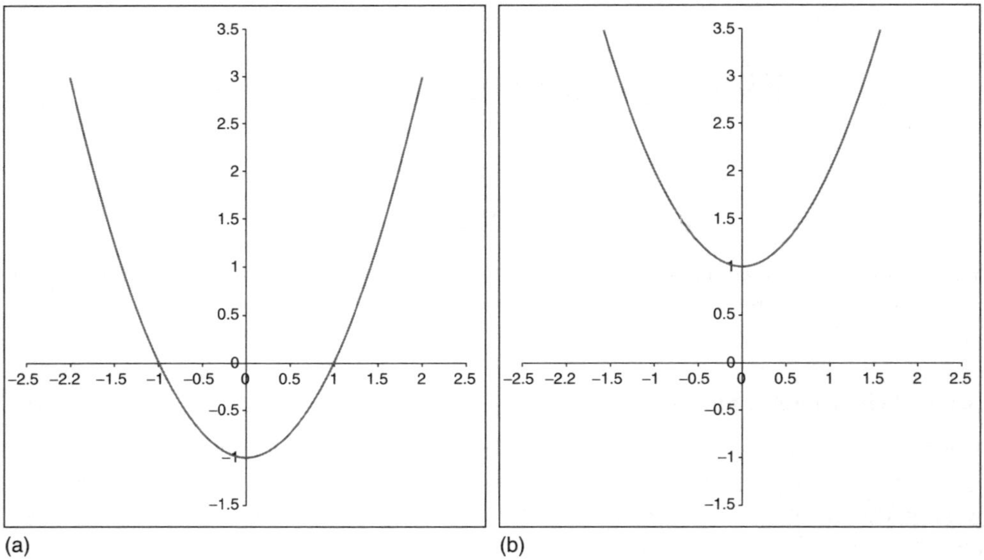

Figure B.1 (a) The function $x^2 - 1$, (b) the function $x^2 + 1$

B.1.2 Imaginary numbers

By definition, the imaginary unit i is one solution of the quadratic equation (B.2) or equivalently

$$x^2 = -1 \tag{B.3}$$

Since there is no real number that squares to any negative real number, we define such a number and assign to it the symbol i. It is important to realize, though, that i is just as well-defined a mathematical construct as the real numbers, despite being less intuitive to study. Real-number operations can be extended to imaginary and complex numbers by treating i as an unknown quantity while manipulating an expression, and then using the definition to replace occurrences of i^2 with -1. Higher integral powers of i can also be replaced with $-i$, 1, i, or -1.

Being a second-order polynomial with no multiple real root, the above equation has two distinct solutions that are equally valid and that happen to be additive inverses of each other. More precisely, once a solution i of the equation has been fixed, the value $-i \neq i$ is also a solution. Since the equation is the only definition of i, it appears that the definition is ambiguous (more precisely, not well-defined). However, no ambiguity results as long as one of the solutions is chosen and fixed as the "*positive i*".

Both imaginary numbers have equal claim to square to -1. If all mathematical textbooks and published literature referring to imaginary or complex numbers were rewritten with $-i$ replacing every occurrence of $+i$ (and therefore every occurrence of $-i$ replaced by $-(-i) = +i$), all facts and theorems would continue to be equivalently valid. The distinction between the two roots of $x^2 + 1 = 0$ with one of them as "positive" is purely a notational relic.

The imaginary unit is sometimes written $\sqrt{-1}$ in advanced mathematics contexts (as well as in less-advanced popular texts); however, great care needs to be taken when manipulating formulas involving radicals. The notation is reserved either for the principal square root function, which is only defined for real $x \geq 0$, or for the principal branch of the complex square root function.

Attempting to apply the calculation rules of the principal (real) square root function to manipulate the principal branch of the complex square root function will produce false results:

$$-1 = i \cdot i = \sqrt{-1} \cdot \sqrt{-1} = \sqrt{(-1) \cdot (-1)} = \sqrt{1} = 1$$

The calculation rule

$$\sqrt{a} \cdot \sqrt{b} = \sqrt{a \cdot b}$$

is only valid for real, non-negative values of a and b. To avoid making such mistakes when manipulating complex numbers, a strategy is never to use a negative number under a square root sign. For instance, rather than writing expressions like $\sqrt{-7}$, one should write $i\sqrt{7}$ instead. That is the use for which the imaginary unit is intended.

B.1.3 The complex plane

Any complex number, z, can be written as

$$z = x + iy$$

where x and y are real numbers and i is the imaginary unit, which has been previously defined. The number x defined by

$$x = \Re(z)$$

is the **real part** of the complex number z, and y, defined by

$$y = \Im(z)$$

is the **imaginary part**.

A complex number z can be viewed as a point or a position vector in a two-dimensional Cartesian coordinate system called the **complex plane** or Argand diagram. The point and hence the complex number z can be specified by Cartesian (rectangular) coordinates. The Cartesian coordinates of the complex number are the **real part** x and the **imaginary part** y, so we can refer to z with the ordered pair (x, y).

Formally, complex numbers can be defined as ordered pairs of real numbers (a, b) together with the operations:

$$(a, b) + (c, d) = (a + c, b + d)$$

$$(a, b) \cdot (c, d) = (ac - bd, bc + ad)$$

So defined, the complex numbers form a field, the complex number field, denoted by $\mathbb{C}$ (a field is an algebraic structure in which addition, subtraction, multiplication and division are defined and satisfy certain algebraic laws; for example, the real numbers form a field).

The real number a is identified with the complex number $(a, 0)$, and in this way the field of real numbers $\mathbb{R}$ becomes a subfield of $\mathbb{C}$. The imaginary unit i can then be defined as the complex number $(0, 1)$, which verifies

$$(a, b) = a \cdot (1, 0) + b \cdot (0, 1) = a + bi$$

$$i^2 = (0, 1) \cdot (0, 1) = (-1, 0) = -1$$

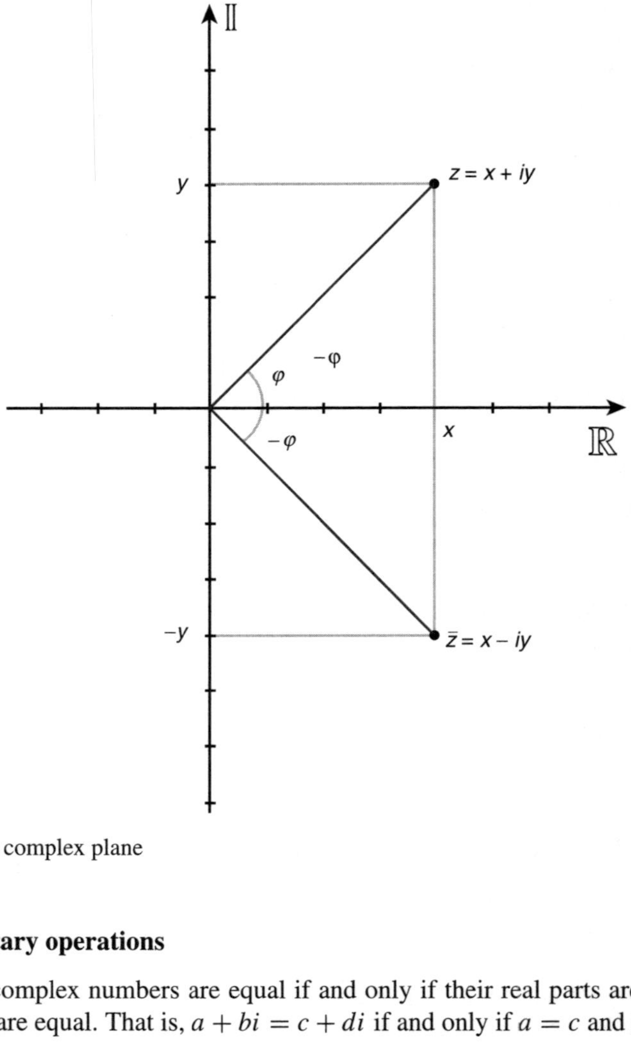

Figure B.2 The complex plane

B.1.4 Elementary operations

Equality. Two complex numbers are equal if and only if their real parts are equal and their imaginary parts are equal. That is, $a + bi = c + di$ if and only if $a = c$ and $b = d$.

Operations. Complex numbers are added, subtracted, multiplied, and divided by formally applying the associative, commutative and distributive laws of algebra, together with the definition $i^2 = -1$:

- *Addition*: $(a + bi) + (c + di) = (a + c) + (b + d)i$
- *Subtraction*: $(a + bi) - (c + di) = (a - c) + (b - d)i$
- *Multiplication*: $(a + bi)(c + di) = ac + bci + adi + bdi^2 = (ac - bd) + (bc + ad)i$
- *Division*: $(a + bi)/(c + di) = \left(\frac{ac+bd}{c^2+d^2}\right) + \left(\frac{bc-ad}{c^2+d^2}\right) i$

Absolute value. The absolute value (or modulus or magnitude) of a complex number z is defined as

$$|z| = \sqrt{a^2 + b^2}$$

1. $|z| = 0$ if and only if $z = 0$
2. $|z + w| \leq |z| + |w|$, (triangle inequality)
3. $|z \cdot w| = |z| \cdot |w|$

for all complex numbers z and w.

Complex Conjugate. The complex conjugate of the complex number $z = a + bi$ is defined to be $a - bi$, written as $\bar{z}$ or $z^\star$. As seen in the previous figure, $\bar{z}$ is the "reflection" of z about the real axis. The following can be checked:

- $\overline{z + w} = \bar{z} + \bar{w}$
- $\overline{z \cdot w} = \bar{z} \cdot \bar{w}$
- $\overline{(z/w)} = \bar{z}/\bar{w}$
- $\bar{\bar{z}} = z$
- $\bar{z} = z$ if and only if z is real
- $|z| = |\bar{z}|$
- $|z|^2 = z \cdot \bar{z}$
- $z^{-1} = \bar{z} \cdot |z|^{-2}$ if z is non-zero.

B.1.5 Polar form

Alternatively to the Cartesian representation $z = a + ib$, the complex number z can be specified by polar coordinates. The polar coordinates are: $r = |z| \geq 0$, called the **absolute value** or **modulus**; and $\phi = \arg(z)$, called the **argument** of z. For $r = 0$ any value of ϕ describes the same number.

To get a unique representation, a conventional choice is to set $\arg(0) = 0$. For $r > 0$ the argument ϕ is unique modulo 2π; that is, if any two values of the complex argument differ by an exact integer multiple of 2π, they are considered equivalent. To get a unique representation, a conventional choice is to limit ϕ to the interval $(-\pi, \pi]$, i.e. $-\pi < \phi \leq \pi$.

The representation of a complex number by its polar coordinates is called the polar form of the complex number.

Conversion from the polar form to the Cartesian form

$$x = r\cos\varphi \qquad y = r\sin\varphi$$

Conversion from the Cartesian form to the polar form

$$r = \sqrt{x^2 + y^2}$$

$$\varphi = \begin{cases} \arctan\left(\frac{y}{x}\right) & \text{if} \quad x > 0 \\ \arctan\left(\frac{y}{x}\right) + \pi \text{ or } -\pi & \text{if} \quad x < 0 \quad \text{and} \quad y \geq 0 \\ \arctan\left(\frac{y}{x}\right) + \pi \text{ or } -\pi & \text{if} \quad x < 0 \quad \text{and} \quad y < 0 \\ +\frac{\pi}{2} & \text{if} \quad x = 0 \quad \text{and} \quad y > 0 \\ -\frac{\pi}{2} & \text{if} \quad x = 0 \quad \text{and} \quad y < 0 \\ \text{undefined} & \text{if} \quad x = 0 \quad \text{and} \quad y = 0 \end{cases}$$

For the the second/third case you can add or subtract π depending on whether you want your answer in positive or negative radians (respectively), even though keeping your radians positive seems to be the convention. The previous formula requires rather laborious case differentiations. However, many programming languages provide a variant of the arctangent function which is often named atan2 and processes the cases internally.

> For example, in Python we have the following definition:
>
> atan2(*y*, *x*). *Return* atan(*y*/*x*), *in radians. The result is between* $-\pi$ *and* π. *The vector in the plane from the origin to point* (*x*, *y*) *makes this angle with the positive X axis. The point of* atan2() *is that the signs of both inputs are known to it, so it can compute the correct quadrant for the angle. For example*, atan(1) *and* atan2(1, 1) *are both* $\pi/4$, *but atan2*(−1, −1) *is* $-3 * \pi/4$.

From the previous equations it is easy to obtain the so-called trigonometric form of a complex number:

$$z = r(\cos\varphi + i\sin\varphi)$$

Using Euler's formula it can also be written as

$$z = r\,\mathrm{e}^{i\varphi}$$

Multiplication, division, exponentiation, and root extraction are much easier in the polar form than in the Cartesian form. Using sum and difference identities it's possible to obtain that

$$r_1\,\mathrm{e}^{i\varphi_1} \cdot r_2\,\mathrm{e}^{i\varphi_2} = r_1\,r_2\,\mathrm{e}^{i(\varphi_1+\varphi_2)}$$

$$\frac{r_1\,\mathrm{e}^{i\varphi_1}}{r_2\,\mathrm{e}^{i\varphi_2}} = \frac{r_1}{r_2}\,\mathrm{e}^{i(\varphi_1-\varphi_2)}$$

Exponentiation with integer exponents; according to De Moivre's formula,

$$\left(r\,\mathrm{e}^{i\varphi}\right)^n = r^n\,\mathrm{e}^{in\varphi}$$

All the roots of any number, real or complex, may be found with a simple algorithm. The nth roots are given by

$$\sqrt[n]{r\,\mathrm{e}^{i\varphi}} = \sqrt[n]{r}\,\mathrm{e}^{i\left(\frac{\varphi+2k\pi}{n}\right)}$$

for $k = 0, 1, 2, \ldots, n-1$, where $\sqrt[n]{r}$ represents the principal nth root of r.

The addition of two complex numbers is just the vector addition of two vectors, and multiplication by a fixed complex number can be seen as a simultaneous rotation and stretching. Multiplication by i corresponds to a counter-clockwise rotation by 90 degrees ($\pi/2$ radians). The geometric content of the equation $i^2 = -1$ is that a sequence of two 90-degree rotations results in a 180-degree (π radians) rotation. Even the fact $(-1)\,(-1) = +1$ from arithmetic can be understood geometrically as the combination of two 180-degree turns.

B.2 FUNCTIONS OF COMPLEX VARIABLES

B.2.1 Definitions

A complex function is a function in which the independent variable and the dependent variable are both complex numbers. More precisely, a complex function is a function whose domain Ω is a subset of the complex plane and whose range is also a subset of the complex plane. For any complex function, both the independent variable and the dependent variable may be separated into real and imaginary parts:

$$z = x + iy$$

and

$$w = f(z) = u(z) + iv(z)$$

where $x, y \in \mathbb{R}$, and $u(z)$, $v(z)$, are real-valued functions.

In other words, the components of the function $f(z)$,

$$u = u(x, y)$$

and

$$v = v(x, y)$$

can be interpreted as real-valued functions of the two real variables, x and y. However this class of functions is too general for our purposes. We are interested only in functions which are differentiable with respect to the complex variable z, a restriction which is much stronger than the condition that u and v be differentiable with respect to x and y. Therefore, one of our first tasks in the study of complex function theory will be to determine the necessary and sufficient conditions for a complex function to have a derivative with respect to the complex variable z.

B.2.2 Analytic functions

Single-valued functions (of a complex variable) which have derivatives throughout a region of the complex plane, are called *analytic functions*.

Just as in real analysis, a "smooth" complex function $w = f(z)$ may have a derivative at a particular point in its domain Ω. In fact, the definition of the derivative

$$f'(z) = \frac{dw}{dz} = \lim_{h \to 0} \frac{f(z+h) - f(z)}{h}$$

is analogous to the real case, with one very important difference. **In real analysis, the limit can only be approached by moving along the one-dimensional number line. In complex analysis, the limit can be approached from any direction in the two-dimensional complex plane**.

If this limit, the derivative, exists for every point z in Ω, then $f(z)$ is said to be differentiable on Ω.

This is a much more powerful result than the analogous theorem that can be proved for real-valued functions of real numbers. In the calculus of real numbers, we can construct a function $f(x)$ that has a first derivative everywhere, but for which the second derivative does not exist at one or more points in the function's domain. But in the complex plane, if a function $f(z)$ is differentiable in a neighbourhood it must also be infinitely differentiable in that neighbourhood. The theory of analytic functions contains a number of amazing theorems, and they all result from this stringent initial requirement that the function possesses "isotropic" derivatives.

Example B.2.1 *Verify that the function*

$$w = z^2 = (x + iy)^2 = x^2 - y^2 + 2ixy$$

is analytic everywhere in the complex plane

Let us write the derivative at z_0 in the form

$$f'(z_0) = \lim_{\Delta z \to 0} \frac{f(z_0 + \Delta z) - f(z_0)}{\Delta z}$$

For $f(z) = z^2$ we have

$$f'(z_0) = \lim_{\Delta z \to 0} \frac{(z_0 + \Delta z)^2 - z_0^2}{\Delta z} = \lim_{\Delta z \to 0} (2z_0 + \Delta z) = 2z_0$$

a result which is clearly independent of the path along with $\Delta z \to 0$, so $f(z) = z^2$ is differentiable and analytic everywhere.

Example B.2.2 *Verify if the function*

$$f(z) = z^\star = x - iy$$

is analytic in some region of the complex plane.

Using the same definition as before, we can write

$$f'(z_0) = \lim_{\Delta z \to 0} \frac{z_0^\star + \Delta z^\star - z_0^\star}{\Delta z} = \lim_{\Delta z \to 0} \frac{\Delta z^\star}{\Delta z}$$

Now if $\Delta z \to 0$ along the real axis, then $\Delta z = \Delta x$ and $\Delta z^\star = \Delta x^\star = \Delta x$, so $f'(z_0) = +1$. However, if Δz approaches zero along the imaginary y-axis, then $\Delta z = i\Delta y$, so $\Delta z^\star = -i\Delta y = -\Delta z$, so $f'(z_0) = -1$. Since at any point z_0 the limit as $z \to z_0$ depends on the direction of approach, the function is not differentiable or analytic anywhere.

B.2.3 Cauchy–Riemann conditions

We now determine the necessary and sufficient conditions for a function of complex variables to be differentiable at a point.

First we assume that

$$f(z) = u(z) + iv(z)$$

is differentiable for some point z_0, so

$$\lim_{h \to 0} \frac{f(z_0 + h) - f(z_0)}{h} = f'(z_0)$$

If this limit exists, then it may be computed by taking the limit as $h \to 0$ along the real axis or imaginary axis; in either case it should give the same result. Approaching along the real axis, one finds

$$\lim_{h \to 0} \frac{f(z_0 + h) - f(z_0)}{h} = \frac{\partial f}{\partial x}(z_0) = \frac{\partial u}{\partial x}(z_0) + i\frac{\partial v}{\partial x}(z_0)$$

On the other hand, approaching along the imaginary axis,

$$\lim_{h\to 0}\frac{f(z_0+ih)-f(z_0)}{ih}=\lim_{h\to 0}\left[-i\frac{f(z_0+ih)-f(z_0)}{h}\right]=-i\frac{\partial f}{\partial y}(z_0)=\frac{\partial v}{\partial y}(z_0)-i\frac{\partial u}{\partial y}(z_0)$$

But by assumption of differentiability, these two limits must be equal. Therefore equating real and imaginary parts, we have

$$\frac{\partial u}{\partial x}=\frac{\partial v}{\partial y} \tag{B.4}$$

$$\frac{\partial v}{\partial x}=-\frac{\partial u}{\partial y} \tag{B.5}$$

Equations (B.4) and (B.5) are known as the Cauchy–Riemann equations. They give a *necessary* condition for differentiability, the sufficient conditions for the differentiability of $f(z)$ at z_0 are, first, that the Cauchy–Riemann equations hold there and, second, that the first partial derivatives of $u(x, y)$ and $v(x, y)$ exist and are continuous at z_0. The reader is referred to the literature for the proof.

By differentiating this system of two partial differential equations, first with respect to x, and then with respect to y, we can easily show that

$$\frac{\partial^2 u}{\partial x^2}+\frac{\partial^2 u}{\partial y^2}=0$$

$$\frac{\partial^2 v}{\partial x^2}+\frac{\partial^2 v}{\partial y^2}=0$$

or, in another common notation,

$$u_{xx}+u_{yy}=v_{xx}+v_{yy}=0$$

In other words, the real and imaginary parts of a differentiable function of a complex variable are harmonic functions because they satisfy Laplace's equation.

Example B.2.3 *Consider the function z^3. We have*

$$z^3=(x^3-3xy^2)+i(3x^2y-y^3)=u+iv$$

So

$$\frac{\partial u}{\partial x}=3x^2-3y^2=\frac{\partial v}{\partial y}$$

$$\frac{\partial v}{\partial x}=6xy=-\frac{\partial u}{\partial y}$$

Thus the Cauchy–Riemann conditions hold everywhere. Since the partial derivatives are continuous, the function z^3 is in fact analytic everywhere.

A function which is analytic in the entire complex plane is said to be an *entire* function.

B.2.4 Multi-valued functions

Up to this point we have implicitly assumed a property for a generic function of a complex number, that is, if we pick any point z_0 in the complex plane and follow any path from z_0

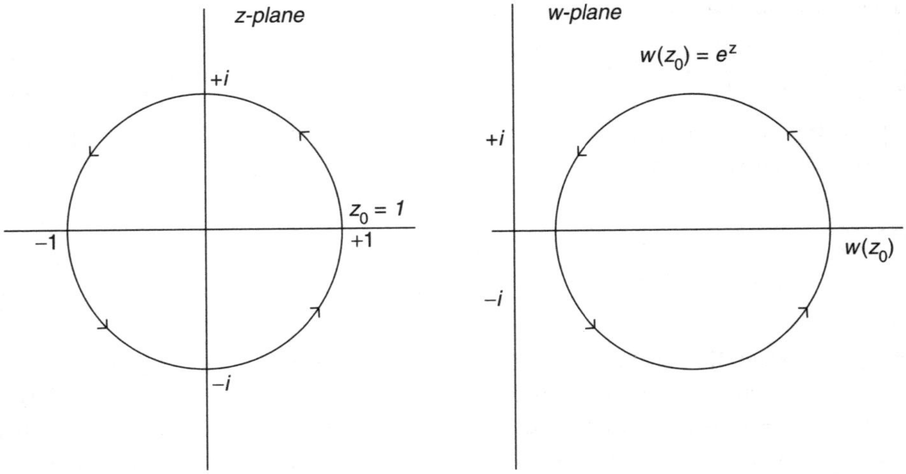

Figure B.3 A circular contour in the z-plane about the origin and its mapping by the function $w(z) = e^z$

through the plane back to z_0, then the value of the function changes continuously along the path, returning to its original value at z_0.

For example, suppose that we consider the function $f(z) = e^z$ and start at the point $z_0 = 1$, encircling the origin in the z-plane counter-clockwise along the unit circle. Figure B.3 shows the circular path in the z-plane and the corresponding path in the f-plane. We note that both paths are closed, which is just the geometrical statement of the fact that if we start at a point z_0 where the function has the value $f(z_0)$, then, when we move along a closed curve back to z_0, the functional values also follow a smooth path back to $f(z_0)$.

However, if we look at another simple function, that is, the square root, we will see that things do not go so smoothly. Let us write:

$$f(z) = \sqrt{x + iy}$$

As we have previously seen, we can rewrite this function in polar form as

$$f(z) = \sqrt{z} = \sqrt{r}e^{i\theta/2} = \sqrt{r}[\cos(\theta/2) + i\sin(\theta/2)]$$

Using this definition, let us vary z along the same path chosen in Figure B.3, starting at $r = 1, \theta = 0$. After making a complete circle around the origin in the z-plane we arrive at the point $w = -1$ in the f-plane and not at $w = +1$. In fact we have

$$f(r = 1, \theta = 2\pi) = \sqrt{1}[\cos(\pi) + i\sin(\pi)] = -1$$

In order to get back to $w = +1$, we must let θ go from 2π to 4π; that is, make the circular trip in the z-plane one more time.

Actually this is not the best way to describe the situation; we do not want to think of tracing the circular path in the original z-plane a second time, but rather of tracing an identical circular path in a *different* z-plane; this corresponds to the fact that, in the first circuit, θ went from 0 to 2π whereas, in the second circuit, it went from 2π to 4π.

In the case of $\sqrt{z}$ we need two planes, usually referred to as *Riemann sheets*, to characterize the values of $f(z)$ in a single-valued manner.

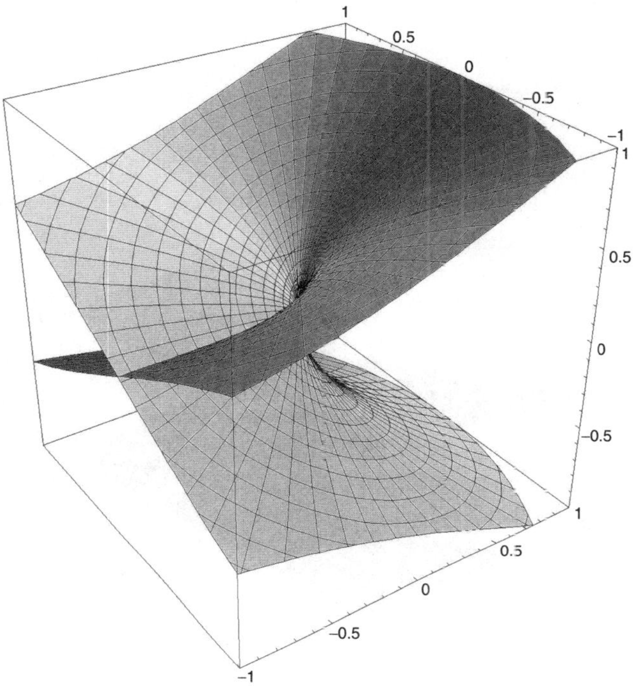

Figure B.4 The Riemann surface for the function $\sqrt{z}$ (from Wikipedia "Complex Square Root" entry). Reproduced with permission of Jan Homann

It is important to note that the path in the z-plane of Figure B.3 encloses the origin. If we choose a closed path which neither encloses the origin nor intersects the positive real axis, then we also obtain a closed path in the f-plane.

It is readily seen that the difficulties described above for $f(z) = \sqrt{z}$ will persist for any path beginning on the positive real axis and returning to the original point along a path enclosing the origin. Thus, if we wish to consider $f(z) = \sqrt{z}$ in the simple fashion that we used for e^z then we conclude that $f(z) = \sqrt{z}$ is not continuous along the positive real axis and is not analytic there. However, to avoid this conclusion we may say that when we come back to the real axis after a circuit of 2π radians, we transfer continuously onto the second Riemann sheet. If we go around $z = 0$ once more on the second sheet, when we return towards the positive real axis we transfer continuously back to the first Riemann sheet. Thus the two sheets can be imagined to be *cut* along the positive real axis and joined in the manner illustrated in Figure B.4. With this convention, the function $f(z) = \sqrt{z}$ is seen to be single-valued everywhere and analytic everywhere except at the origin. Thus the origin is a singular point for $f(z) = \sqrt{z}$. In general, suppose that we have a singular point z_0 of some function $f(z)$ and a path starting at z_1 which encircles z_0. If we must sweep through an angle greater than 2π in order to return to the original value at z_1, then z_0 is called a *branch point* of $f(z)$ and the cut that emanates from this point is called a *branch cut*.

It should be noticed that the choice of the real axis as the branch cut for $f(z) = \sqrt{z}$ was entirely arbitrary. Any other ray, say $\theta = \theta_0$, will serve equally well, the only thing that is not arbitrary is the choice of $z = 0$ as a branch point.

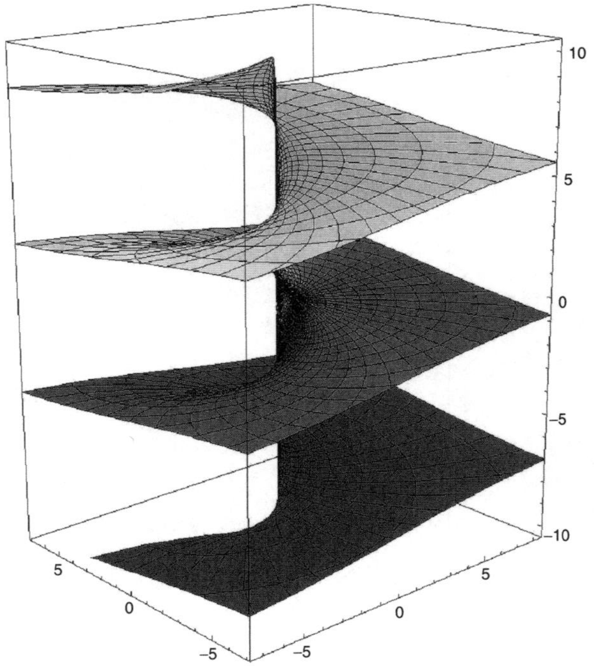

Figure B.5 The Riemann surface for the function $\ln z$ (from Wikipedia "Complex Logarithm Root" entry). Reproduced with permission of Jan Homann

As another example of a multi-valued function, we consider the logarithm (Figure B.5). Again using $z = re^{i\theta}$ we define

$$\log(z) = \ln(r) + i\theta \tag{B.6}$$

With the logarithm, the multi-valuedness difficulties described above are all the more striking since no matter how many times one encircles the origin starting, say, at some point on the positive real axis, one would *never* return to the original value of the logarithm. The logarithm increases by $2\pi i$ on each circuit, thus an infinite number of Riemann sheets, each one joined to the one below it by means of a cut along the positive real axis, is necessary to turn $\log(z)$ into a single-valued function. When this is done $\log(z)$ is analytic everywhere except at $z = 0$ where we assign the value $-\infty$ on all sheets.

C
Complex Integration

C.1 DEFINITIONS

Let t be a real parameter ranging from t_A to t_B, and let $z = z(t)$ be a curve, or *contour* C in the complex plane with endpoints $A = z(t_A)$, $B = z(t_B)$. Now we mark off a number of points t_i between t_A and t_B and approximate the curve by a series of straight lines drawn from each $z(t_i)$ to $z(t_{i+1})$.

To define the integral of a function f of a complex variable, we form the quantity

$$\lim_{|\Delta z_i|\to 0} = \sum_{i=0}^{n} f(z_i)\Delta z_i \equiv \int_C f(z)\,\mathrm{d}z$$

where $\Delta z_i = z(t_{i+1}) - z(t_i)$ and $f(z_i)$ is the function evaluated at a point z_i on C between $z(t_{i+1})$ and $z(t_i)$. The sum is evaluated in the limit of an arbitrarily fine partition of the range through which the real parameter t moves while generating the contour from A to B: that is, as $n \to \infty$, or, what is the same thing, in the limit of arbitrarily small $|z_i|$ for all i.

Writing $f(z) = u(x, y) + iv(x, y)$ and $\mathrm{d}z = \mathrm{d}x + i\,\mathrm{d}y$ we have

$$\int_C f(z)\,\mathrm{d}z = \int_C (u\,\mathrm{d}x - v\,\mathrm{d}y) + i\int_C (u\,\mathrm{d}y + v\,\mathrm{d}x)$$

We can also write this in parametric form. If

$$\mathrm{d}x = x'(t)\,\mathrm{d}t, \quad \mathrm{d}y = y'(t)\,\mathrm{d}t$$

we have

$$\int_C f(z)\,\mathrm{d}z = \int_{t_A}^{t_B} \left(u\frac{\mathrm{d}x}{\mathrm{d}t} - v\frac{\mathrm{d}y}{\mathrm{d}t}\right)\mathrm{d}t + i\int_{t_A}^{t_B} \left(u\frac{\mathrm{d}y}{\mathrm{d}t} + v\frac{\mathrm{d}x}{\mathrm{d}t}\right)\mathrm{d}t$$

For a given contour C running from A to B, we define the opposite contour, written as $-C$ to be the same curve but traversed from B to A. The integral of $f(z)$ along $-C$ is clearly given by the above equation but with t_A and t_B interchanged. Thus

$$\int_C = -\int_{-C}$$

It also follows that

$$\int_{C_1} + \int_{C_2} = \int_{C_1+C_2}$$

If C is a closed curve that does not intersect itself, we shall always interpret $\oint$ to mean the integral taken counter-clockwise along the closed contour C.

Example C.1.1 *Let us integrate the function $f(z) = z^\dagger$ counter-clockwise around the unit circle centred at the origin. The values of z on this curve are given by $z = \mathrm{e}^{i\theta}$ $\theta = 0, 2\pi$.*

Therefore

$$I = \oint z^\dagger \, \mathrm{d}z = \int_0^{2\pi} \mathrm{e}^{-i\theta} \, \mathrm{e}^{i\theta} \, \mathrm{d}\theta = 2\pi i$$

Example C.1.2 *Consider the function $f(z) = 1/z$, and let the contour C be the unit circle about* 0*, which can be parameterized by* e^{it}*, with t in* $[0, 2\pi)$*. Substituting, we find*

$$\oint_C f(z) \, \mathrm{d}z = \int_0^{2\pi} \frac{1}{\mathrm{e}^{it}} i \, \mathrm{e}^{it} \, \mathrm{d}t = i \int_0^{2\pi} \mathrm{e}^{-it} \, \mathrm{e}^{it} \, \mathrm{d}t = i \int_0^{2\pi} \mathrm{d}t = i(2\pi - 0) = 2\pi i$$

No integral around the closed contour is zero. The reason, as we shall see, is that $z^\dagger$ is not analytic anywhere and therefore not within C, and z^{-1} is not analytic at $z = 0$ which is within C. Both these examples are explained by the Cauchy–Goursat theorem.

C.2 THE CAUCHY–GOURSAT THEOREM

Definition C.2.1 *An open subset U of $\mathbb{C}$ is said to be* simply connected *if U has no "holes"; for instance, every open disk $U = z : |z - z_0| < r$ qualifies.*

The theorem is usually formulated for closed paths as follows:

Theorem C.2.1 (*Chauchy*) *Let U be an open subset of $\mathbb{C}$ which is simply connected, let $f : U \to \mathbb{C}$ be an analytic function with $f'(z)$ continuous throughout this region, and let C be a contour in U whose start point is equal to its end point. Then,*

$$\oint_C f(z) \, \mathrm{d}z = 0$$

Proof. Let us consider the following identity

$$\oint_C f(z) \, \mathrm{d}z = \oint_C (u \, \mathrm{d}x - v \, \mathrm{d}y) + i \oint_C (u \, \mathrm{d}y + v \, \mathrm{d}x)$$

to evaluate the two line integrals on the right, we use Green's theorem for line integrals. It states that if the derivatives of P and Q are continuous functions within and on a closed contour C, then

$$\oint_C (P \, \mathrm{d}x + Q \, \mathrm{d}y) = \int \int_S \left(\frac{\partial Q}{\partial x} - \frac{\partial P}{\partial y} \right) \mathrm{d}x \, \mathrm{d}y$$

where S is the surface bounded by C. By hypothesis $f'(z)$ is continuous, so the first partial derivatives of u and v are also continuous; then Green's theorem yields

$$\oint_C (u \, \mathrm{d}x - v \, \mathrm{d}y) + i \oint_C (u \, \mathrm{d}y + v \, \mathrm{d}x) \\ = \int \int_S \left(\frac{\partial u}{\partial y} + \frac{\partial v}{\partial x} \right) \mathrm{d}x \, \mathrm{d}y + i \int \int_S \left(\frac{\partial u}{\partial x} - \frac{\partial v}{\partial y} \right) \mathrm{d}x \, \mathrm{d}y$$

But since the Cauchy–Riemann equations hold, the integrands above all vanish, therefore

$$\oint_C f(z)\,\mathrm{d}z = 0 \qquad \text{(QED)}$$

The condition that U be simply connected is crucial; consider

$$C(t) = \mathrm{e}^{it} \quad t \in [0, 2\pi]$$

which traces out the unit circle and then the contour integral

$$\oint_C \frac{1}{z}\,\mathrm{d}z$$

As we have seen in the previous example, its contour integral is non-zero: the Cauchy integral theorem does not apply here since $f(z) = 1/z$ is not defined (and certainly not analytic) at $z = 0$.

One important consequence of the theorem is that contour integrals of analytic functions on simply connected domains can be computed in a manner familiar from the fundamental theorem of real calculus: let U be a simply connected open subset of $\mathbb{C}$, let $f : U \to \mathbb{C}$ be a holomorphic function, and let C be a piecewise continuously differentiable contour in U with start point A and end point B, then

$$\int_C f(z)\,\mathrm{d}z = F(b) - F(a)$$

As was shown by Goursat, Cauchy's integral theorem can be proved assuming only that the complex derivative $f'(z)$ exists everywhere in U without requiring continuity. This is because any function which is analytic in a region *necessarily* has a continuous derivative. In fact an analytic function has derivatives of all orders and therefore all its derivatives are continuous, the continuity of the nth derivative being a consequence of the existence of the derivative of order $n + 1$. But it is possible to establish this result on higher derivatives *only after* one shows that the continuity of $f'(z)$ is not needed in the proof of Cauchy's theorem. The relaxation of this hypotheses is therefore of utmost importance, and it is Goursat's result that really distinguishes the theory of integration of a function of complex variable from the theory of line integrals in the real plane.

Theorem C.2.2 (*Chauchy–Goursat*) *Let U be an open subset of $\mathbb{C}$ which is simply connected, let $f : U \to \mathbb{C}$ be an analytic function and let C be a contour in U whose start point is equal to its end point. Then,*

$$\oint_C f(z)\,\mathrm{d}z = 0$$

The proof of the theorem is more involved than the previous one and we refer the interested reader to the literature.

C.3 CONSEQUENCES OF CAUCHY'S THEOREM

The Cauchy integral theorem leads to the Cauchy integral formula and the residue theorem.

Theorem C.3.1 *Suppose U is an open subset of the complex plane $\mathbb{C}$, and as usual $f : U \to \mathbb{C}$ is an analytic function, and the disk $D = \{z : |z - z_0| < r\}$ is completely contained*

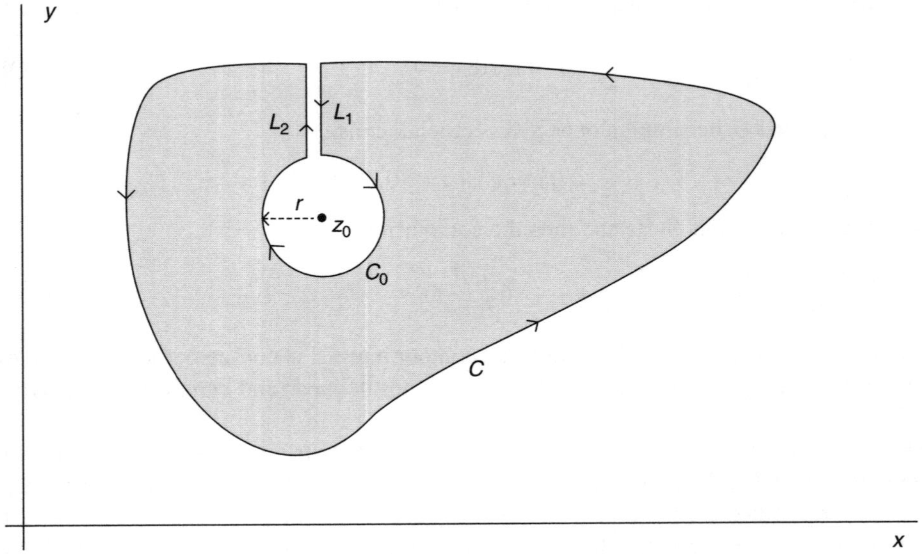

Figure C.1 Chauchy integral theorem

in U. Let C be the circle forming the boundary of D. Then for every 'a' in the interior of D we have:

$$f(a) = \frac{1}{2\pi i} \oint_C \frac{f(z)}{z-a}\,\mathrm{d}z$$

where the contour integral is to be taken counter-clockwise.

The proof of this statement uses the Cauchy integral theorem and, just like that theorem, only needs f to be complex differentiable. It is worth following the proof in order to become acquainted with complex integral calculus.

Proof. Let us consider Figure C.1: inside the contour C we draw a circle C_0 of radius r about z_0 and consider the contour formed by the circle C_0, the line C and the two straight line segments L_1 and L_2, which lie arbitrarily close to each other. Let us call this entire contour C'. Now consider

$$\oint_{C'} \frac{f(z)}{z-z_0}\,\mathrm{d}z = \oint_C \frac{f(z)}{z-z_0}\,\mathrm{d}z + \int_{L_1} \frac{f(z)}{z-z_0}\,\mathrm{d}z + \oint_{C_0} \frac{f(z)}{z-z_0}\,\mathrm{d}z + \int_{L_2} \frac{f(z)}{z-z_0}\,\mathrm{d}z$$

Inside C', $\frac{f(z)}{z-z_0}$ is analytic, so by the Cauchy–Goursat theorem

$$\oint_{C'} \frac{f(z)}{z-z_0}\,\mathrm{d}z = 0$$

Now, as we bring the line segments L_1 and L_2 arbitrarily close together,

$$\int_{L_1} \frac{f(z)}{z-z_0}\,\mathrm{d}z \to -\int_{L_2} \frac{f(z)}{z-z_0}\,\mathrm{d}z$$

since the lines are traversed in opposite directions. Thus, in this limit we have

$$\oint_{C'} \frac{f(z)}{z - z_0}\,\mathrm{d}z = 0 = \oint_C \frac{f(z)}{z - z_0}\,\mathrm{d}z + \oint_{C_0} \frac{f(z)}{z - z_0}\,\mathrm{d}z$$

so that

$$\oint_C \frac{f(z)}{z - z_0}\,\mathrm{d}z = -\oint_{C_0} \frac{f(z)}{z - z_0}\,\mathrm{d}z$$

At this point we note that C_0 is traversed in a clockwise direction, since it is considered as a contour in its own right i.e. not just as a part of C'. Let us therefore define $C_0' = -C_0$ so that C_0' is a counter-clockwise contour, then we may write

$$\oint_C \frac{f(z)}{z - z_0}\,\mathrm{d}z = \oint_{C_0'} \frac{f(z)}{z - z_0}\,\mathrm{d}z = f(z_0)\oint_{C_0'} \frac{1}{z - z_0}\,\mathrm{d}z + \oint_{C_0'} \frac{f(z) - f(z_0)}{z - z_0}\,\mathrm{d}z$$

We now use the fact that C_0' is a circle to write $z - z_0 = r\,\mathrm{e}^{i\theta}$ on C_0', thus the first integral on the right becomes

$$\oint_{C_0'} \frac{1}{z - z_0}\,\mathrm{d}z = \int_0^{2\pi} \frac{ir\,\mathrm{e}^{i\theta}}{r\,\mathrm{e}^{i\theta}}\,\mathrm{d}\theta = 2\pi i$$

for all $r > 0$ within C. A Cauchy formula will therefore be established if we can show that

$$\oint_{C_0'} \frac{f(z) - f(z_0)}{z - z_0}\,\mathrm{d}z = 0$$

for some choice of the contour C_0'. The continuity of $f(z)$ at z_0 tells us that, for all $\epsilon > 0$, there exists a δ such that if $|z - z_0| \le \delta$, then $|f(z) - f(z_0)| \le \epsilon$. So, by taking $r = \delta$, we satisfy the condition $|z - z_0| \le \delta$ which in turn implies that

$$\left|\oint_{C_0'} \frac{f(z) - f(z_0)}{z - z_0}\,\mathrm{d}z\right| \le \oint_{C_0'} \frac{|f(z) - f(z_0)|}{|z - z_0|}\,\mathrm{d}z < \frac{\epsilon}{\delta}(2\pi\delta) = 2\pi\epsilon$$

Thus by taking r small enough but still greater than zero, the absolute value of the integral can be made smaller than any pre-assigned number, implying that:

$$\oint_C \frac{f(z)}{z - z_0}\,\mathrm{d}z = 2\pi i f(z_0)$$

This result means, among the other things, that if a function is analytic within and on a contour C, its value at every point inside C is determined by its values on the bounding curve C.

One may replace the circle C with any closed rectifiable curve in U which doesn't have any self-intersections and which is oriented counter-clockwise. The formulas remain valid for any point z_0 from the region enclosed by this path.

One can then deduce from the formula that f must actually be infinitely often continuously differentiable, with

$$f^{(n)}(z_0) = \frac{n!}{2\pi i}\oint_C \frac{f(z)}{(z - z_0)^{n+1}}\,\mathrm{d}z$$

Some call this identity Cauchy's differentiation formula. A proof of this last identity is a by-product of the proof that holomorphic functions are analytic.

An important consequence of the Cauchy's integral formula is the following:

Theorem C.3.2 (*Liouville's theorem*) *If $f(z)$ is entire and $|f(z)|$ is bounded for all values of z, then $f(z)$ is a constant.*

Proof. From Cauchy's integral formula, taking the derivative of both members, we have that

$$f'(z_0) = \frac{1}{2\pi i} \oint_C \frac{f(z)}{(z-z_0)^2} \mathrm{d}z$$

if we take C to be the circle $|z - z_0| = r_0$, then

$$|f'(z_0)| \leq \left| \frac{1}{2\pi i} \right| \oint_{C_0} \frac{|f(z)|}{|(z-z_0)^2|} |\mathrm{d}z| < \frac{1}{2\pi r_0^2} M 2\pi r_0 = \frac{M}{r_0}$$

where $|f(z)| < M$ within and on C_0. Therefore $|f'(z_0)| < M/r_0$, and we may take r_0 as large as we like because $f(z)$ is entire. So taking r_0 large enough, we can make $|f'(z_0)| < \epsilon$ for any pre-assigned ϵ. That is $|f'(z_0)| = 0$, which implies that $f'(z_0) = 0$ for all z_0 so $f(z_0) = \text{constant}$.

In particular, from Liouville's theorem we can conclude that if we have a function $f(z)$ that is analytic in the entire complex plane and is such that $|f(z)| \to 0$ as $|z| \to \infty$ in the entire complex plane, then this function is identically zero in the entire plane.

C.4 PRINCIPAL VALUE

Let us begin by considering a function $f(z)$ that is analytic in the upper half of the complex plane and is such that $|f(z)| \to 0$ as $|z| \to \infty$ in the upper half plane. Now consider the contour integral

$$\oint_C \frac{f(z)}{z-\alpha} \mathrm{d}z$$

where C is the contour shown in Figure C.2 and α is real. By assumption, $f(z)$ is analytic within and on C; so is $1/(z-\alpha)$. Thus

$$\oint_C \frac{f(z)}{z-\alpha} \mathrm{d}z = 0$$

Let us break this integral as follows:

$$\oint_C \frac{f(z)}{z-\alpha} \mathrm{d}z = \int_{-R}^{\alpha-\delta} \frac{f(x)}{x-\alpha} \mathrm{d}x + \int_{S_\delta} \frac{f(z)}{z-\alpha} \mathrm{d}z + \int_{\alpha+\delta}^{+R} \frac{f(x)}{x-\alpha} \mathrm{d}x + \int_{S_R} \frac{f(z)}{z-\alpha} \mathrm{d}z = 0$$

Here δ is the radius of the small semicircle S_δ centred at $x = \alpha$ and R is the radius of the large semicircle S_R centred at the origin, as shown in Figure C.2. The radius δ can be chosen as small as we please, and R can be chosen as large as we like. In the limit of arbitrarily small δ, the quantity

$$\int_{-R}^{\alpha-\delta} \frac{f(x)}{x-\alpha} \mathrm{d}x + \int_{\alpha+\delta}^{+R} \frac{f(x)}{x-\alpha} \mathrm{d}x$$

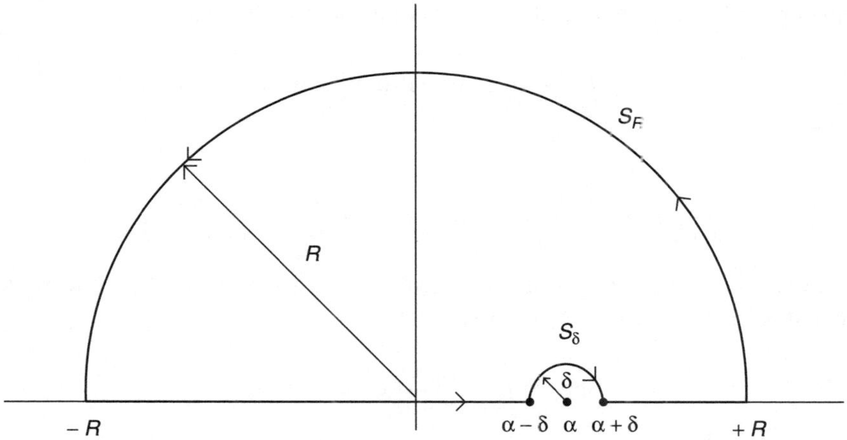

Figure C.2 The contour, C, used to obtain equation (C.1) . The radius, R, of the semicircle, S_R, may be made as large as necessary and the radius, δ, of the semicircle, S_δ, may be made as small as we please

is called the *principal-value integral* of $f(x)/(x-\alpha)$ and is denoted by

$$P\int_{-R}^{+R}\frac{f(x)}{x-\alpha}\,\mathrm{d}x$$

Now along the large semicircle S_R we set $z = R\,\mathrm{e}^{i\theta}$, so that

$$\int_{S_R}\frac{f(z)}{z-\alpha}\,\mathrm{d}z = i\int_0^{\pi}\frac{f(R\,\mathrm{e}^{i\theta})}{R\,\mathrm{e}^{i\theta}-\alpha}\,R\,\mathrm{e}^{i\theta}\,\mathrm{d}\theta$$

But

$$|R\,\mathrm{e}^{i\theta}-\alpha| = [R^2+\alpha^2-2R\alpha\cos\theta]^{1/2} \geq [R^2+\alpha^2-2R\alpha]^{1/2} = |R-\alpha|$$

so we can write

$$\left|\int_{S_R}\frac{f(z)}{z-\alpha}\,\mathrm{d}z\right| \leq \frac{R}{|R-\alpha|}\int_0^{\pi}|f(R\,\mathrm{e}^{i\theta})|\,\mathrm{d}\theta$$

But as $R\to\infty$ $|f(z)|\to 0$ and $R/(R-\alpha)\to 1$. Therefore the integral over the semicircle of radius R can be made arbitrarily small by choosing R sufficiently large. Thus we may write:

$$\lim_{R\to\infty} P\int_{-R}^{+R}\frac{f(x)}{x-\alpha}\,\mathrm{d}x = -\int_{S_\delta}\frac{f(z)}{z-\alpha}\,\mathrm{d}z = -f(\alpha)\int_{S_\delta}\frac{1}{z-\alpha}\,\mathrm{d}z - \int_{S_\delta}\frac{f(z)-f(\alpha)}{z-\alpha}\,\mathrm{d}z$$

where we have added and subtracted the term

$$\int_{S_\delta}\frac{f(\alpha)}{z-\alpha}\,\mathrm{d}z$$

Setting

$$z-\alpha = \delta\,\mathrm{e}^{i\theta}$$

in the first integral on the right-hand side of this equation, we find that

$$-f(\alpha)\int_{S_\delta}\frac{1}{z-\alpha}\,\mathrm{d}z = -\mathrm{i}f(\alpha)\int_{\pi}^{0}\mathrm{d}\theta = i\pi f(\alpha)$$

Thus

$$\lim_{R\to\infty} P\int_{-R}^{+R}\frac{f(x)}{x-\alpha}\,\mathrm{d}x = i\pi f(\alpha) - \int_{S_\delta}\frac{f(z)-f(\alpha)}{z-\alpha}\,\mathrm{d}z$$

Since $f(z)$ is continuous at $z=\alpha$, the argument used in deriving Cauchy's integral formula tell us that this last integral over S_δ vanishes. Hence

$$\lim_{R\to\infty} P\int_{-R}^{+R}\frac{f(x)}{x-\alpha}\,\mathrm{d}x = i\pi f(\alpha)$$

For the sake of brevity we write this simply as

$$P\int_{-R}^{+R}\frac{f(x)}{x-\alpha}\,\mathrm{d}x = i\pi f(\alpha) \tag{C.1}$$

where $f(x)$ is a complex-valued function of a real variable.

The principal-value integral can be seen as a way to avoid singularities on a path of integration: one integrates to within δ of the singularity in question, skips over the singularity and begins integrating again a distance δ beyond the singularity.

This prescription is also very useful in the one-dimensional real analysis where it enables one to make sense of such integrals as:

$$\int_{-R}^{+R}\frac{\mathrm{d}x}{x}$$

One would like this integral to be zero, since we are integrating an odd function over a symmetric domain. However, unless we insert a P in front of this integral, the singularity at the origin makes the integral meaningless. Following the prescription for principal-value integrals we can easily evaluate the above integral, we have

$$P\int_{-R}^{+R}\frac{\mathrm{d}x}{x} = \lim_{\delta\to 0}\left[\int_{-R}^{-\delta}\frac{\mathrm{d}x}{x} + \int_{\delta}^{+R}\frac{\mathrm{d}x}{x}\right]$$

In the first integral on the right-hand side, set $x=-y$. Then

$$P\int_{-R}^{+R}\frac{\mathrm{d}x}{x} = \lim_{\delta\to 0}\left[\int_{R}^{\delta}\frac{\mathrm{d}y}{y} + \int_{\delta}^{+R}\frac{\mathrm{d}x}{x}\right]$$

The sum of the two integrals inside the bracket is obviously zero since

$$\int_a^b = -\int_b^a$$

thus

$$P\int_{-R}^{+R}\frac{\mathrm{d}x}{x} = 0$$

Example C.4.1 *Let us evaluate the following integral*

$$P \int_{-R}^{+R} \frac{\mathrm{d}x}{x-a}$$

where $-R < a < R$.

Answer: *First of all we write the integral in the form*

$$P \int_{-R}^{+R} \frac{\mathrm{d}x}{x-a} = \lim_{\delta \to 0} \left[\int_{-R}^{a-\delta} \frac{\mathrm{d}x}{x-a} + \int_{a+\delta}^{+R} \frac{\mathrm{d}x}{x-a} \right]$$

Setting $x = -y$ *in the first integral on the right-hand side, we find that*

$$\begin{aligned} P \int_{-R}^{+R} \frac{\mathrm{d}x}{x-a} &= \lim_{\delta \to 0} \left[\int_{R}^{\delta} \frac{\mathrm{d}y}{y+a} + \ln(R-a) - \ln \delta \right] \\ &= \lim_{\delta \to 0} [\ln \delta - \ln(R+a) + \ln(R-a) - \ln \delta] \end{aligned}$$

thus

$$P \int_{-R}^{+R} \frac{\mathrm{d}x}{x-a} = \ln \left(\frac{R-a}{R+a} \right), \quad -R < a < R$$

C.5 LAURENT SERIES

We now come to one of the most important applications of the Cauchy–Goursat theorem, namely the possibility of expanding an analytic function in a power series. The main result may be stated as follows:

Theorem C.5.1 *If* $f(z)$ *is analytic throughout the annular region between and on the concentric circles* C_1 *and* C_2 *centred at* $z = a$ *and of radii* r_1 *and* $r_2 < r_1$ *respectively, then there exists a unique series expansion in terms of positive and negative powers of* $(z-a)$,

$$f(z) = \sum_{k=0}^{\infty} a_k (z-a)^k + \sum_{k=1}^{\infty} b_k (z-a)^{-k}$$

where

$$\begin{aligned} a_k &= \frac{1}{2\pi i} \oint_{C} \frac{f(z)}{(z-a)^{k+1}} \mathrm{d}z \\ b_k &= \frac{1}{2\pi i} \oint_{C_2} (z-a)^{k-1} f(z) \, \mathrm{d}z \end{aligned}$$

Proof. Let there be two circular contours C_2 and C_1, with the radius of C_1 larger than that of C_2. Let z_0 be at the centre of C_1 and C_2, and z be between C_1 and C_2. Now create a cut line C_c between C_1 and C_2, and integrate around the path $C = C_1 + C_c - C_2 - C_c$, so that the plus and minus contributions of C_c cancel one another, as illustrated in Figure C.3. Since

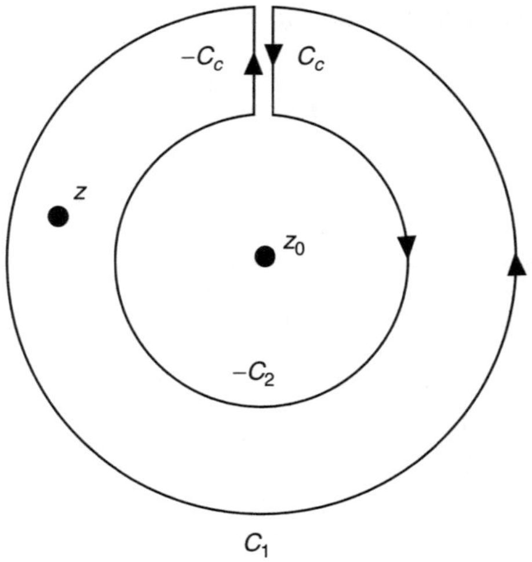

Figure C.3 Complex integral contour used for the proof of unicity of Laurent Series

$f(z)$ is analytic within and on C, from the Cauchy integral formula,

$$\begin{aligned}
f(z) &= \frac{1}{2\pi i}\int_C \frac{f(z')}{z'-z}\,dz' = \frac{1}{2\pi i}\int_{C_1} \frac{f(z')}{z'-z}\,dz' \\
&+ \frac{1}{2\pi i}\int_{C_c} \frac{f(z')}{z'-z}\,dz' - \frac{1}{2\pi i}\int_{C_2} \frac{f(z')}{z'-z}\,dz' - \frac{1}{2\pi i}\int_{C_c} \frac{f(z')}{z'-z}\,dz' \\
&= \frac{1}{2\pi i}\int_{C_1} \frac{f(z')}{z'-z}\,dz' - \frac{1}{2\pi i}\int_{C_2} \frac{f(z')}{z'-z}\,dz' \qquad \text{(C.2)}
\end{aligned}$$

since contributions from the cut line in opposite directions cancel out. Now

$$\begin{aligned}
f(z) &= \frac{1}{2\pi i}\int_{C_1} \frac{f(z')}{(z'-z_0)-(z-z_0)}\,dz' \\
&- \frac{1}{2\pi i}\int_{C_2} \frac{f(z')}{(z'-z_0)-(z-z_0)}\,dz' \\
&= \frac{1}{2\pi i}\int_{C_1} \frac{f(z')}{z'-z_0}\left[1-\frac{z-z_0}{z'-z_0}\right]^{-1}dz' \\
&- \frac{1}{2\pi i}\int_{C_2} \frac{f(z')}{z-z_0}\left[\frac{z'-z_0}{z-z_0}-1\right]^{-1}dz' \\
&= \frac{1}{2\pi i}\int_{C_1} \frac{f(z')}{z'-z_0}\left[1-\frac{z-z_0}{z'-z_0}\right]^{-1}dz' \\
&+ \frac{1}{2\pi i}\int_{C_2} \frac{f(z')}{z-z_0}\left[1-\frac{z'-z_0}{z-z_0}\right]^{-1}dz' \qquad \text{(C.3)}
\end{aligned}$$

For the first integral, $|z' - z_0| > |z - z_0|$. For the second, $|z' - z_0| < |z - z_0|$. Now use the Taylor expansion (valid for $|t| < 1$)

$$\frac{1}{1-t} = \sum_{n=0}^{\infty} t^n$$

to obtain

$$\begin{aligned}
f(z) &= \frac{1}{2\pi i}\int_{C_1} \frac{f(z')}{z'-z_0}\sum_{n=0}^{\infty}\left(\frac{z-z_0}{z'-z_0}\right)^n dz' + \frac{1}{2\pi i}\int_{C_2}\frac{f(z')}{z-z_0}\sum_{n=0}^{\infty}\left(\frac{z'-z_0}{z-z_0}\right)^n dz' \\
&= \frac{1}{2\pi i}\sum_{n=0}^{\infty}(z-z_0)^n\int_{C_1}\frac{f(z')}{(z'-z_0)^{n+1}}\,dz' + \frac{1}{2\pi i}\sum_{n=0}^{\infty}(z-z_0)^{-n-1}\int_{C_2}(z'-z_0)^n f(z')\,dz' \\
&= \frac{1}{2\pi i}\sum_{n=0}^{\infty}(z-z_0)^n\int_{C_1}\frac{f(z')}{(z'-z_0)^{n+1}}\,dz' + \frac{1}{2\pi i}\sum_{n=1}^{\infty}(z-z_0)^{-n}\int_{C_2}(z'-z_0)^{n-1} f(z')\,dz'
\end{aligned} \tag{C.4}$$

where the second term has been re-indexed. Re-indexing again,

$$\begin{aligned}
f(z) = {} & \frac{1}{2\pi i}\sum_{n=0}^{\infty}(z-z_0)^n\int_{C_1}\frac{f(z')}{(z'-z_0)^{n+1}}\,dz' \\
& + \frac{1}{2\pi i}\sum_{n=-\infty}^{-1}(z-z_0)^n\int_{C_2}\frac{f(z')}{(z'-z_0)^{n+1}}\,dz'
\end{aligned} \tag{C.5}$$

Since the integrands, including the function $f(z)$, are analytic in the annular region defined by C_1 and C_2, the integrals are independent of the path of integration in that region. If we replace paths of integration C_1 and C_2 by a circle C of radius r with $r_1 \le r \le r_2$, then

$$\begin{aligned}
f(z) &= \frac{1}{2\pi i}\sum_{n=0}^{\infty}(z-z_0)^n\int_{C}\frac{f(z')}{(z'-z_0)^{n+1}}\,dz' \\
&\quad + \frac{1}{2\pi i}\sum_{n=-\infty}^{-1}(z-z_0)^n\int_{C}\frac{f(z')}{(z'-z_0)^{n+1}}\,dz' \\
&= \frac{1}{2\pi i}\sum_{n=-\infty}^{\infty}(z-z_0)^n\int_{C}\frac{f(z')}{(z'-z_0)^{n+1}}\,dz' \\
&= \sum_{n=-\infty}^{\infty} a_n(z-z_0)^n
\end{aligned} \tag{C.6}$$

Generally, the path of integration can be any path γ that lies in the annular region and encircles z_0 once in the positive (counter-clockwise) direction.

The complex residues a_n are therefore defined by

$$a_n = \frac{1}{2\pi i}\int_{\gamma}\frac{f(z')}{(z'-z_0)^{n+1}}\,dz'$$

C.6 COMPLEX RESIDUE

The constant a_{-1} in the Laurent series

$$f(z) = \sum_{n=-\infty}^{\infty} a_n(z - z_0)^n$$

of $f(z)$ about a point z_0 is called the residue of $f(z)$. If f is analytic at z_0, its residue is zero, but the converse is not always true (for example, $1/z^2$ has residue 0 at $z = 0$ but is not analytic at $z = 0$).

The residue of a function f at a point z_0 may be denoted $\text{Res}_{z=z_0}(f(z))$. Two basic examples of residues are given by $\text{Res}_{z=0}1/z = 1$ and $\text{Res}_{z=0}1/z^n = 0$ for $n > 1$.

The residue of a function f around a point z_0 is also defined by

$$\text{Res}_{z_0} f = \frac{1}{2\pi i} \oint_\gamma f\,\text{d}z$$

where γ is a counter-clockwise simple closed contour, small enough to avoid any other poles of f. In fact, any counter-clockwise path with contour-winding number 1 which does not contain any other pole gives the same result by the Cauchy integral formula. Figure C.4 shows a suitable contour for which to define the residue of function, where the poles are indicated as black dots.

The residues of a function $f(z)$ may be found without explicitly expanding into a Laurent series as follows. If $f(z)$ has a pole of order m at z_0, then $a_n = 0$ for $n < -m$ and $a_{-m} \neq 0$. Therefore,

$$f(z) = \sum_{n=-m}^{\infty} a_n(z - z_0)^n = \sum_{n=0}^{\infty} a_{-m+n}(z - z_0)^{(-m+n)}$$

$$(z - z_0)^m f(z) = \sum_{n=0}^{\infty} a_{-m+n}(z - z_0)^n$$

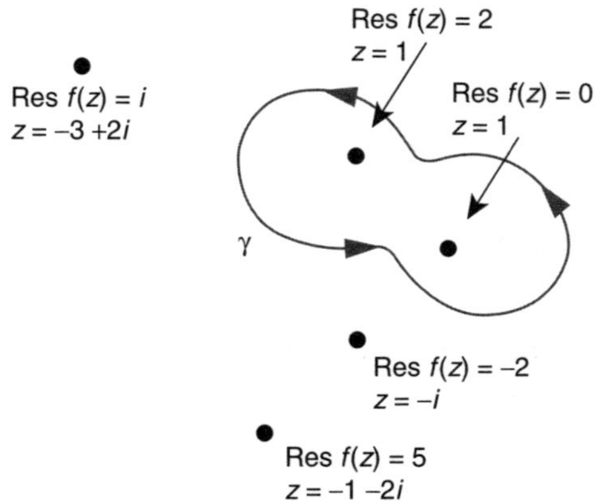

Figure C.4 Complex integral contour for the example in section C.7

$$\begin{aligned}\frac{\mathrm{d}}{\mathrm{d}z}\left[(z-z_0)^m f(z)\right] &= \sum_{n=0}^{\infty} n a_{-m+n}(z-z_0)^{(n-1)} \\ &= \sum_{n=1}^{\infty} n a_{-m+n}(z-z_0)^{(n-1)} \\ &= \sum_{n=0}^{\infty}(n+1)a_{-m+n+1}(z-z_0)^n \end{aligned} \tag{C.7}$$

$$\begin{aligned}\frac{\mathrm{d}^2}{\mathrm{d}z^2}\left[(z-z_0)^m f(z)\right] &= \sum_{n=0}^{\infty} n(n+1)a_{-m+n+1}(z-z_0)^{(n-1)} \\ &= \sum_{n=1}^{\infty} n(n+1)a_{-m+n+1}(z-z_0)^{(n-1)} \\ &= \sum_{n=0}^{\infty}(n+1)(n+2)a_{-m+n+2}(z-z_0)^n \end{aligned} \tag{C.8}$$

Iterating,

$$\begin{aligned}\frac{\mathrm{d}^{m-1}}{\mathrm{d}z^{m-1}}\left[(z-z_0)^m f(z)\right] &= \sum_{n=0}^{\infty}(n+1)(n+2)\ldots(n+m-1)a_{n-1}(z-z_0)^n \\ &= (m-1)!a_{-1} + \sum_{n=1}^{\infty}(n+1)(n+2)\ldots(n+m-1)a_{n-1}(z-z_0)^{n-1} \end{aligned} \tag{C.9}$$

So

$$\lim_{z\to z_0}\frac{\mathrm{d}^{m-1}}{\mathrm{d}z^{m-1}}\left[(z-z_0)^m f(z)\right] = \lim_{z\to z_0}(m-1)!a_{-1} + 0 = (m-1)!a_{-1}$$

and the residue is

$$a_{-1} = \frac{1}{(m-1)!}\frac{\mathrm{d}^{m-1}}{\mathrm{d}z^{m-1}}\left[(z-z_0)^m f(z)\right]_{z=z_0}$$

The residues of a holomorphic function at its poles characterize a great deal of the structure of a function, appearing for example in the amazing residue theorem of contour integration.

C.7 RESIDUE THEOREM

Let there exist an analytic function $f(z)$ whose Laurent series is given by

$$f(z) = \sum_{n=-\infty}^{\infty} a_n(z-z_0)^n$$

and integrate term by term using a closed contour γ encircling z_0,

$$\begin{aligned}\oint_\gamma f(z)\,\mathrm{d}z &= \sum_{n=-\infty}^{\infty} a_n \oint_\gamma (z-z_0)^n\,\mathrm{d}z \\ &= \sum_{n=-\infty}^{-2} a_n \oint_\gamma (z-z_0)^n\,\mathrm{d}z + a_{-1}\oint_\gamma (\mathrm{d}z)/(z-z_0) \\ &\quad +\sum_{n=0}^{\infty} a_n \oint_\gamma (z-z_0)^n\,\mathrm{d}z \end{aligned} \tag{C.10}$$

The Cauchy integral theorem requires that the first and last terms vanish, so we have

$$\oint_\gamma f(z)\,\mathrm{d}z = a_{-1}\oint_\gamma (\mathrm{d}z)/(z-z_0)$$

where a_{-1} is the complex residue. Using the contour $z=\gamma(t)=\mathrm{e}^{it}+z_0$ gives

$$\oint_\gamma (\mathrm{d}z)/(z-z_0) = \int_0^{2\pi} (i\,\mathrm{e}^{it}\,\mathrm{d}t)/(\mathrm{e}^{it}) = 2\pi i$$

so we have

$$\oint_\gamma f(z)\,\mathrm{d}z = 2\pi i a_{-1}$$

If the contour γ encloses multiple poles, then the theorem gives the general result

$$\oint_\gamma f(z)\,\mathrm{d}z = 2\pi i \sum_{a\in A} \mathrm{Res}_{z=a_i} f(z)$$

where A is the set of poles contained inside the contour. This amazing theorem therefore says that the value of a contour integral for any contour in the complex plane depends only on the properties of a few very special points inside the contour.

Figure C.4 shows an example of the residue theorem applied to the illustrated contour γ and the function

$$f(z) = \frac{3}{(z-1)^2} + \frac{2}{(z-i)} - \frac{2}{(z+i)} + i\frac{1}{(z+3-2i)} + \frac{5}{(z+1+2i)}$$

Only the poles at 1 and i are contained in the contour, and have residues of 0 and 2, respectively. The values of the contour integral is therefore given by

$$\oint_\gamma f(z)\,\mathrm{d}z = 2\pi i(0+2) = 4\pi i$$

Example C.7.1 *Consider again the integral*

$$\int_{-\infty}^{\infty} \frac{1}{(x^2+1)^2}\,\mathrm{d}x$$

Now we are going to solve it using the residue approach. Consider the complex-valued function

$$f(z) = \frac{1}{(z^2+1)^2}$$

The Laurent series of $f(z)$ about i, the only singularity we need to consider, is

$$f(z) = \frac{-1}{4(z-i)^2} + \frac{-i}{4(z-i)} + \frac{3}{16} + \frac{i}{8}(z-i) + \frac{-5}{64}(z-i)^2 + \cdots$$

It is clear by inspection that the residue is $-i/4$, so, by the residue theorem, we have

$$\oint_C f(z)\,\mathrm{d}z = \oint_C \frac{1}{(z^2+1)^2}\,\mathrm{d}z = 2\pi i\,\mathrm{Res}_{z=i} f = 2\pi i(-i/4) = \frac{\pi}{2}$$

C.8 JORDAN'S LEMMA

Jordan's lemma shows the value of the integral

$$I = \int_{-\infty}^{\infty} f(x)\,\mathrm{e}^{iax}\,\mathrm{d}x$$

along the infinite upper semicircle and with $a > 0$ is 0 for "nice" functions which satisfy

$$\lim_{R\to\infty} |f(R\,\mathrm{e}^{i\theta})| = 0$$

Thus, the integral along the real axis is just the sum of complex residues in the contour. The lemma can be established using a contour integral I_R that satisfies

$$\lim_{R\to\infty} |I_R| \le \pi/a \lim_{R\to\infty} \epsilon(R) = 0$$

To derive the lemma, write

$$x = R\,\mathrm{e}^{i\theta} = R(\cos\theta + i\sin\theta)$$
$$\mathrm{d}x = iR\,\mathrm{e}^{i\theta}\,\mathrm{d}\theta$$

and define the contour integral

$$I_R = \int_0^{\pi} f(R\,\mathrm{e}^{i\theta})\mathrm{e}^{iaR\cos\theta - aR\sin\theta} iR\,\mathrm{e}^{i\theta}\,\mathrm{d}\theta$$

Then

$$\begin{aligned} |I_R| &\le R\int_0^{\pi} |f(R\,\mathrm{e}^{i\theta})||\mathrm{e}^{iaR\cos\theta}||\mathrm{e}^{-aR\sin\theta}||i||\mathrm{e}^{i\theta}|\,\mathrm{d}\theta \\ &= R\int_0^{\pi} |f(R\,\mathrm{e}^{i\theta})|\mathrm{e}^{-aR\sin\theta}\,\mathrm{d}\theta = 2R\int_0^{\pi/2} |f(R\,\mathrm{e}^{i\theta})|\mathrm{e}^{-aR\sin\theta}\,\mathrm{d}\theta \end{aligned} \tag{C.11}$$

Now, if $\lim_{R\to\infty} |f(R\,\mathrm{e}^{i\theta})| = 0$, choose an ϵ such that $|f(R\,\mathrm{e}^{i\theta})| \le \epsilon$, so

$$|I_R| \le 2R\epsilon \int_0^{\pi/2} \mathrm{e}^{-aR\sin\theta}\,\mathrm{d}\theta$$

But, for θ in $[0, \pi/2]$,

$$\frac{2}{\pi}\theta \le \sin\theta$$

so

$$|I_R| \leq 2R\epsilon \int_0^{\pi/2} \mathrm{e}^{-2aR\theta/\pi}\, \mathrm{d}\theta = 2\epsilon R \frac{1-\mathrm{e}^{-aR}}{2aR/\pi} = \frac{\pi\epsilon}{a}(1-\mathrm{e}^{-aR}) \qquad \text{(C.12)}$$

As long as $\lim_{R\to\infty} |f(z)| = 0$, Jordan's lemma

$$\lim_{R\to\infty} |I_R| \leq \pi/a \lim_{R\to\infty} \epsilon(R) = 0$$

then follows.

D

Vector Spaces and Function Spaces

D.1 DEFINITIONS

A *vector space* over the set of complex number $\mathbb{C}$ is a set of elements V called vectors, which satisfy the following axioms:

1. There exists an operation $(+)$ on the vectors such that
   - if a,b and $c \in V$ then $a + (b + c) = (a + b) + c$ (*associativity*);
   - there exists an identity element $0 \in V$ such that for all $a \in V$, $a + 0 = 0 + a = a$;
   - for every $a \in V$ there exists an inverse element in V denoted $-a$, such that $a + (-a) = (-a) + a = 0$.
2. For every $\alpha \in \mathbb{C}$ and $x \in V$ there exists a vector $\alpha x \in V$; furthermore:
   - $\alpha(\beta x) = (\alpha\beta)x$;
   - $1(x) = x$, for all $x \in V$;
   - $\alpha(x + y) = \alpha x + \alpha y$;
   - $(\alpha + \beta)x = \alpha x + \beta x$.

An example is the set of all complex numbers, where we interpret $x + y$ and αx as ordinary complex numerical addition and multiplication is a complex vector space. Another example of a complex vector space is the set P of all polynomials in a real variable t with complex coefficients, provided that we interpret vector addition and scalar multiplication as the ordinary addition of two polynomials and the multiplication of a polynomial by a complex number. The 0 vector in P is the polynomial which is identically zero; it is worth noting that P is not a finite-dimensional vector space.

For the sake of completeness we shall also recall the following definitions:

Definition D.1.1 *A mapping f: $V \to W$ from a complex vector space to another is said to be antilinear (or conjugate-linear or semilinear) if*

$$f(ax + by) = a^{\dagger} f(x) + b^{\dagger} f(y)$$

for all a, b in $\mathbb{C}$ and all x, y in V.

Definition D.1.2 *A sesquilinear form on a complex vector space V is a map $V \times V \to \mathbb{C}$ that is linear in one argument and antilinear in the other. Specifically a map $\varphi : V \times V \to \mathbb{C}$ is sesquilinear if*

$$\varphi(x + y, z + w) = \varphi(x, z) + \varphi(x, w) + \varphi(y, z) + \varphi(y, w) \quad \text{(D.1)}$$

$$\varphi(ax, by) = a^{\dagger} b\, \varphi(x, y) \quad \text{(D.2)}$$

for all $x, y, z, w \in V$ and all $a, b \in \mathbb{C}$.

A word of clarification is in order concerning the above definition. Conventions differ as to which argument should be linear. We take the first to be conjugate-linear and the second to be linear. This convention is used by essentially all physicists and originates in *Dirac's bra–ket* notation in quantum mechanics. The opposite convention is perhaps more common in mathematics but is not universal.

Many important results from different fields of mathematics can be attained when functions are viewed as vectors in an appropriately defined vector space. This kind of representation produces a number of additional considerations concerning the attempt to represent a function as a linear combination of some given set of functions, i.e. the problem of series expansions.

Addition of the two vectors f_1 and f_2 in function space is defined according to the following rule

$$(f_1 + f_2)(x) = f_1(x) + f_2(x)$$

and multiplication by a complex scalar α is defined as

$$(\alpha f)(x) = \alpha f(x)$$

All the typical questions of analysis such as those of convergence therefore become relevant. Of course we cannot afford to analyse properly these subjects in this book, so we refer the interested reader to the appropriate bibliography.

Example D.1.1 *When we define a function vector space, i.e. a vector space whose elements are functions, we have to specify the property of the function set. For example, a very important function space is that of complex-valued functions of a real variable x defined on the closed interval $[a, b]$ and which are square integrable, i.e. functions for which*

$$\int_a^b |f(x)|^2\, dx$$

exists and is finite. We shall show that the set of square integrable functions form a vector space. This space is called L^2.

The only possible difficulty in showing that these operations satisfy the various axioms that define a vector space is establishing closure. In the case at hand, are the sums and scalar multiples of square integrable functions also square integrable? The answer is yes and so the space is in fact a vector space. We may in fact prove closure of the sum:

$$\begin{aligned} |f_1 + f_2|^2 &= |f_1|^2 + |f_2|^2 + f_1^\dagger f_2 + f_1 f_2^\dagger \\ &= |f_1|^2 + |f_2|^2 + 2\,\mathrm{Re}\,(f_1^\dagger f_2) \\ &\le |f_1|^2 + |f_2|^2 + 2|f_1^\dagger f_2| \\ &\le |f_1|^2 + |f_2|^2 + 2|f_1||f_2| \end{aligned} \tag{D.3}$$

Also

$$0 \le (|f_1| - |f_2|)^2 = |f_1|^2 + |f_2|^2 - 2|f_1||f_2|$$

so

$$|f_1|^2 + |f_2|^2 \ge 2|f_1||f_2|$$

we use this last inequality to replace $2|f_1||f_2|$ in equation (D.3) with something larger, thereby preserving the inequality. Thus the inequality

$$0 \le |f_1 + f_2|^2 \le 2|f_1|^2 + 2|f_2|^2$$

holds at every point in $[a, b]$. *Integrating over both sides we obtain that square integrability of* f_1 *and* f_2 *ensures square integrability of their sum.*

D.2 INNER PRODUCT SPACE

In mathematics, an inner product space is a vector space of arbitrary (possibly infinite) dimensions with the additional structure of an inner product. This additional structure associates, to each pair of vectors in the space, a scalar quantity known as the *inner product* (also called a *scalar product* and *dot product*) of the vectors. Inner products allow the rigorous introduction and generalization of intuitive geometrical notions such as the angle between vectors or length of vectors in spaces of any dimensionality. It also provides the means to define orthogonality between vectors (zero scalar product). Inner product spaces generalize Euclidean spaces (with the dot product as the inner product) and are very important in functional analysis.

Let us concentrate our attention, as usual, on the field of complex numbers $\mathbb{C}$. Formally, an inner product space is a vector space V over the field $\mathbb{C}$ together with a positive-definite sesquilinear form, called, as expected, the inner product. For real vector spaces, this is actually a positive-definite symmetric bilinear form. Thus the inner product is a map

$$\langle \cdot, \cdot \rangle : V \times V \to \mathbb{C}$$

satisfying the following axioms for all $x, y, z \in V$, $a, b \in \mathbb{C}$:

- *Conjugate symmetry*:

$$\langle x, y \rangle = \langle y, x \rangle^{\dagger}$$

This condition implies that $\langle x, x \rangle \in \mathbb{R}$, because $\langle x, x \rangle = \langle x, x \rangle^{\dagger}$.
- *Anti-linearity* in the first variable:

$$\langle ax, y \rangle = a^{\dagger} \langle x, y \rangle$$
$$\langle x + y, z \rangle = \langle x, z \rangle + \langle y, z \rangle$$

- *Linearity* in the second variable:

$$\langle x, by \rangle = b \langle x, y \rangle \tag{D.4}$$

By combining these with *conjugate symmetry*, we get:

$$\langle x, by \rangle = b \langle x, y \rangle$$
$$\langle x, y + z \rangle = \langle x, y \rangle + \langle x, z \rangle$$

so $\langle \cdot, \cdot \rangle$ is a sesquilinear form.
- *Positivity*:

$$\langle x, x \rangle > 0 \quad \text{for all } x \neq 0$$

- *Definiteness*:

$$\langle x, x \rangle = 0 \Rightarrow x = 0$$

The property of an inner product space V that

$$\langle x+y, z\rangle = \langle x, z\rangle + \langle y, z\rangle$$
$$\langle x, y+z\rangle = \langle x, y\rangle + \langle x, z\rangle$$

is called additivity.

Example D.2.1 *A trivial example is given by real numbers with the standard multiplication as the inner product*

$$\langle x, y\rangle = xy$$

More generally any Euclidean space $\mathbb{R}^n$ with the dot product is an inner product space

$$\langle (x_1, \ldots, x_n), (y_1, \ldots, y_n)\rangle := \sum_{i=1}^{n} x_i y_i = x_1 y_1 + \cdots + x_n y_n$$

The general form of an inner product on $\mathbb{C}^n$ is given by:

$$\langle \mathbf{x}, \mathbf{y}\rangle := \mathbf{y}^{\dagger}\mathbf{M}\mathbf{x}$$

with M any symmetric positive-definite matrix, and $y^{\dagger}$ the conjugate transpose of y. For the real case this corresponds to the dot product of the results of directionally differential scaling of the two vectors, with positive scale factors and orthogonal directions of scaling. Apart from an orthogonal transformation, it is a weighted-sum version of the dot product, with positive weights.

Inner product spaces have a naturally defined norm

$$\|x\| = \sqrt{\langle x, x\rangle}$$

This is well defined because of the non-negativity axiom of the definition of inner product space. The norm is thought of as the length of the vector x. Directly from the axioms, one can prove the *Cauchy–Schwarz* inequality:

Theorem D.2.1 *For x, y elements of V*

$$|\langle x, y\rangle| \leq \|x\| \cdot \|y\|$$

holds with equality if and only if x and y are linearly dependent.

This is one of the most important inequalities in mathematics. Its short proof should be noticed. First, it is trivial in the case $y = 0$. Thus we may concentrate on $\langle y, y\rangle$ as non-zero. Now, just let

$$\lambda = \langle y, y\rangle^{-1}\langle x, y\rangle$$

and it follows that

$$0 \leq \langle x - \lambda y, x - \lambda y\rangle = \langle x, x\rangle - \langle y, y\rangle^{-1}|\langle x, y\rangle|^2$$

and the result simply shows up by multiplying out.

The geometric interpretation of the inner product in terms of angle and length motivates much of the geometric terminology that we use in regard to these spaces. In particular, we will say that non-zero vectors x, y of V are orthogonal if and only if their inner product is zero.

D.3 TOPOLOGICAL VECTOR SPACES

A topological vector space is one of the basic structures investigated in functional analysis. As the name suggests, the space blends a topological structure with the algebraic concept of a vector space. The elements of topological vector spaces are typically functions, and the topology is often defined so as to capture a particular notion of convergence of sequences of functions. Hilbert spaces and Banach spaces are well-known examples.

Let us first recall the definition of a *topological space*. A topological space is a set S in which a collection τ of subsets (called *open sets*) is specified by the following properties:

- S is open;
- $\emptyset$ is open;
- the intersection of any two open sets is open;
- the union of every collection of open sets is open.

Such a collection τ is called a *topology on* S and is often denoted by (S, τ).

Suppose now that τ is a topology on a vector space X such that

- every point of X is a closed set, and
- the vector space operations are continuous with respect to τ.

Under these conditions, τ is said to be a *vector topology* on X and X is a *topological vector space*. The second point means that addition and multiplication are continuous with respect to τ. As far as the addition is concerned, this means that the mapping

$$(x, y) \rightarrow x + y$$

of the Cartesian product $X \times X$ into X is such that if $x_i \in X$ for $i = 1, 2$ and if V is a neighbourhood of $x_1 + x_2$ there should exist neighbourhoods V_i of x_i such that $V_1 + V_2 \subset V$. Similarly, the assumption that scalar multiplication is continuous means that the mapping

$$(\alpha, x) \rightarrow \alpha x$$

of $\mathbf{C} \times X$ into X is continuous, i.e. if $x \in X$, α is a scalar, and V is a neighbourhood of αx, then for some $r > 0$ and some neighbourhood W of x we have $\beta W \subset V$ whenever $|\beta - \alpha| < r$. In particular, topological vector spaces are uniform spaces and one can thus talk about completeness, uniform convergence and uniform continuity. The vector space operations of addition and scalar multiplication are actually uniformly continuous. Because of this, every topological vector space can be completed and is thus a dense linear subspace of a complete topological vector space.

D.4 FUNCTIONALS AND DUAL SPACE

Let us consider now a map from a vector space to the field underlying the vector space. In other words, this is an application that takes functions as its argument or input and returns a scalar. In mathematics such an object is usually called a *functional*. Its use goes back to the calculus of variations where one searches for a function which minimizes a certain functional. In functional analysis, the functional is also used in a broader sense as a mapping from an arbitrary linear vector space into the underlying scalar field (usually, real or complex numbers). A special kind of such functionals, linear functionals, gives rise to the study of dual spaces.

There are two types of dual spaces: the algebraic dual space, and the continuous dual space. The algebraic dual space is defined for all vector spaces. When defined for a topological vector space there is a subspace of this dual space, corresponding to continuous linear functionals, which constitutes a continuous dual space.

D.4.1 Algebraic dual space

Given a vector space V over the field $\mathbb{C}$, we define the dual space $V^\star$ to be the set of all linear functionals on V, i.e. scalar-valued linear maps on V (in this context, a "scalar" is a member of the base-field $\mathbb{C}$). $V^\star$ itself becomes a vector space over $\mathbb{C}$ under the following definition of addition and scalar multiplication:

$$(\varphi + \psi)(x) = \varphi(x) + \psi(x)$$
$$(a\varphi)(x) = a\varphi(x)$$

for all φ, $\psi \in V^\star$, $a \in \mathbb{C}$ and x in V.

The pairing of a functional φ in the dual space $V^\star$ and an element x of V is often denoted by an angular bracket, such as

$$\varphi(x) = [\varphi, x] \quad \text{or} \quad \varphi(x) = \langle \varphi, x \rangle$$

D.4.2 Continuous dual space

When dealing with topological vector spaces, one is typically only interested in the continuous linear functionals from the space into the base field. This gives rise to the notion of the "continuous dual space", which is a linear subspace of the algebraic dual space $V^\star$, denoted V'. For any finite-dimensional normed vector space or topological vector space, such as Euclidean n-space, the continuous dual and the algebraic dual coincide. This is, however, false for a infinite-dimensional normed space. In topological contexts $V^\star$ may sometimes be used just for the continuous dual space and the continuous dual may just be called the dual.

E
The Fast Fourier Transform

E.1 DISCRETE FOURIER TRANSFORM

The discrete Fourier transform (DFT) is one of the specific forms of Fourier analysis. As such, it transforms one function into another, which is called the frequency domain representation, or simply the DFT of the original function (which is often a function in the time domain). The DFT requires an input function that is a finite sequence of real or complex numbers, and for this reason it is ideal for processing information stored in computers. In particular, the DFT is widely employed in signal processing and related fields to analyse the frequencies contained in a sampled signal, to solve partial differential equations, and to perform other operations such as convolutions. The DFT can be computed efficiently in practice using a fast Fourier transform (FFT) algorithm.

Since FFT algorithms are so commonly employed to compute the DFT, the two terms are often used interchangeably in colloquial settings, although there is a clear distinction: "DFT" refers to a mathematical transformation, regardless of how it is computed, while "FFT" refers to any one of several efficient algorithms for the DFT.

The sequence of N complex numbers $x_0, \ldots, x_{N-1}$ is transformed into the sequence of N complex numbers $X_0, \ldots, X_{N-1}$ by the DFT according to the formula

$$X_k = \sum_{n=0}^{N-1} x_n \, e^{\frac{2\pi i}{N} kn}, \qquad k = 0, \ldots, N-1$$

where $e^{\frac{2\pi i}{N}}$ is a primitive Nth root of unity.

The inverse discrete Fourier transform (IDFT) is given by

$$x_n = \frac{1}{N} \sum_{k=0}^{N-1} X_k \, e^{-\frac{2\pi i}{N} kn}, \qquad n = 0, \ldots, N-1$$

Note that the normalization factor multiplying the DFT and the IDFT (here 1 and $1/N$) and the signs of the exponents are merely conventions, and differ in some treatments. The only requirements of these conventions are that the DFT and the IDFT have opposite-sign exponents and that the product of their normalization factors is $1/N$. A normalization of $1/\sqrt{N}$ for both the DFT and the IDFT makes the transforms unitary, which has some theoretical advantages, but it is often more practical in numerical computation to perform the scaling all at once, as above (and a unit scaling can be convenient in other ways).

The vectors $e^{\frac{2\pi i}{N} kn}$ form an orthogonal basis over the set of N-dimensional complex vectors:

$$\sum_{n=0}^{N-1} \left(e^{\frac{2\pi i}{N} kn} \right) \left(e^{-\frac{2\pi i}{N} k'n} \right) = N \delta_{kk'}$$

where $\delta_{kk'}$ is the Kronecker delta. This orthogonality condition can be used to derive the formula for the IDFT from the definition of the DFT, and is equivalent to the unitarity property below.

If the expression that defines the DFT is evaluated for all integers k instead of just for $k = 0, \ldots, N-1$, then the resulting infinite sequence is a periodic extension of the DFT, periodic with period N.

The periodicity can be shown directly from the definition:

$$\sum_{n=0}^{N-1} x_n \, e^{\frac{2\pi i}{N}(k+N)n} = \sum_{n=0}^{N-1} x_n \, e^{\frac{2\pi i}{N}kn} \, e^{2\pi i n} = \sum_{n=0}^{N-1} x_n \, e^{\frac{2\pi i}{N}kn}$$

where we have used the fact that $e^{-2pi} = 1$. In the same way it can be shown that the IDFT formula leads to a periodic extension.

E.2 FAST FOURIER TRANSFORM

A fast Fourier transform (FFT) is an efficient algorithm used to compute the discrete Fourier transform (DFT) and its inverse. FFTs are of great importance to a wide variety of applications, from digital signal processing and solving partial differential equations to algorithms for the quick multiplication of large integers.

By far the most common FFT is the Cooley–Tukey algorithm. This is a divide and conquer algorithm that recursively breaks down a DFT of any composite size into many smaller DFTs, along with $O(N)$ multiplications by complex roots of unity. This method (and the general idea of an FFT) was made popular by a publication of J.W. Cooley and J.W. Tukey in 1965, but it was later discovered that those two authors had independently reinvented an algorithm known to Carl Friedrich Gauss around 1805 (and subsequently rediscovered several times in limited forms).

A fundamental question of longstanding theoretical interest is: What is the computational cost required to calculate the Discrete Fourier Transform of a function composed of N points? Up to halfway through the 1960s the answer was: Let us define the complex number

$$W_n = e^{2\pi i/N} \tag{E.1}$$

then the DFT can be rewritten in the form

$$X_k = \sum_{n=0}^{N-1} W^{nk} x_n \tag{E.2}$$

In other terms, the vector x_n is multiplied for a matrix whose (n, k)-element is equal to W raised to the product nk. As a result, the matrix product produces a vector whose elements are the points of the DFT. This complex multiplication (plus a few operations necessary for producing the powers of W) evidently requires N^2 operations. Therefore, the DFT appears to be a process of order N^2. Actually, this conclusion is false since, as we shall see, the DFT can be calculated with a process of order $N \log_2 N$.

The FFT algorithm is based on a previous analysis by Danielson and Lanczos. In 1942, they showed that a DFT of length N could be rewritten as the sum of two DFTs of length $N/2$, the first made up by the points in even position in the starting vector, and the second by the points in odd position. The demonstration of this is very simple:

$$
\begin{aligned}
X_k &= \sum_{n=0}^{N-1} x_n \, e^{2\pi i n k/N} \\
&= \sum_{n=0}^{N/2-1} x_{2n} \, e^{2\pi i (2n)k/N} + \sum_{n=0}^{N/2-1} x_{2n+1} \, e^{2\pi i (2n+1)k/N} \\
&= \sum_{n=0}^{N/2-1} x_{2n} \, e^{2\pi i n k/(N/2)} - W^k \sum_{n=0}^{N/2-1} x_{2n+1} \, e^{2\pi i n k/(N/2)} \\
&= X_k^e + W_N^k X_k^o
\end{aligned}
\tag{E.3}
$$

where X_k^e is the kth component of the Fourier transform (of length $N/2$) formed by the even components of the original signal, while X_k^o is formed from the odd components. Each of the two sub-Fourier transforms is periodic with period $N/2$.

The most interesting thing about this result is that the procedure can be used recursively. In fact, we can apply the previous procedure to calculate the two DFTs of length $N/2$ decomposing each of these in two DFT (this turn of length $N/4$) made up by taking from X_k^e and X_k^o the points in even and odd position. If the initial number of points is a power of 2 (and we will always stick to this case), in the end this recursive procedure will produce a set of k DFTs composed of a single point, and this will exactly happen after $\log_2 N$ steps. Since it is mandatory to understand this point well, it is worthwhile to make a simple example. Let us consider a function composed by 8 points; application to this set of the Danielson–Lanczos algorithm (1942) allows us to write

$$
\begin{aligned}
X_k &= X_k^e + W_8^k X_k^o \\
&= (X_k^{ee} + W_4 X_k^{eo}) + W_8^k (X_k^{oe} + W_4^k X_k^{oo}) \\
&= [(X_k^{eee} + W_2^k X_k^{eeo}) + W_4^k (X_k^{eod} + W_2^k X_k^{eoo})] \\
&\quad + [(X_k^{oee} + W_2^k X_k^{oeo}) + W_4^k (X_k^{ood} + W_2^k X_k^{ooo})]
\end{aligned}
\tag{E.4}
$$

The various final quantities are single points of the original function. So, leaving out of consideration for a moment the computation of phase factors, we see that the first action to perform, in order to compute the FFT, is to sort the original data into a new order. As we can see, the final order is obtained by reversing the binary expression of the number which shows the position of a point in the departure string (*bit-reversal*). It is quite easy to understand the reason for this if we realize that the successive subdivisions of the data into even and odd are tests of successive low-order (less significant) bits of n.

From a computational point of view, the most interesting thing to notice about the first phase of the FFT algorithm is that, in order to calculate the new position, it is not necessary to make any conversion from decimal to binary, or vice versa. Let us see why.

To begin with, notice that since the sorting is obtained by exchanging couples of numbers, the computation cost is of order $N/2$ and not N. Furthermore, all the even numbers of the first half, expressed with $\log_2 N$ digits, have a 0 digit in both the first and the last position (see Table E.1 as an example for 16 numbers). Therefore their bit-reversed mapping will be in the first half too. As regards the odd numbers, on the contrary, from their binary expression we can infer that each of them will be exchanged with an even number of the second half (Table D.1 is always kept for reference).

Table E.1 Decimal–binary conversion table

Original position	Binary expression	Final position	Binary expression
0	0000	0	0000
1	0001	8	1000
2	0010	4	0100
3	0011	12	1100
4	0100	2	0010
5	0101	10	1010
6	0110	6	0110
7	0111	14	1110
8	1000	1	0001
9	1001	9	1001
10	1010	5	0101
11	1011	13	1101
12	1100	3	0011
13	1101	11	1011
14	1110	7	0111
15	1111	15	1111

Note that for all the following considerations it is essential that each number must always be expressed using all the digits at our disposal; for instance, if one has a succession composed of the 64 first integer numbers, the decimal number 5 must always be written in the form 00101 and not as a 101. Although this is equivalent from any other point of view, the two forms are not equivalent for our present discussion. In fact, we can realize easily that the opposite of the first expression turns out to be 101000 (which is equal to 40 in the decimal system) while in the second case the number remains unchanged!

The number 1 will always be mapped in $N/2$ independently on N. In fact, as we have said, to represent N numbers in binary notation we need $l \log_2 N$ digits so the bit-reverse of 1 will be $100\ldots.000$, that is $N/2$. Obviously the number 2, which in binary notation is $0000\ldots.10$, will be mapped in $N/4$, and so on.

We can also easily show that

(1) *if we know the bit-reversed of a generic even number, it is possible to immediately find the bit-reversed of the following odd number*

and, vice versa,

(2) *if we know the bit-reversed of an odd number, it is possible to obtain directly the bit-reversed of the following even number.*

Let us begin with the first case.

Any even number, expressed in binary notation, has 0 as least significant bit (lsb). Clearly the immediately next odd number is only different in having the lsb equal to 1. Therefore, if we have the bit-reversed of any even number, the bit-reversed of the following odd number is obtained simply by adding $100\ldots..0$, that is $N/2$. Let us see an example: consider the bit-reversed of the number 4 in a set of 16 numbers

$$4 = 0100 \rightarrow 0010 = 2 \tag{E.5}$$

the next odd number, 5, will be mapped in the number 10 (decimal), in fact

$$5 = 0101 \rightarrow 1010 = 0010 + 1000 = 2 + 16/2 = 2 + 8 = 10 \tag{E.6}$$

So, if we know the bit-reversed of a generic even number, say j, the bit-reversed of the immediately next odd number will be $j + N/2$. It is worthwhile to notice than we do not need to make any conversion from binary to decimal, or vice versa!

Obtaining the bit-reversed value of an even number given the bit-reversed value of the preceding odd number, is also very easy. Begin with noticing how one obtains (always in binary notation obviously) an even number starting from the previous odd number... obviously adding 1! This is trivial, but let us see this simple sum in some detail; we take, for instance, any odd number

$$100010111010101111$$

and add 1 according to the binary arithmetic rules

$$\begin{aligned} &100010111010101111 + \\ &000000000000000001 \\ = {} &100010111010110000 \end{aligned} \tag{E.7}$$

As we can see, this operation is equivalent to replacing all the consecutive least significant 1's with the same number of 0's and in replacing the first digit 0, which came immediately after such sequence of 1's, with a digit 1. This simple observation gives us the solution of our problem. In fact, given the bit-reversed value of an odd number, we can obtain the bit-reversed value of the following even number replacing all the consecutive most significant 1's with an equal number of 0's and replacing the first 0 immediately next with a 1.

Consider a practical example that will show how this procedure can be implemented with decimal operations only. We wish to know the bit-reversed mapping of a generic odd number (of a set of 64 numbers), for example,

$$37 = 100111 \rightarrow 111001 = 57 \tag{E.8}$$

The replacement of the first most significant digit 1 with a 0 is equivalent to subtracting the $N/2$ number from the assigned number; therefore, in our case it is necessary to subtract $32 = 64/2$ from 57 obtaining

$$111001 - 100000 = 57 - 32 = 011001 = 25 \tag{E.9}$$

We must still replace the two significant 1's and the 0 digit in third position. To replace the first 1 is, in turn, equivalent to subtracting $N/4$, in fact

$$011001 - 10000 = 25 - 16 = 001001 = 9 \tag{E.10}$$

then, to replace the second 1, we must subtract $N/8$, obtaining

$$001001 - 001000 = 9 - 8 = 000001 = 1 \tag{E.11}$$

Now we add $N/16$ to obtain the number we are looking for

$$000001 + 000100 = 1 + 4 = 000101 = 5 \tag{E.12}$$

It is easy to verify that the bit-reversed value of 38 is 5:

$$38 = 101000 \rightarrow 000101 = 5 \tag{E.13}$$

Therefore we can summarize the procedure as follows:

Let us start from a generic number j; in the first step we subtract the quantity $m = N/2$ in order to replace the first digit 1, then we check if the new number obtained is greater or less than m. If it is greater than m, this means that the digit immediately next to the replaced 1 is also 1 and therefore the process must be repeated, this time subtracting $m = m/2$. The process continues until the obtained j number is less than m which, in turn, is progressively divided by 2. When we reach this point, m is added to j.

This algorithm can be implemented easily with the following loop:

```
m = N/2
DO WHILE j > m AND m ≥ 2
   j = j - m
   m = m / 2
LOOP
j = j + m
```

As we can see, we start the loop only if $j > m$; this allows us to compute both the even numbers bit-reversal and the odd ones with the same code. In fact the initial value of m is $N/2$, so, if $j > m$ this means that the number previously reversed was an odd number (otherwise it would be left inside the first half, remember!) so we must find the position of the next even number and therefore we will apply the last procedure described. On the contrary, if $j < m$ this means that the previous reversed number was an even number and then the mapping of the current number (that is an odd one) is obtained simply by adding $N/2$.
Now we are going to see how to implement the phase factor calculation; for this purpose we recall the result by Danielson and Kivelson

$$X_k = \sum_{n=0}^{N-1} x_n \, e^{2\pi ink/N} = X_k^e + W_N^k X_k^o \tag{E.14}$$

where

$$W_N^k = \exp\left[2\pi i \frac{k}{N}\right] \tag{E.15}$$

The fundamental key to save computation time is to exploit the periodicity of the exponential functions in order to avoid redundant operations. In fact, both X_k^e and X_k^o are periodic functions in the interval $0 \leq k < N/2$ since

$$\exp\left[2\pi i \frac{n}{N/2}\left(\frac{N}{2} + k\right)\right] = \exp\left[2\pi in\right] \exp\left[2\pi i \frac{nK}{N/2}\right] = \exp\left[2\pi i \frac{nK}{N/2}\right] \tag{E.16}$$

because $\exp(-2\pi in) = 1 \; \forall \, n \in \mathbf{Z}$. From this we obtain

$$W_N^{k+N/2} = -W_N^k \tag{E.17}$$

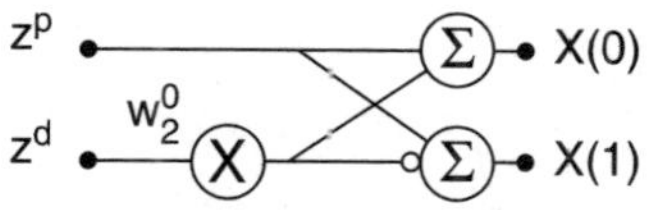

Figure E.1 Formal diagram corresponding to equation E.20

So we do not need to compute the phase factor for the second half of the function points. We can write

$$X_k = X_k^e + W_N^k X_k^o, \quad 0 \leq k < N/2 \tag{E.18}$$

$$X_k = X_k^e - W_N^k X_k^o, \quad N/2 \leq k < N-1 \tag{E.19}$$

therefore the problem of calculating the DFT of N points reduces one to calculating two DFTs of $N/2$ points with multiplicative phase factors, respectively, equal to W_N^k and $-W_N^k$.

As we have already seen, this procedure can be iterated until we reach the simplest case, which is to calculate the DFT of two single points, say z^e and z^o. In this case the DFT will be simply given by the points

$$\begin{aligned} X_0 &= z^e + W_2^0 z^o \\ X_1 &= z^e - W_2^0 z^o \end{aligned} \tag{E.20}$$

The computational process can be described by a diagram, as shown in Figure E.1.

The next step will have four points, in this case we can write the set as:

$$\begin{aligned} X_0 &= z_0^e + W_4^0 z_0^o \\ X_1 &= z_1^e + W_4^1 z_1^o \\ X_2 &= z_0^e - W_4^0 z_0^o \\ X_3 &= z_1^e - W_4^1 z_1^o \end{aligned} \tag{E.21}$$

This set of equations can be represented by the diagram in Figure E.2.

Finally, in the case of eight points, we obtain the diagram shown in Figure E.3.

As we can see, the fundamental schema is always the same. In the general case where the input variables are complex, the general scheme is equivalent to the following set of equations

$$\begin{aligned} \text{Re}(C) &= \text{Re}(A) + \text{Re}(B)\cos\theta + \text{Im}(B)\sin\theta \\ \text{Im}(C) &= \text{Im}(A) + \text{Im}(B)\cos\theta - \text{Re}(B)\sin\theta \\ \text{Re}(D) &= \text{Re}(A) - \text{Re}(B)\cos\theta - \text{Im}(B)\sin\theta \\ \text{Im}(D) &= \text{Im}(A) - \text{Im}(B)\cos\theta + \text{Re}(B)\sin\theta \end{aligned} \tag{E.22}$$

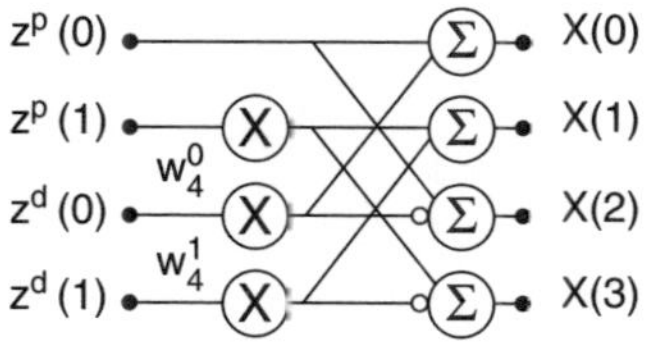

Figure E.2 Formal diagram corresponding to equation E.21

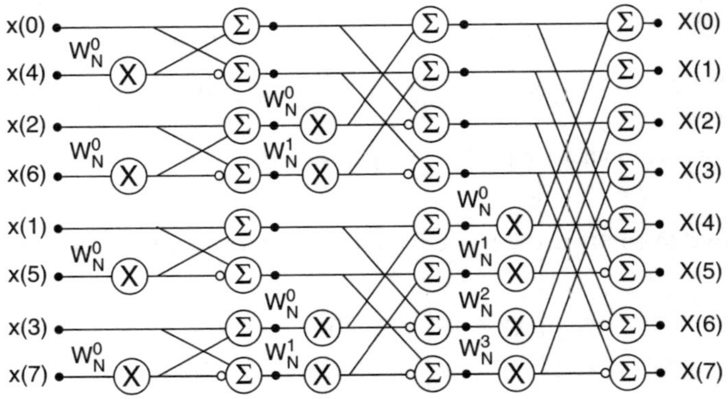

Figure E.3 Formal diagram corresponding to 8 points

If we note that

$$W_N^{k+1} = \exp\left[2\pi i \frac{k+1}{N}\right] = \exp\left[2\pi i \frac{1}{N}\right] W_N^k \tag{E.23}$$

we easily obtain

$$\begin{aligned} \mathrm{Re}(W_N^{k+1}) &= \mathrm{Re}(W_N^k)\cos\left(\frac{2\pi}{N}\right) + \mathrm{Im}(W_N^k)\sin\left(\frac{2\pi}{N}\right) \\ \mathrm{Im}(W_N^{k+1}) &= \mathrm{Re}(W_N^k)\sin\left(\frac{2\pi}{N}\right) - \mathrm{Im}(W_N^k)\cos\left(\frac{2\pi}{N}\right) \end{aligned} \tag{E.24}$$

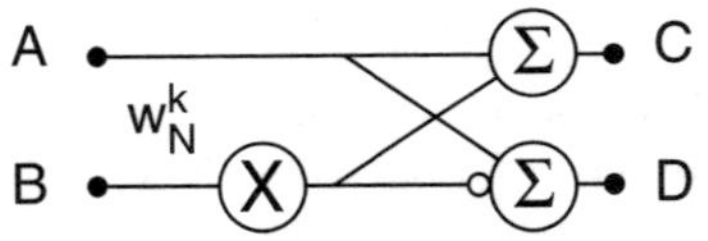

Figure E.4 All previous schemes are combinations of this basic scheme

F
The Fractional Fast Fourier Transform

The typical problem that we might want to tackle with the use of FFT is the computation of an infinite sum of the type:

$$p(x) = \sum_{n=-\infty}^{+\infty} \hat{p}_n \, \mathrm{e}^{-i2\pi n x \Delta}$$

Recall that $\Delta = 1/2X_c$, where X_c is the spatial cutoff, that is a value beyond which (say $|x| > X_c$) the function $p(x)$ can be considered negligibly small.

From the original infinite sum we pass to the finite series

$$p_N(x) = \sum_{n=-N/2}^{N/2} \hat{p}_n \, \mathrm{e}^{-i2\pi n x \Delta}$$

and, as far as the values

$$x_m = \frac{m}{N\Delta}, \qquad 0 \le m < N$$

are concerned, we can compute very efficiently the N-number $p_N(x_m)$ (efficiently means $O(N \log(N))$) using the FFT.

The convenience of the FFT introduces some inflexibility, namely, the highest resolution we can achieve is given by:

$$\delta x = \frac{1}{N\Delta} = \frac{2X_c}{N}$$

and this sometimes is just too coarse. We always have the option to increase the number N of Fourier modes, but this has a cost. The alternative, which should always be weighted with care, is to resort to the fractional FFT.

Let's decide that the spacing we want for the set x_m is given by:

$$x_m = m\theta$$

this would require computing

$$p_N(m\theta) = \sum_{n=-N/2}^{N/2} \hat{p}_n \, \mathrm{e}^{-i2\pi n m \theta \Delta} =: \sum_{n=-N/2}^{N/2} \hat{p}_n \, \mathrm{e}^{-i2\pi n m \eta}, \quad \eta = \theta\Delta \tag{F.1}$$

and the problem that we have to solve consists in devising an efficient way to compute the sum (F.1) for an arbitrary real value of η.

Since

$$e^{-i2\pi nm\eta} = e^{i\pi(n-m)^2\eta}\, e^{-i\pi n^2\eta}\, e^{-i2\pi m^2\eta}$$

equation (F.1) can be written as:

$$p_N(m\theta) = e^{-i\pi m^2\eta} \sum_{n=-N/2}^{N/2} e^{i\pi(n-m)^2\eta}\, e^{-i\pi n^2\eta}\, \hat{p}_n$$

If we define:

$$\begin{aligned} f_m &:= p_N\left(\left(m - \frac{N}{2}\right)\theta\right) e^{i\pi(m-N/2)^2\eta} \\ q_n &:= e^{-i\pi(n-N/2)^2\eta}\, \hat{p}_{n-N/2} \\ T_{nm} &:= e^{i\pi(n-m)^2\eta} \end{aligned}$$

in matrix notation, equation (F.1) becomes:

$$\mathbf{f} = \mathbf{T}\mathbf{q}$$

The matrix **T** has a peculiar form, it is in fact only a function of the difference between the two indices:

$$T_{nm} = T(n-m)$$

and such a matrix is a well-known object in the computational literature. It is known as a Toepliz matrix.

We will now take a brief detour in the world of Toepliz matrices.

F.1 CIRCULAR MATRIX

Before we tackle Toepliz matrices we must describe another special kind of matrix, that is a circular matrix.

A circular matrix **C** is a matrix of the form:

$$\mathbf{C} = \begin{pmatrix} c_0 & c_{N-1} & c_{N-2} & \cdots & c_1 \\ c_1 & c_0 & c_{N-1} & \cdots & c_2 \\ & & \cdots & & \\ c_{N-1} & c_{N-2} & \cdots & & c_0 \end{pmatrix}$$

The matrix **C** is fully specified by its first column:

$$\mathbf{c}^1 = \begin{pmatrix} c_0 \\ c_1 \\ \cdots \\ c_{N-1} \end{pmatrix}$$

and the generic element C_{ij} can be written in the form:

$$C_{ij} = g(j-i), \quad g(m) = g(m+N)$$

In particular:

$$g(m) = \mathbf{c}^1_m, \quad 0 \leq m < N$$

Theorem F.1.1 *The N functions f_n defined by:*

$$f_n(j) := \exp\left(\frac{i2\pi nj}{N}\right)$$

are eigenfunctions of any circular matrix **C**.

Eigenvalues are given by:

$$\lambda_n = \sum_{j=0}^{N-1} \mathbf{c}^1_j f_n(j)$$

Proof. From the definition:

$$\sum_{j=0}^{N-1} C_{ij} f_n(j) = \sum_{j=0}^{N-1} g(j-i) f_n(j)$$

a change of variable gives us:

$$\sum_{j=0}^{N-1} C_{ij} f_n(j) = \sum_{j=-i}^{N-1-i} g(j) f_n(j+i)$$

From the definition of the functions $f_n(i)$ we have $f_n(i+j) = f_n(i) f_n(j)$ and

$$\sum_{j=0}^{N-1} C_{ij} f_n(j) = f_n(i) \sum_{j=-i}^{N-1-i} g(j) f_n(j)$$

The sum on the r.h.s. involves a periodic function of period N. Necessarily it does not depend on which particular window of N elements we sum it. More precisely we observe that

$$S[m] := \sum_{j=m}^{N-1+m} g(j) f_n(j)$$

is independent of m, therefore:

$$\sum_{j=-i}^{N-1-i} g(j) f_n(j). = \sum_{j=0}^{N-1} g(j) f_n(j) = \lambda_n$$

and we conclude that

$$\sum_{j=0}^{N-1} C_{ij} f_n(j) = \lambda_n f_n(i)$$

□

If we use the symbol $\mathcal{F}$ to denote also the discrete Fourier transform (hoping that from the context it will always be clear whether we are looking at a discrete or a continuous transform), from the definition of $f_n(j)$ we can write:

$$[\mathcal{F}c][n] = \sum_{i=0}^{N-1} f_n(i)c_i, \quad [\mathcal{F}b][i] = \sum_{n=0}^{N-1} f_n(i)b_n$$

As usual, let $f^\dagger$ be the complex conjugate, then:

$$[\overline{\mathcal{F}}c][n] = \sum_{i=0}^{N-1} f_n^\dagger(i)c_i, \quad [\overline{\mathcal{F}}b][i] = \sum_{n=0}^{N-1} f_n^\dagger(i)b_n$$

Since

$$\sum_{n=0}^{N-1} f_n(i) f_n^\dagger(j) = \begin{cases} N & i = j \\ 0 & i \neq j \end{cases}$$

we have

$$C_{ij} = \sum_{n=0}^{N-1} \lambda_n f_n^\dagger(i) f_n(j) \tag{F.2}$$

where

$$\boldsymbol{\lambda} = \mathcal{F}\mathbf{c}^1$$

F.1.1 Matrix vector multiplication

A matrix vector multiplication, where a circular matrix in involved, can be performed very efficiently, exploiting the decomposition (F.2). We want to compute:

$$x_i = \sum_{j=0}^{N-1} C_{ij} v_j$$

Using the decomposition (F.2) we get:

$$x_i = \sum_{n=0}^{N-1} f_n^\dagger(i)\lambda_n \sum_{j=0}^{N-1} f_n(j)v_j \tag{F.3}$$

The quantity λ, as we have seen, is the Fourier transform of the defining vector $\mathbf{c}^1$, and both sums appearing in equation (F.3) can be computed with the discrete fast Fourier transform.

In a more compact notation, equation (F.3) can be written as:

$$\mathbf{x} = \overline{\mathcal{F}}(\,[\mathcal{F}\mathbf{c}^1][\mathcal{F}\mathbf{v})]\,)$$

This calls for three fast Fourier transforms and one pointwise vector multiplication for an asymptotically computational complexity of $O(N \log(N))$.

F.2 TOEPLIZ MATRIX

A Toepliz matrix $\mathbf{T}$ is a matrix of the form:

$$\mathbf{T} = \begin{pmatrix} t_0 & t_{-1} & t_{-2} & \cdots & t_{-(N-1)} \\ t_1 & t_0 & t_{-1} & \cdots & t_{-(N-2)} \\ & & \cdots & & \\ t_{N-1} & t_{N-2} & \cdots & & t_0 \end{pmatrix}$$

The matrix $\mathbf{T}$ is fully specified by its first column

$$\mathbf{t}^1 = \begin{pmatrix} t_0 \\ t_1 \\ \cdots \\ t_{N-1} \end{pmatrix}$$

and its first row:

$$\mathbf{t}_1 = \left(t_0, t_{-1}, \ldots, t_{-(N-1)}\right)$$

and the generic element T_{ij} can be written in the form

$$T_{ij} = t(j - i), \quad 0 \leq i, j < N$$

F.2.1 Embedding in a circular matrix

Let's consider a column vector

$$\mathbf{r}_i = \begin{cases} t_i & 0 \leq i < N \\ 0 & N \leq i \leq N + Q \\ t_{-[(2N+Q)-i]} & N + Q < i < 2N + Q \end{cases} \qquad 2N + Q = 2^M$$

with M the smallest integer such that $2^M \geq 2N$.

In the next step we build the circular matrix $\mathbf{C}(r)$ based on $\mathbf{r}$. We want to show that the $N \times N$ top left corner is the original Toepliz matrix.

Let's compute $\mathbf{C}(r)_{ij}, \quad 0 \leq i, j < N$; that is, the top left corner of $\mathbf{C}(r)$. For $i \geq j$ we have:

$$\mathbf{C}(r)_{ij} = r(i - j) = t(i - j)$$

while, for $i < j$,

$$\begin{aligned} \mathbf{C}(r)_{ij} &= r(i - j) \\ &= r(-(j - i)) \\ &= r(2N + Q - (j - i)) \end{aligned}$$

Clearly

$$1 \leq j - i \leq N - 1$$

therefore

$$N + Q < 2N + Q - (j - i) < 2N + Q$$

and

$$r(2N+Q-(j-i)) = t_{-[2N+Q-(2N+Q-(j-i))]} = t(i-j)$$

The top left corner of $\mathbf{C}(r)_{ij}$ is therefore the original Toepliz matrix. If we are interested in computing the matrix vector product:

$$\mathbf{z} = \mathbf{T}\mathbf{x}.$$

we can compute:

$$\begin{pmatrix} \mathbf{z} \\ \mathbf{u} \end{pmatrix} = \mathbf{C}(\mathbf{r}) \begin{pmatrix} \mathbf{x} \\ \mathbf{0} \end{pmatrix}$$

F.2.2 Applications to pricing

Let's recall that the basic pricing formula is based on the fundamental sum:

$$d_N(k,\alpha) = \sum_{n=-N/2}^{+N/2} \mathrm{e}^{-2\pi i n k \Delta} \phi_X(n\Delta - \alpha) \frac{1-(-1)^n}{n\Delta}$$

(see equation (1.30)), where $\Delta = 1/2X_c$ is the cutoff and $k = \log(B(t,T)K/S_t)$ is related to the strike K. As we have seen in the text, if we can get away with computing strikes k_n evenly spaced,

$$k_n = n\frac{2X_c}{N}$$

we can handsomely solve the problem using the FFT algorithm. In most applications of the Fourier transform methodology to finance, we have a sequence of prices corresponding to a set of strikes $\{\overline{k}_n\}$. Most of the strikes $\overline{k}$ will have to be interpolated from the available strike k_n and the interpolation will be more precise the narrower is the step size. To gain some insight, let's look at a situation with $N = 1024$ and $X_c = 8$. The default strike resolution is:

$$\Delta k = \frac{2X_c}{N} = 0.0078$$

If we decide to use the fractional FFT, we can use any desired spacing provided we cover the whole range $[\overline{k}_m = \min\{\overline{k}\}, \overline{k}_M = \max\{\overline{k}\}]$. This last requirements demands that

$$\frac{N}{2}\theta > \log\left(\frac{k_M}{S_t}\right)$$
$$-\frac{N}{2}\theta < \log\left(\frac{k_m}{S_t}\right)$$

If the range of strikes to match runs from, say, 0.8 through 1.2, we end up with the constraint:

$$\theta \geq \frac{2}{N}\max(\log(k_M/S_t), -\log(k_m/S_t)) = 0.00043$$

which would presumably produce a much more accurate result.

F.3 SOME NUMERICAL RESULTS

We present some comparison between the interpolation performed with the FFT and the fractional FFT. In what follows the notation **CN-Call, AN-Put**, etc., will correspond to the following payoff:

Symbol	*Payoff*
AN-Call	$S_T\mathbf{1}[S_T > K]$
CN-Call	$\mathbf{1}[S_T > K]$
Call	$[S_T - K]^+$
AN-Put	$S_T\mathbf{1}[S_T < K]$
CN-Put	$\mathbf{1}[S_T < K]$
Put	$[K - S_T]^+$

What we have done is to compare all of the above payoff for a set of strikes equally spaced. We have computed the non-interpolated payoff using the Fourier transform algorithm for the exact strikes in question. And this would represent for us a result as close as possible to the true result. Then, keeping fixed the number of Fourier modes (120) we have computed the same payoffs, extrapolating the results from the set of "Fourier Strikes" and "Fractional Fast Fourier Strikes".

The row labelled "err" reports the sum of the square of the differences (for each payoff) between the interpolation method and the non-interpolated results. As you can see, the accuracy gained by resorting to the fractional transform is about two orders of magnitude higher. The fractional transform is, on average, four times as slow as the direct FFT. Whenever these two extra digits are relevant, the fractional method is by all means the method of choice, given that the number of modes needed in the FFT to gain these two orders of magnitude is well above four. Of course we realize that a pricing accurate to six places is hardly an issue.

F.3.1 The Variance Gamma model

The first set of results have been obtained with the Variance Gamma model that we have run at the, by now usual, parameters (see Tables F.1, F.2 and F.3)

$$\sigma = 0.4390, \quad \theta = -0.7030, \quad \nu = 0.0286$$

Table F.1 FT: Variance Gamma model

K	AN-Call	CN-Call	Call	AN-Put	CN-Put	Put
0.8000	0.8029	0.4985	0.3044	0.1971	0.2625	0.0654
0.8500	0.7637	0.4892	0.2744	0.2363	0.3193	0.0830
0.9000	0.7228	0.4760	0.2468	0.2772	0.3801	0.1029
0.9500	0.6810	0.4595	0.2215	0.3190	0.4442	0.1252
1.0000	0.6388	0.4404	0.1984	0.3612	0.5108	0.1496
1.0500	0.5969	0.4195	0.1774	0.4031	0.5793	0.1762
1.1000	0.5557	0.3973	0.1584	0.4443	0.6491	0.2048
1.1500	0.5156	0.3743	0.1413	0.4844	0.7196	0.2352
1.2000	0.4769	0.3511	0.1258	0.5231	0.7904	0.2673

Table F.2 FFT: Variance Gamma model

K	AN-Call	CN-Call	Call	AN-Put	CN-Put	Put
0.8000	0.8029	0.4985	0.3044	0.1971	0.2625	0.0654
0.8500	0.7636	0.4891	0.2745	0.2364	0.3195	0.0831
0.9000	0.7227	0.4758	0.2470	0.2773	0.3803	0.1031
0.9500	0.6810	0.4593	0.2217	0.3190	0.4444	0.1253
1.0000	0.6388	0.4403	0.1985	0.3612	0.5109	0.1497
1.0500	0.5969	0.4195	0.1775	0.4031	0.5793	0.1763
1.1000	0.5558	0.3972	0.1586	0.4442	0.6492	0.2050
1.1500	0.5158	0.3743	0.1416	0.4842	0.7197	0.2355
1.2000	0.4773	0.3511	0.1262	0.5227	0.7904	0.2676
Err	1.5201e-04	1.1368e-04	1.8572e-04	1.5201e-04	1.1368e-04	1.8572e-04

Table F.3 Fractional FFT: Variance Gamma model

K	AN-Call	CN-Call	Call	AN-Put	CN-Put	Put
0.8000	0.8029	0.4985	0.3044	0.1971	0.2625	0.0654
0.8500	0.7637	0.4892	0.2744	0.2363	0.3193	0.0830
0.9000	0.7228	0.4760	0.2468	0.2772	0.3801	0.1029
0.9500	0.6810	0.4595	0.2215	0.3190	0.4442	0.1252
1.0000	0.6388	0.4404	0.1984	0.3612	0.5108	0.1497
1.0500	0.5969	0.4195	0.1774	0.4031	0.5793	0.1762
1.1000	0.5557	0.3973	0.1584	0.4443	0.6491	0.2048
1.1500	0.5156	0.3743	0.1413	0.4844	0.7196	0.2352
1.2000	0.4769	0.3511	0.1258	0.5231	0.7904	0.2673
Err	5.5190e-07	1.2694e-06	1.2721e-06	5.5190e-07	1.2694e-06	1.2721e-06

Table F.4 FT: Heston model

K	AN-Call	CN-Call	Call	AN-Put	CN-Put	Put
0.8000	0.8222	0.5206	0.3015	0.1778	0.2403	0.0625
0.8500	0.7854	0.5153	0.2701	0.2146	0.2932	0.0786
0.9000	0.7461	0.5052	0.2409	0.2539	0.3509	0.0970
0.9500	0.7048	0.4909	0.2140	0.2952	0.4128	0.1176
1.0000	0.6621	0.4729	0.1892	0.3379	0.4783	0.1405
1.0500	0.6186	0.4519	0.1666	0.3814	0.5468	0.1654
1.1000	0.5747	0.4286	0.1462	0.4253	0.6178	0.1925
1.1500	0.5311	0.4034	0.1276	0.4689	0.6905	0.2216
1.2000	0.4881	0.3770	0.1110	0.5119	0.7644	0.2525

Table F.5 FFT: Heston model

K	AN-Call	CN-Call	Call	AN-Put	CN-Put	Put
0.8000	0.8222	0.5206	0.3015	0.1778	0.2403	0.0625
0.8500	0.7853	0.5151	0.2702	0.2147	0.2934	0.0787
0.9000	0.7460	0.5049	0.2411	0.2540	0.3512	0.0972
0.9500	0.7047	0.4906	0.2141	0.2953	0.4130	0.1178
1.0000	0.6621	0.4728	0.1893	0.3379	0.4784	0.1406
1.0500	0.6186	0.4519	0.1667	0.3814	0.5469	0.1655
1.1000	0.5747	0.4284	0.1464	0.4253	0.6180	0.1927
1.1500	0.5312	0.4032	0.1280	0.4688	0.6907	0.2219
1.2000	0.4883	0.3769	0.1114	0.5117	0.7646	0.2529
Err	1.0773e-04	1.8344e-04	2.0095e-04	1.0773e-04	1.8344e-04	2.0095e-04

F.3.2 The Heston model

The second set of results concern the Heston model (Tables F.4, F.5 and F.6), computed at the parameters:

$$\lambda = 1.4810, \quad \overline{v} = 0.1575, \quad \eta = 0.2560, \quad v_0 = 0.2104, \quad \rho = -0.8941$$

Table F.6 Fractional FFT: Heston model

K	AN-Call	CN-Call	Call	AN-Put	CN-Put	Put
0.8000	0.8222	0.5206	0.3015	0.1778	0.2403	0.0625
0.8500	0.7854	0.5153	0.2701	0.2146	0.2932	0.0786
0.9000	0.7461	0.5052	0.2409	0.2539	0.3509	0.0970
0.9500	0.7048	0.4909	0.2140	0.2952	0.4128	0.1176
1.0000	0.6621	0.4729	0.1892	0.3379	0.4783	0.1405
1.0500	0.6186	0.4519	0.1666	0.3814	0.5469	0.1654
1.1000	0.5747	0.4286	0.1462	0.4253	0.6178	0.1925
1.1500	0.5311	0.4034	0.1276	0.4689	0.6905	0.2216
1.2000	0.4881	0.3770	0.1110	0.5119	0.7644	0.2525
Err	7.0673e-07	1.8538e-06	1.3055e-06	7.0673e-07	1.8538e-06	1.3055e-06

G Affine Models: The Path Integral Approach

G.1 THE PROBLEM

Here we focus on the computation of the expectation of

$$\exp\left(-\Lambda \int_0^T \nu_t \, \mathrm{d}t\right)$$

within the CIR model. While this is by now standard textbook knowledge, for the sake of completeness we report it in full, using an unusual technique.

Let us consider the stochastic process defined in the risk-neutral measure:

$$\mathrm{d}X_t = \mu_t \, \mathrm{d}t + \sigma_t \, \mathrm{d}W_t \tag{G.1}$$

where we have set

$$\mu_t = \mu(t, X_t), \quad \sigma_t = \sigma(t, X_t)$$

As usual we split the time interval $T - t$ into N intervals of length δt such that

$$N\delta t = T - t$$

the transition probability density is given by

$$\begin{aligned} p_{\delta t} = p(x_{t+\delta t}, t + \delta t | x_t, t) &= \frac{\mathrm{d}x_{t-\delta t}}{\sqrt{2\pi\sigma_t^2 \delta t}} \exp\left\{-\frac{1}{2\sigma_t^2 \delta t}\left[x_{t+\delta t} - x_t - \mu_t \delta t\right]^2\right\} \\ &\equiv \frac{\mathrm{d}x_{t-\delta t}}{\sqrt{2\pi\sigma_t^2 \delta t}} \exp(\mathcal{L}(t)\delta t) \end{aligned}$$

with an obvious definition for the function $\mathcal{L}(t)$.

The transition probability $(x_t, t \to x_T, T)$ is the N-fold convolution

$$p(x_T, T | x_t, t) = I_N(\{\mathrm{d}x\}) = p_{\delta t}^{*N}$$

Let us define:

$$\begin{aligned} \epsilon &= \delta t \\ x_n &= x_{t+n\epsilon} \\ \mu_n &= \mu(t + n\epsilon, x_n) \\ \sigma_n &= \sigma(t + n\epsilon, x_n) \\ \mathcal{L}_n &= \mathcal{L}(t + n\epsilon) \end{aligned}$$

We are interested in computing

$$\phi_X(\{f\}) \equiv E\left[\exp\left(-\int_t^T f(s)x(s)\,\mathrm{d}s\right)\bigg| x_t = x(t)\right] \tag{G.2}$$

where $f(t)$ is a measurable function.

Let's define the N-measure I_N as:

$$I_N(\{x\},\{f\}) \equiv I_N(\{\mathrm{d}x\})\exp\left(-\sum_{n=1}^{N} f_n x_n \epsilon\right)$$

Consistent with our notation, we can write:

$$E\left[\exp\left(-\sum_{n=1}^{N} f_n X_n \epsilon\right)\bigg| X_t = x(t)\right] = \int I_N(\{\mathrm{d}x\},\{f\})$$

It turns out to be convenient to compute a more general expression:

$$\overline{I} = \int I_N(\{\mathrm{d}x\},\{k\})\exp(\alpha_N - \beta_N x_N)$$

We can single out the terms contributing to the integral over x_N and write:

$$\overline{I} = \int I_{N-1}(\{\mathrm{d}x\},\{f\})\int\left[\frac{\mathrm{d}x_N}{\sqrt{2\pi\sigma_{N-1}^2\epsilon}}\right]\mathrm{e}^{(-[\mathcal{L}_{N-1}\epsilon + f_N x_N \epsilon] + \alpha_N - \beta_N x_N)}$$

The integral

$$G_N = \int\left[\frac{\mathrm{d}x_N}{\sqrt{2\pi\sigma_{N-1}^2\epsilon}}\right]\mathrm{e}^{-\mathcal{L}_{N-1}\epsilon + \alpha_N - (\beta_N + f_N\epsilon)x_N}$$

is a Gaussian integral, whose result is

$$G_N = \exp\left(\alpha_N - \gamma_N(x_{N-1} + \mu_{N-1}\epsilon) + \frac{\epsilon}{2}\gamma_N^2\sigma_{N-1}^2\right)$$

where

$$\gamma_q \equiv \beta_q + f_q\epsilon$$

We confine ourselves to affine form for the process parameters:

$$\mu_n = a_n + b_n x_n \tag{G.3}$$

$$\sigma_n^2 = c_n + d_n x_n \tag{G.4}$$

where a_n, b_n, c_n, d_n may depend on t but NOT on x.

Then:

$$G_N = \exp(\alpha_{N-1} - \beta_{N-1}x_{N-1})$$

where

$$\alpha_{N-1} = \alpha_N - \gamma_N a_{N-1}\epsilon + \frac{\epsilon}{2}\gamma_N^2 c_{N-1}$$
$$\beta_{N-1} = \beta_N + \epsilon f_N + \epsilon\gamma_N b_{N-1} - \frac{\epsilon}{2}\gamma_N^2 d_{N-1}$$

It follows that

$$\bar{I} = \int I_{N-1}(\{\mathrm{d}x\}, \{f\}) \exp(\alpha_{N-1} - \beta_{N-1}x_{N-1})$$

and we are clearly faced with a recursive behaviour where the generic nth term would satisfy

$$\frac{\alpha_{n-1} - \alpha_n}{\epsilon} = -\gamma_n a_{n-1} + \frac{1}{2}\gamma_n^2 c_{n-1}$$
$$\frac{\beta_{n-1} - \beta_n}{\epsilon} = f_n + \gamma_n^2 b_{n-1} - \frac{1}{2}\gamma_n^2 d_{n-1}$$

sending $\epsilon \to 0, \quad \alpha_n \to A_T(t, \{f\}), \quad \beta_n = B_T(t, \{f\})$ we get:

$$\mathrm{d}X_s = [a(s) + b(s)X_s]\,\mathrm{d}s + \sigma_s\,\mathrm{d}W_s, \quad s > t, \quad X_t = x(t)$$
$$\sigma_s^2 = c_s + \mathrm{d}_s X_s$$

the following equation must hold:

$$\phi_X(\{f\}, t) \equiv E\left[\exp\left(-\int_t^T f(s)X(s)\,\mathrm{d}s\right)\bigg|\mathcal{F}_t\right] = \exp\Big(A_T(t, \{f\}) - B_T(t, \{f\})x(t)\Big)$$

where A_T and B_T are the unique solution of the p.d.e.:

$$\begin{aligned}
\frac{\mathrm{d}A_T(t, \{f\})}{\mathrm{d}t} &= a(t)B_T(t, \{f\}) - \frac{1}{2}B_T^2(t, \{f\})c(t) \\
\frac{\mathrm{d}B_T(t, \{f\})}{\mathrm{d}t} &= -f(t) - b(t)B_T(t, \{f\}) + \frac{1}{2}B_T^2(t, \{f\})\,\mathrm{d}(t) \\
A_T(T, \{f\}) &= B_T(T, \{f\}) = 0
\end{aligned} \tag{G.5}$$

G.2 SOLUTION OF THE RICCATI EQUATIONS

From the comparison of equation (7.6) with equation (G.2) we get

$$f(t) = \Lambda$$

Comparing equations (7.4), (G.1), (G.3) and (G.4) we get

$$a(t) = \kappa\theta, \quad b(t) = -\kappa, \quad c(t) = 0, \quad d(t) = \eta^2$$

and we have to solve

$$\begin{aligned}
\frac{\mathrm{d}A_T(t, \Lambda)}{\mathrm{d}t} &= \kappa\theta B_T(t, \Lambda) \\
\frac{\mathrm{d}B_T(t, \Lambda)}{\mathrm{d}t} &= -\Lambda - \kappa B_T(t, \Lambda) + \frac{\eta^2}{2}B_T^2(t, \Lambda) \\
A_T(T, \Lambda) &= B_T(T, \Lambda) = 0
\end{aligned} \tag{G.6}$$

The solution for $B_T(t,\Lambda)$ that fulfils the boundary condition $B_T(T,\Lambda)=0$ is given by:

$$\int_{B_T(t,\Lambda)}^{0}\frac{\mathrm{d}x}{x^2+(2\kappa/\eta^2)x-(2\Lambda/\eta^2)}=\frac{\eta^2}{2}(T-t)$$

Let $z_\pm$ be the roots of the equation:

$$x^2+\frac{2\kappa}{\eta^2}x-\frac{2\Lambda}{\eta^2}=0,\quad \text{that is}\quad z_{m/p}=\begin{cases}-\frac{\gamma+\kappa}{\eta^2} & z_m<0\\ \frac{\gamma-\kappa}{\eta^2} & z_p>0\end{cases}$$

where

$$\gamma=\sqrt{\kappa^2+2\Lambda\eta^2}$$

and:

$$\frac{1}{x^2+(2\kappa/\eta^2)x-(2\Lambda/\eta^2)}=-\frac{\eta^2}{2\gamma}\left[\frac{1}{x-z_m}-\frac{1}{x-z_p}\right]$$

The solution is therefore obtained by solving the integral equation

$$\int_{B_T(t,\Lambda)}^{0}\frac{\mathrm{d}x}{x-z_p}-\int_{B_T(t,\Lambda)}^{0}\frac{\mathrm{d}x}{x-z_m}=\gamma(T-t)$$

It follows that

$$-\left(\frac{z_m}{z_p}\right)\left(\frac{z_p-B_T(t,\Lambda)}{B_T(t,\Lambda)-z_m}\right)=\mathrm{e}^{-\gamma(T-t)}$$

Let us introduce the quantity

$$g=\frac{z_p}{z_m}=-\frac{\gamma-\kappa}{\gamma+\kappa}$$

then

$$B_T(t,\Lambda)=z_p\frac{1-\mathrm{e}^{-\gamma(T-t)}}{1-g\,\mathrm{e}^{-\gamma(T-t)}} \tag{G.7}$$

A simple manipulation displays an alternative form for $\beta(t)$ which is highly suitable to compute $A_T(t,\Lambda)$:

$$B_T(t,\Lambda)=z_p-\frac{2}{\eta^2}\frac{\mathrm{d}}{\mathrm{d}t}\log\left(1-g\,\mathrm{e}^{-\gamma(T-t)}\right)$$

it is now a simple matter to compute $A_T(t,\Lambda)$:

$$\begin{aligned}A_T(t,\Lambda)&=-\kappa\theta z_p(T-t)+\frac{2\kappa\theta}{\eta^2}\log\left(\frac{1-g}{1-g\,\mathrm{e}^{-\gamma(T-t)}}\right)\\&=-\lambda\bar{v}\left[z_p(T-t)-\frac{2}{\eta^2}\log\left(\frac{1-g}{1-g\,\mathrm{e}^{-\gamma(T-t)}}\right)\right]\end{aligned}$$

where we have used the mapping defined in equation (7.5). This completes our computation for the characteristic function of the Heston model.

Bibliography

Ané, T. and Geman, H. (2000) Order flow, transactions clock and normality of asset returns. *Journal of Finance*, **55**, 2259–2284.

Applebaum, D. (2009) *Lévy Processes and Stochastic Calculus*. Cambridge University Press.

Asmussen, S. and Rosinski, J. (2001) Approximations of small jumps of Lévy processes with a view towards simulations. *Journal of Applied Probability*, **38**, 482–493.

Bachelier, L. (1900) *Theorie de la Speculation*. Gauthier-Villard, Paris.

Bakshi, G. and Madan, D. (2000) Spanning and derivative securities evaluation. *Journal of Financial Economics*, **55** (2), 205–238.

Bakshi, G., Chao, C. and Chen, Z. (1997) Empirical performance of alternative option pricing models. *Journal of Finance*, **52**, 2003–2049.

Barndorff-Nielsen, O.E. (1998) Processes of Normal Inverse Gaussian type. *Finance and Stochastics*, **2** (1), 41–68.

Barndorff-Nielsen, O.E. and Shephard, N. (2001) Non-Gaussian Ornstein–Uhlenbeck based models and some of their uses in financial economics. *Journal of the Royal Statistical Society*, Series B, **63**, 167–241.

Bertoin, J. (1996) *Lévy processes*. Cambridge University Press.

Billingsley, P. (1986) *Probability and Measure* (2nd edition). John Wiley & Sons, Inc., New York.

Björk, T. (1998) *Arbitrage Theory in Continuous Time*. Oxford University Press, Inc., New York.

Black, F. and Scholes, M. (1973) The pricing of options and corporate liabilities. *Journal of Political Economy*, **81**, 637–654.

Bouziane, M. (2008) *Pricing Interest Rate Derivatives: A Fourier Transform Approach*. Springer, Berlin.

Bracewell, R. (1965) *The Fourier Transform and Its Applications*. McGraw-Hill, New York.

Breeden, D.T. and Litzenberger, R.H. (1978) Prices of state-contingent claims implicit in option prices. *Journal of Business*, **51**, 621–651.

Brigo, D. and Mercurio, F. (2006) *Interest Rate Models. Theory and Practice* (2nd edition). Springer Finance.

Carr, P. and Madan, D. (1999) Option valuation using the Fast Fourier Transform. *Journal of Computational Finance*, **2**, 61–73.

Carr, P. and Wu, L. (2003) Finite moment log-stable process and option pricing. *Journal of Finance*, **58**, 753–777.

Carr, P. and Wu, L. (2003) What type of process underlies options? A simple robust test. *Journal of Finance*, **58**, 2581–2610.

Carr, P. and Wu, L. (2004) Time changed Lévy processes and option pricing. *Journal of Financial Economics*, **17**, 113–141.

Carr, P., Geman, H., Madan, D. and Yor, M. (2002) The fine structure of asset returns: An empirical investigation. *Journal of Business*, **75** (2), 305–332.

Carr, P., Geman, H., Madan, D. and Yor, M. (2003) Stochastic volatility for Lévy processes. *Mathematical Finance*, **13**, 345–382. .

Carr, P., Geman, H., Madan, D. and Yor, M. (2004) From local volatility to local Lévy processes. *Quantitative Finance*, **4**, 581–588.

Carr, P., Geman, H., Madan, D. and Yor, M. (2005) Pricing options on realized variance. *Finance and Stochastics*, **9**, 453–475.

Carr, P., Geman, H., Madan, D. and Yor, M. (2007) Self-decomposability and option pricing. *Mathematical Finance*, **17** (1), 31–57.

Clark, P. (1973) A subordinated stochastic process with finite variance for speculative prices. *Econometrica*, **41**, 135–155.

Cont, R. and Tankov P. (2004) *Financial Modelling With Jump Processes*. Chapman & Hall.

Cooley, J.W. and Tukey, J.W. (1965) An algorithm for the machine calculation of complex Fourier series. *Mathematics of Computation*, **19** (April), 297.

Cox, J. and Ross, S. (1985) The valuation of options for alternative stochastic processes. *Journal of Financial Economics*, **3**, 144–156.

Cox, J., Ingersoll, J. and Ross, S. (1985) A theory of the term structure of interest rates. *Econometrica*, **53**, 385–408.

Dambis, K.E. (1965) On the decomposition of continuous submartingales. *Theory of Probability and Applications*, **10**, 401–410.

Danielson, G.C. and Lanczos, C. (1942) Some improvements in practical Fourier analysis and their application to X-ray scattering from liquids. *Journal of the Franklin Institute*, **233** (4), 365–380; and **233** (5), 435–452.

Delbaen, F. and Schachermayer, W. (1998) The fundamental theorem of asset pricing for unbounded stochastic processes. *Mathematische Annalen*, **312**, 215–250.

Dubins, L.E. and Schwarz, G. (1965) On continuous martingales. *Proceeding of National Academy of Sciences USA*, **53**, 913–916.

Duffie, D. (2001) *Dynamic Asset Pricing Theory* (3rd edition). Princeton University Press.

Duffie, D. and Kan, R. (1996) A yield factor model for interest rates. *Mathematical Finance*, **6** (4), 379–406.

Duffie, D., Pan, J. and Singleton, K. (2000) Transform analysis and option pricing for affine jump-diffusions. *Econometrica*, **68** (6), 1343–1376.

Eberlein, E. (2001) Application of generalized hyperbolic Lévy motions to finance. In O.E. Barndorff-Nielsen, T. Mikosch, and S. Resnick (eds), *Lévy Processes: Theory and Applications* (pp. 319–337). Birkhäuser Verlag.

Eberlein, P., Keller, U. and Prause, K. (1998) New insight into smile, mispricing and value at risk. *Journal of Business*, **71**, 371–406.

Embrechts, P., Kluppenberg, P. and Mikosch, T. (1997) *Modeling Extremal Event for Insurance and Finance*. Springer, Berlin.

Engle, R.F. (ed.) (1996) *ARCH Selected Readings*. Oxford University Press.

Fama, E.F. (1965) Efficient capital markets: A review of theory and empirical work. *Journal of Finance*, **25** (2), 383–417.

Fama, E.F. (1965) The behaviour of asset prices. *Journal of Business*, **38**, 34–105.

Feller, W. (1968) *An Introduction to Probability Theory and Its Applications,* Vol. I. John Wiley & Sons, Inc., New York.

Feller, W. (1971) *An Introduction to Probability Theory and Its Applications,* Vol. II. John Wiley & Sons, Inc., New York.

Feng, L. and Linetsky, V. (2008) Pricing discretely monitored barrier options and defaultable bonds in Lévy process models: A Fast Hilbert Transform approach. *Mathematical Finance*, **18** (3), 337–384.

Follmer, H. and Schweitzer, M. (1991) Hedging of contingent claims under incomplete information. In M.H.A. Davis and R.J. Elliot (eds), *Applied Stochastic Analysis, Stochastics Monograph* 5 (pp. 389–414). Gordon Breach, London and New York.

Gatheral, J. (2007) *The Volatility Surface*. John Wiley & Sons, Ltd, Chichester, UK.

Geman, H. (1989) *The importance of the forward risk neutral probability in a stochastic approach of interest rates*. Working Paper, ESSEC.

Geman, H. (2002), Pure jump Lévy processes for asset price modelling, *Journal of Banking and Finance*, **26** (7), 1297–1316.

Geman, H. and Yor, M. (1993) Bessel processes, Asian options and perpetuities. *Mathematical Finance*, **2** (4), 349–375.

Geman, H., Madan, D. and Yor, M. (2001) Time changes for Lévy processes. *Mathematical Finance*, **11** (1), 79–96.

Gil-Pelaez, J. (1951) A note on the inversion theorem. *Biometrika*, **38** (4), 481–482.

Glasserman, P. (2003) *Monte Carlo Methods in Financial Engineering*. Springer.

Harrison, J.M. and Kreps, D. (1979) Martingales and arbitrage in multiperiod security markets, *Journal of Economic Theory*, **2**, 381–408.

Harrison, J.M. and Pliska, S.R. (1981) Martingales and stochastic integrals in the theory of continuous trading. *Stochastic Processes and Applications*, **11**, 215–260.

Heston, S.L. (1993) A closed form solution for options with stochastic volatility with applications to bond and currency options. *Review of Financial Studies*, **6**, 327–343.

Huang, J.Z. and Wu, L. (2004) Specification analysis of option pricing models based on time changed Lévy processes. *Journal of Finance*, **59** (3), 1405–1440.

Hull, J. and White, A. (1998) Value at risk when daily changes in market variables are not normally distributed. *Journal of Derivatives*, **5** (3), 9–19.

Hull, J. and White, A. (1987) The pricing of options on assets with stochastic volatility. *Journal of Finance*, **42**, 281–300.

Ingersoll, J.E. (2000) Digital contracts: Simple tools for pricing complex derivatives. *Journal of Business*, **73** (1), 62–88.

Jamshidian, F. (1989) An exact bond option pricing formula. *Journal of Finance*, **44**, 205–209.

Jeanblanc, M., Pitman, J. and Yor, M. (2001) Self-similar processes with independent increments associated with Lévy and Bessel processes. *Stochastic Processes and Applications*, **100**, 223–232.

Karlin, S. and Taylor, H.M. (1975) *A First Course in Stochastic Processes*. Academic Press.

Karlin, S. and Taylor, H.M. (1981) *A Second Course in Stochastic Processes*. Academic Press.

Kendall, M. and Stuart, A. (1977) *The Advanced Theory of Statistics* (4th edition). Griffin, London.

Khintchine, A.Y. (1938) *Limit Laws of Sums of Independent Random Variables*. ONTI, Moscow, Russia.

Kingman, J. (1993) *Poisson Processes*. Volume 3 of Oxford University Studies in Probability. Oxford University Press, New York.

Konikov, A.Y. and Madan, D. (2002) Option pricing using variance gamma Markov chains. *Review of Derivative Research*, **5**, 81–115.

Kyprianou, A.E. (2006) *Introductory Lectures on Fluctuation of Lévy Processes with Applications*. Springer.

Lee, R.W. (2004) Option pricing by transform methods: Extensions, unification, and error control. *Journal of Computational Finance*, **7** (3), 51–86.

Lévy, B. (1937) *Théorie de l'Addition des Variables Aléatoires*. Gauthier-Villars, Paris.

Lewis, A.L. (2000) *Option Valuation Under Stochastic Volatility*. Finance Press.

Lewis, A.L. (2001) *A simple option pricing formula for general jump diffusion and other exponential Lévy processes*. Manuscript. Envision Financial System and OptionCity.net.

Madan, D. and Milne, F. (1991) Option pricing with VG martingale components. *Mathematical Finance*, **1**, 39–55.

Madan, D. and Seneta, E. (1990) The Variance Gamma (VG) model for share market returns. *Journal of Business*, **63**, 511–524.

Madan, D., Carr, P. and Chang, E. (1998) The Variance Gamma process and option pricing. *European Finance Review*, **2**, 79–105.

Mandelbrot, B.B. (1963) The variation of certain speculative prices. *Journal of Business*, XXXVI, 392–417.

Merton, R.C. (1973) Theory of rational option pricing. *Bell Journal of Economics and Management*, **4**, 141–183.

Merton, R.C. (1976) Option pricing when underlying returns are discontinuous. Theory of rational option pricing. *Journal of Financial Economics*, **3**, 125–144.

Monroe, I. (1978) Processes that can be embedded in Brownian motion. *Annals of Applied Probability*, **6** (1) 42–56.

Musiela, M. and Rutkowski, M. (2005) *Martingale Methods in Financial Modelling* (2nd edition). Springer Finance.

Ross, S.A. (1976) The arbitrage theory of capital asset pricing. *Journal of Economic Theory*, **13**, 341–360.

Samorodnitsky, G. and Taqqu, M. (1994) *Stable Non-Gaussian Random Processes*. Chapmann and Hall, New York.

Samuelson, P.A. (1963) Proof that properly anticipated prices fluctuate randomly. *Industrial Management Review*, **6**, 41–50.

Samuelson, P.A. (1973a) Mathematics of speculative price. *SIAM Review*, **15** (1), 1–42.
Samuelson, P.A. (1973b) Proof that properly discounted present values of assets fluctuate randomly. *Bell Journal of Economics and Management*, **4** (2), 369–374.
Sato, K. (1991) Self-similar processes with independent increments. *Probability Theory and Related Fields*, **89**, 285–300.
Sato, K. (1999) *Lévy processes and Infinitely Divisible Distributions*. Cambridge University Press.
Schoutens, W. (2003), *Lévy Processes in Finance: Pricing Financial Derivatives*. John Wiley & Sons, Inc., New York.
Schwartz, L. (1961) *Méthode mathématiques pour le sciences physiques*. Herman and Cie, Paris.
Shao, J. (1999) *Mathematical Statistics*. Springer-Verlag, New York.
Vershik, A. and Yor, M. (1995) *Mulitplicativité du processus gamma et étude asymptotique des lois stable d'indice α, lorsque α tends vers 0*. Preprint. Laboratoire de Probabilitées, Université Paris VI.
Winkel, M. *Lévy Processes and Finance*. Lecture Notes (Oxford).
Zemanian, A.H. (1987) *Distribution Theory and Transform Analysis*. Dover.
Zhu, J. (1987) *Modular Option Pricing of Options*. Springer, Berlin.

Index

Index compiled by Terry Halliday